ROUTLEDGE HANDBOOK
OF WATER ECONOMICS
AND INSTITUTIONS

Growing scarcity of freshwater worldwide brings to light the need for sound water resource modeling and policy analysis. While a solid foundation has been established for many specific water management problems, combining those methods and principles in a unified framework remains an ongoing challenge. This Handbook aims to expand the scope of efficient water use to include allocation of sources and quantities across uses and time, as well as integrating demand-management with supply-side substitutes.

Socially efficient water use does not generally coincide with private decisions in the real world, however. Examples of mechanisms designed to incentivize efficient behavior are drawn from agricultural water use, municipal water regulation, and externalities linked to water resources. Water management is further complicated when information is costly and/or imperfect. Standard optimization frameworks are extended to allow for coordination costs, games and cooperation, and risk allocation. When operating efficiently, water markets are often viewed as a desirable means of allocation because a market price incentivizes users to move resources from low to high value activities. However, early attempts at water trading have run into many obstacles. Case studies from the United States, Australia, Europe, and Canada highlight the successes and remaining challenges of establishing efficient water markets.

Kimberly Burnett is an Associate Specialist with the University of Hawaii Economic Research Organization, USA.

Richard Howitt is currently Professor Emeritus in Agricultural and Resource Economics at the University of California at Davis, USA.

James A. Roumasset is a Professor in the Department of Economics at the University of Hawaii, Manoa, USA.

Christopher A. Wada is a Research Economist with the University of Hawaii Economic Research Organization, USA.

ROUTLEDGE HANDBOOK OF WATER ECONOMICS AND INSTITUTIONS

Edited by Kimberly Burnett, Richard Howitt, James A. Roumasset and Christopher A. Wada

LONDON AND NEW YORK

First published 2015 by Routledge

2 Park Square, Milton Park, Abingdon, Oxfordshire OX14 4RN
711 Third Avenue, New York, NY 10017

Routledge is an imprint of the Taylor & Francis Group, an informa business

First issued in paperback 2017

British Library Cataloguing in Publication Data
A catalogue record for this book is available from the British Library

Library of Congress Cataloging in Publication Data
Routledge handbook of water economics and institutions /
edited by Kimberly Burnett, Richard Howitt, James A. Roumasset
and Christopher A. Wada.
pages cm
Includes bibliographical references and index.
1. Water-supply – Management. 2. Water-supply – Management –
Case studies. 3. Water-supply – Economic aspects. 4. Water-supply –
Economic aspects – Case studies. 5. Water resources development.
6. Water resources development – Case studies.
I. Burnett, Kimberly M., editor of compilation.
II. Title: Handbook of water economics and institutions.
HD1691.R69 2014
333.91–dc23
2014022742

ISBN: 978-0-415-72856-0 (hbk)
ISBN: 978-1-138-57319-2 (pbk)

Typeset in Bembo
by HWA Text and Data Management, London

CONTENTS

Contents

FIGURES

TABLES

CONTRIBUTORS

Henning Bjornlund
Canada Research Chair in Water
Policy and Management, University of
Lethbridge
Associate Professor, University of South
Australia
Email: henning.bjornlund@uleth.ca

Kimberly Burnett
Research Economist and Associate
Specialist
University of Hawaii Economic Research
Organization
University of Hawaii, Manoa
540 Saunders Hall
2424 Maile Way
Honolulu, HI 96822
Email: kburnett@hawaii.edu

Ujjayant Chakravorty
Department of Economics
Tufts University
114A Braker Hall
8 Upper Campus Rd.
Medford, MA 02155-6722
Email: ujjayant.chakravorty@tufts.edu

Bonnie Colby
Professor
Department of Agricultural and Resource
Economics
PO Box 210023
The University of Arizona
Tucson, AZ 85721-0023
Email: bcolby@ag.arizona.edu

Lin Crase
Professor of Applied Economics and
Director, Centre for Water Policy and
Management
Faculty of Business, Economics and Law
School of Economics
La Trobe University
Wodonga Victoria 3689
Australia
Email: l.crase@latrobe.edu.au

Osiel González Dávila
SOAS, University of London
Thornhaugh Street
Russell Square
London WC1H 0XG, UK
E-mail: osiel.davila@soas.ac.uk

Gonzalo Delacámara
Researcher
IMDEA Water Institute
Parque Científico Tecnológico de la UAH
C/ Punto Net, 4- 2ª planta
28805, Alcalá de Henares
Madrid
Spain
Email: gonzalo.delacamara@imdea.org

Ariel Dinar
Professor of Environmental Economics
and Policy
School of Public Policy
University of California, Riverside
Riverside, CA 92521
Email: ariel.dinar@ucr.edu

George Frisvold
Professor
Department of Agricultural and Resource
Economics
PO Box 210023
The University of Arizona
Tucson, AZ 85721-0023
Email: frisvold@ag.arizona.edu

Renan-Ulrich Goetz
Professor of Economics
Universitat de Girona
Departament d'Economia
Campus Montilivi
E-17071 Girona
Spain
Email: renan.goetz@udg.edu

Carlos Mario Gómez
Professor of Economics
University of Alcalá
Madrid
Spain
Email: mario.gomez@uah.es

Yazhen Gong
Assistant Professor
School of Environment and Natural
Resources
Renmin University of China
No 59, Zhongguancun Street
Beijing
China 100872
Email: ygong.2010@ruc.edu.cn

Ellen Hanak
Senior Fellow
Public Policy Institute of California
500 Washington Street, Suite 600
San Francisco, CA 94111
Email: hanak@ppic.org

Richard Howitt
Professor of Agricultural & Resource
Economics
2120 Social Sciences & Humanities
University of California, Davis
One Shields Avenue
Davis, CA 95616
Email: howitt@primal.ucdavis.edu

Jikun Huang
Director and Professor
Center for Chinese Agricultural Policy
Chinese Academy of Sciences
Datun Rd, Anwai
Beijing 100101
China
E-Mail: jkhuang.ccap@igsnrr.ac.cn

Qiuqiong Huang
Associate Professor
221 Agriculture Building
Department of Agricultural Economics &
Agribusiness
University of Arkansas
Fayetteville, AR 72701
Email: qqhuang@uark.edu

Ray G. Huffaker
Professor
Agricultural and Biological Engineering
Department
University of Florida
281 Frazier Rogers Hall
PO Box 110570
Gainesville, FL 32611-0570
Email: rhuffaker@ufl.edu

Karl Jandoc
Ph.D. Student
Department of Economics
University of Hawaii, Manoa
Saunders Hall 542
2424 Maile Way
Honolulu, HI 96822
Email: kljandoc@hawaii.edu

Ruben Juarez
Assistant Professor of Economics
Department of Economics
University of Hawaii, Manoa
Saunders Hall 542
2424 Maile Way
Honolulu, HI 96822
Email: rubenj@hawaii.edu

Yukio Kinoshita
Associate Professor of Agricultural
Economics
Faculty of Agriculture
Iwate University
18-8, Ueda 3 chome, Morioka
Japan
Email: kinop@iwate-u.ac.jp

K. K. Klein
Professor of Economics
University of Lethbridge
Department of Economics
4401 University Drive
Lethbridge, AB
T1K 3M4
Canada
Email: klein@uleth.ca

Keith Knapp
Professor of Resource Economics
Department of Environmental Sciences
University of California, Riverside
900 University Avenue
Riverside, CA 92521
Email: keith.knapp@ucr.edu

Phoebe Koundouri
Scientific Director, ReSEES: Research
team on Socio-Economic and
Environmental Sustainability
School of Economics
Athens University of Economics and
Business
Greece
Email: pkoundouri@aueb.gr

Gary D. Libecap
Donald Bren Distinguished Professor of
Corporate Environmental Management
Bren School of Environmental Science
& Management and Department of
Economics
University of California, Santa Barbara
4420 Donald Bren Hall
Santa Barbara, CA 93106
Email: glibecap@bren.ucsb.edu

C.-Y. Cynthia Lin
Associate Professor
Department of Agricultural and Resource
Economics
University of California, Davis
One Shields Avenue
Davis, CA 95616
Email: cclin@primal.ucdavis.edu

John Lynham
Assistant Professor
Department of Economics
University of Hawaii, Manoa
Saunders Hall 542
2424 Maile Way
Honolulu, HI 96822
Email: lynham@hawaii.edu

Josefina Maestu
Director, United Nations Office to
Support the International Decade for
Action: "Water for Life" 2005–2015
UN-Water Decade Programme on
Advocacy and Communication
Email: maestu@un.org

Matthew Mealy
FRM Senior Associate
KPMG LLP

Siwa Msangi
Senior Research Fellow
International Food Policy Research
Institute
2033 K St, NW
Washington, DC 20006-1002
Email: s.msangi@cgiar.org

Sue O'Keefe
Associate Professor, School of Economics,
Centre for Water Policy and Management
La Trobe University
Wodonga Victoria 3689
Australia
Email: S.OKeefe@latrobe.edu.au

Lisa Pfeiffer
Economist
NOAA National Marine Fisheries Service
Northwest Fisheries Science Center
2725 Montlake Boulevard East
Seattle, WA 98112
Email: lisa.pfeiffer@noaa.gov

Sittidaj Pongkijvorasin
Assistant Professor of Economics
Chulalongkorn University
Phayathai Road
Bangkok 10330
Thailand
Email: Sittidaj.P@Chula.ac.th

Mark W. Rosegrant
Division Director
Environment and Production Technology
International Food Policy Research
Institute
2033 K St, NW
Washington, DC 20006-1002
Email: m.rosegrant@cgiar.org

James A. Roumasset
Professor
Department of Economics
University of Hawaii, Manoa
Saunders Hall 542
2424 Maile Way
Honolulu, HI 96822
Email: jimr@hawaii.edu

Scott Rozelle
Helen F. Farnsworth Senior Fellow, FSI
Affiliated Faculty, CDDRL
Co-director, REAP
Stanford University
Shorenstein APARC
Encina Hall East, E407
Stanford, CA 94305-6055
Email: rozelle@stanford.edu

Kurt Schwabe
Associate Professor of Environmental
Economics and Policy
Associate Director, Water Science and
Policy Center
Department of Environmental Sciences
University of California, Riverside
900 University Avenue
Riverside, CA 92521
Email: kurt.schwabe@ucr.edu

Nori Tarui
Associate Professor
Department of Economics
University of Hawaii, Manoa
Saunders Hall 542
2424 Maile Way
Honolulu, HI 96822
Email: nori.tarui@hawaii.edu

Yacov Tsur
Ruth Ochberg Professor
Department of Agricultural Economics
and Management
Robert H. Smith Faculty of Agriculture,
Food and Environment
The Hebrew University of Jerusalem
P.O. Box 12, Rehovot 76100
Israel
Email: tsur@agri.huji.ac.il

Christopher A. Wada
Research Economist
University of Hawaii Economic Research
Organization
University of Hawaii, Manoa
540 Saunders Hall
2424 Maile Way
Honolulu, HI 96822
Email: cawada@hawaii.edu

Jinxia Wang
Professor and Deputy Director
Center for Chinese Agricultural Policy
Chinese Academy of Sciences
Datun Rd, Anwai
Beijing 100101
China
Email: jxwang.ccap@igsnrr.ac.cn

Sarah Wheeler
Senior Research Fellow – CRMA
University of South Australia Business
School
School of Commerce
City West Campus
Office WL3-52
Adelaide, South Australia
Australia
Email: Sarah.Wheeler@unisa.edu.au

Àngels Xabadia
Associate Professor
Universitat de Girona
Departament d'Economia
Campus Montilivi
E-17071 Girona
Spain
Email: angels.xabadia@udg.edu

PART I

Principles and overview

1

GLOBAL OUTLOOK FOR WATER SCARCITY, FOOD SECURITY, AND HYDROPOWER

Mark W. Rosegrant

Introduction

Water is essential for growing food; for household water uses, including drinking, cooking, and sanitation; as a critical input into industry; for tourism and cultural purposes; and in sustaining the earth's ecosystems. But this essential resource is under serious threat. Increasing national, regional, and seasonal water scarcities in much of the world pose severe challenges for national governments, the international development community, and, ultimately, for individual water users. The challenges of growing water scarcity are heightened by a) increasing costs of developing new water; b) rising energy prices that increase costs to deliver clean groundwater but also generate interest in hydropower dams; c) degradation of soils in irrigated areas; d) depletion of groundwater; e) water pollution and degradation of water-related ecosystems; and f) wasteful use of already developed supplies, often encouraged by subsidies and distorted incentives that influence water use.

With global population projected to rise to 9 billion in 2050, farmers need to increase food production to assure its availability for the growing populace. In order to enhance agricultural production, sufficient agricultural inputs—such as land, water, fertilizer, pesticide, seed quality (or high-yielding varieties), and management technique—should be accessible to the farmers. Of all these agricultural inputs, water availability is vital for crop growth, especially in arid and semi-arid regions of the world. The multiplicity of uses of water leads to competition among all agriculture, environment, energy, industry and domestic users. Agriculture is the highest user of freshwater among all sectors, accounting for 71 percent of water withdrawals, followed by industrial use at 20 percent, and domestic use (including household, drinking water, and sanitation) at 9 percent (Wada et al. 2011).

Growing water scarcity and water quality constraints are a major challenge to future outcomes in food security, especially since agriculture is expected to remain the largest user of freshwater resources in all regions of the world for the foreseeable future, despite rapidly growing industrial and domestic demand. As non-agricultural demand for water increases, water will be increasingly transferred from irrigation to other uses in many regions. In addition, the reliability of the agricultural water supply will decline without significant improvements in water management policies and investments. The intensifying sectoral competition and water scarcity problems, along with declining reliability of agricultural water supply, will put downward pressure on food supplies and continue to generate concerns for global food security.

This chapter provides an overview of the challenges for future water resources and discusses policy reforms to address these challenges as they relate to food production and food security. Over the next four decades, water scarcity is projected to grow and have significant consequences for food production. An assessment presented in this chapter compares potential outcomes for water scarcity and food security in 2050 generated by two different paths. The first path is a baseline scenario called Business As Usual (BAU), which assumes continuation of current trends and existing plans in agricultural and water policies and investments in agricultural productivity growth. The second is a constructed alternative path called the Green Growth Scenario, which assumes more efficient global water use, earlier adoption of second generation biofuels, and other changes consistent with more sustainable resource use.

One feature of the Green Growth Scenario is an assumed increase in the use of hydropower for energy generation and irrigation water use. Hydropower is examined in detail in this chapter for its potential to meet energy challenges while expanding irrigation water supplies and food production potential, thereby enhancing global food security. For several reasons outlined below, hydropower has become a relatively forgotten part of the water security picture that deserves renewed attention, particularly given simultaneous concerns with energy supplies, water scarcity, and general food security. Unlike a major competing form of renewable energy—biofuels—hydropower does not consume energy nor reduce food availability. As a result, development of hydropower could complement food production by developing structures (and power) that also provide irrigation water and support its distribution for growing food crops. Hydropower can have environmental and social impacts that need to be considered as well.

The last section reviews more general strategies for addressing challenges for food security stemming from water scarcity, including investment in infrastructure and water supply, water management and policy reform, and improved crop productivity and water use.

Water scarcity and food security

Rising global population will intensify competition for water. More than one-third of the global population—approximately 2.4 billion people—already live in water-scarce regions, i.e., river basins with annual water withdrawals greater than 40 percent of total renewable water, and 22 percent of the world's gross domestic product (GDP) (US$9.4 trillion at 2000 prices) is produced in water-short areas (Ringler et al. 2015). Moreover, 39 percent of global grain production is not sustainable in terms of water use, and expected levels of water productivity in the coming years will not be sufficient to reduce risks and ensure sustainability. By 2050, just over one-half of the global population (52 percent or 4.8 billion people), 49 percent of global grain production, and 45 percent

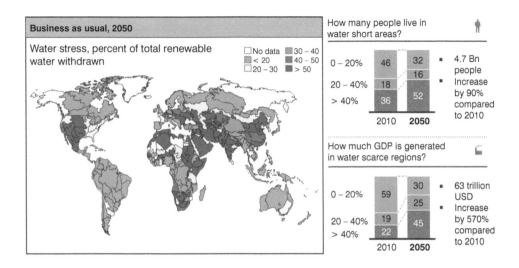

Figure 1.1 Projected water stress level in different regions of the world by 2050 (source: adapted from Ringler et al. 2015)
Note: Water scarce areas are river basins with annual water withdrawals greater than 40 percent of total renewable water.

(US$63 trillion) of total GDP will be at risk due to water stress, which will likely impact investment decisions, increase operation costs and affect the competitiveness of certain regions (Figure 1.1).

For China and India and many other rapidly-developing countries, water scarcity has already started and is already widespread—in these two countries alone 1.4 billion people live in areas of high water stress today (Ringler et al. 2015). In 2005 per capita water availability in the most populous countries—China and India—was fairly low, at 1,691 and 1,101 cubic meters respectively. In contrast, per capita water availability in Brazil (ranked fifth in terms of population) was 32,525 cubic meters and in Russia (ranked seventh in terms of population), 28,259 cubic meters. As a result of demographic changes in China and India, water availability is expected to decline to 1,507 and 856 cubic meters per capita respectively by 2030.

Furthermore, the ultimate outcome of climate change and its effect on water availability are not known with precision. Unknowns include geographic location, direction of change (less/more precipitation), degree of change in precipitation (low/high), change in precipitation intensity (low/high), and timing (within next five years or over multiple decades). Shifting precipitation patterns and warming temperatures could increase water scarcity in some regions while other areas may experience increased soil-moisture availability, which could expand opportunities for agricultural production (Malcolm et al. 2012; Rosegrant, Ringler and Zhu 2014).

Although water resources may improve in some areas, the World Bank expects climate change will make it more difficult to manage the world's water because it affects the entire water cycle (World Bank 2009, p. 137). Warming speeds up the hydrological cycle, increasing precipitation in some areas and for the world as a whole. Nonetheless, the report concludes that increased evaporation will make drought more prevalent across wide swaths of the world by the middle of the twenty-first century, as shown

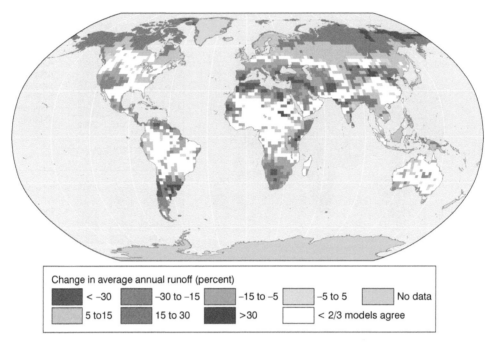

Change in average annual runoff (percent)

■ < –30	■ –30 to –15	■ –15 to –5	☐ –5 to 5	☐ No data
☐ 5 to15	■ 15 to 30	■ >30	☐ < 2/3 models agree	

Figure 1.2 Percentage change in average annual runoff across the regions of the world (sources: Milly et al. 2008; Milly et al. 2005 in World Bank 2010)

in Figure 1.2. Many of these regions are already net food importers, such as much of Africa, the Middle East, and Central America. Other areas where average annual runoff is expected to decline sharply include major agricultural producing regions, such as much of Europe, and parts of South America, North America, and Australia. The average number of consecutive dry days could increase by up to 20 days in many of these regions (Figure 1.2). Moreover, intensity of precipitation is expected to rise in almost all areas, regardless of whether total precipitation is decreasing or increasing. Increased intensity will likely pose challenges for agricultural and other users of water when trying to capture and manage available water supplies.

International Model for Policy Analysis of Agricultural Commodities and Trade Scenario Assessment

In the coming decades, increasing water scarcity is expected to contribute to a slowdown in agricultural growth and rising food prices. A scenario assessment is used here to determine whether enhancements in key aspects of water use could make significant improvements in food and water security. The analysis uses the International Model for Policy Analysis of Agricultural Commodities and Trade (IMPACT) model: a partial equilibrium, multi-commodity, multi-country model which generates projections of global food supply, demand, trade, and prices.[1] IMPACT covers over 46 crops and livestock commodities and it includes 115 countries/regions where each country is linked to the rest of the world through international trade and 281 food producing units (grouped according to political boundaries and major river basins). Demand is a function of prices, income, and population growth. Crop production is determined by crop and input prices, the rate of productivity growth, and water availability.

The analytical starting point is a BAU scenario for agriculture and water, which assumes a continuation of current trends and existing plans in agricultural and water policies and investments in agricultural productivity growth. Population projections are the "Medium" variant population growth rate projections from the Population Statistics division of the United Nations (UN 2011). The population numbers for the countries in the UN data have been aggregated to be in sync with IMPACT's regional definitions which comprise 115 economic regions. GDP projections are estimated by the authors, drawing upon the Millennium Ecosystem Assessment (Millennium Ecosystem Assessment 2005).

Under the BAU scenario, food prices rise, and there is only a slight improvement in food security because of the growth in demand and the constraints on both crop productivity (and area) and expansion of livestock production. BAU projects rising prices for cereals between 2010 and 2050, with the highest projected increase for maize (55 percent), followed by wheat (40 percent) and rice (33 percent). Price increases for millet, sorghum, and other grains range between 10 percent and 31 percent. For meat prices, the greatest increase between 2010 and 2050 is for pork (54 percent), followed by poultry (48 percent), beef (20 percent), and lamb (2 percent).

Because of the rising prices for major commodities, the number of persons at risk of hunger remains relatively high but declines modestly in developing countries, from 875 million in 2010 to 700 million in 2050, and in South Asia from approximately 320 million to 210 million. For Sub-Saharan Africa, though, the number of persons at risk of hunger increases in the baseline, given higher cereal prices and rapid population growth.

In order to assess how food security outcomes might change given alternative public investments and policies, the Green Growth Scenario for 2050 is developed to present a view of the world where economic growth and development is in harmony with general sustainability and environmental goals. The scenario emphasizes the importance of sustainable development to achieve economic growth through a set of drivers, ranging from high income growth to earlier adoption of second generation biofuels, switching to higher use of hydropower, and increased water use efficiency. Specifically, the scenario is constructed with the following changes relative to the BAU scenario:

- To represent a world with enhanced economic growth, the scenario applies higher GDP growth rates to all countries in IMPACT.
- To represent sustainability in the scenario, the efficiency in the use of water resources is increased in three sectors—irrigation (agricultural sector), domestic, and industrial. Increased water use efficiency lowers the water use in these sectors, where water is a critical component, thus, underlining the importance of sustainable use of natural resources.
- To illustrate faster expansion of the biofuels sector in the Green Growth Scenario, significant production of second generation biofuels start 5 years earlier than assumed in the BAU scenario (i.e., 2025 rather than 2030), thus lowering demand for agricultural feedstocks for the first generation biofuels.

- To incorporate the impact of higher fertilizer prices in the world market and sustainable development goals, the growth rate of fertilizer prices is increased by 25 percent, which captures the effect of lower fertilizer use per hectare.
- The scenario does not incorporate any effect from climate change on the agricultural sector (assumes perfect mitigation). Hence, crop area and yields in IMPACT remains unaffected by climate change for this scenario.
- The impact of enhanced agricultural research and development (R&D) and improved water productivity is reflected in an increase in growth rates for crop yields with respect to land area and water use.

Scenario results for the year 2050—water use

Results from the scenario analysis indicate major impacts on the water sector. The Water Simulation Module of the IMPACT model is used to simulate water allocation and uses by sector over the period 2000–2050. The combined effect of both direct and indirect factors contributes to the differences in water use between BAU and the Green Growth Scenario for 2050. To reflect the direct water-saving effects in the Green Growth Scenario, water use efficiencies are raised in the domestic, industrial, and irrigation sectors relative to BAU. Second, indirect water use consequences are channeled through changes in irrigated crop areas caused by higher income growth, earlier adoption of new biofuels technology, higher crop and livestock productivity growth, and higher fertilizer prices which are specified in the Green Growth Scenario. The next section on food prices and food security incorporates these efficiency gains, with implications for reducing water demand per unit of food production and for food security.

For all regions, total consumptive water uses decrease in the Green Growth Scenario relative to the BAU scenario, though the magnitude of change differs by region. The largest percent reductions are found in Europe and Central Asia, and Developed Europe, while the smallest percent reductions are found for Middle East and North Africa, and South Asia. Figure 1.3 presents the percent changes in total water consumption due to efficiency gains, and Figure 1.4 shows the percent changes in irrigation water consumption due to efficiency gains.

The small reduction in total water consumption in South Asia reflects an *increase* in irrigation water consumption in South Asia under the Green Growth Scenario (Figure 1.4). In this region, the growth in water use efficiency in the domestic and industrial sectors reduces water consumption in these two sectors, and, as a result, more water is left for irrigation. In the the Middle East and North Africa region, total water consumption also declines only marginally under the Green Growth Scenario because the gain in irrigation efficiency for this region is significantly smaller than any other regions, given that its irrigation efficiency is initially at a high level. Irrigation water consumption increases in these two regions because, with higher efficiencies, domestic and industrial sectors consume less water and the "saved water" is consequently used in irrigation for which a large gap exists between demand and supply.

A measure of the level of water scarcity in irrigation is irrigation water supply reliability (IWSR), defined as the ratio of irrigation water supply to demand on an annual basis (Rosegrant, Cai, and Cline 2002). Figure 1.5 shows the results for 2000 and 2050. In general, Europe and Central Asia, South Asia and, to some extent, East Asia and Pacific

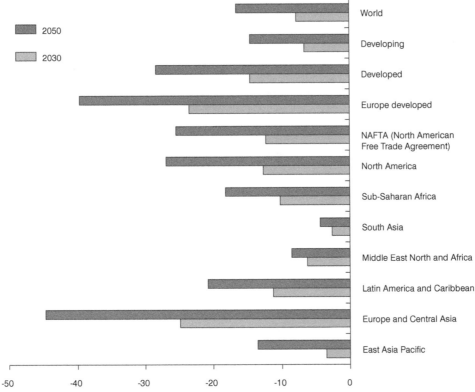

Figure 1.3 Percentage change in total water consumption in Green Growth Scenario compared to BAU, 2030 and 2050 (source: IFPRI IMPACT Model, 2012 Simulations.)

have the most serious irrigation water shortages, as indicated by their low IWSR values. Under the BAU scenario, their IWSR values decrease over time and reach fairly low levels by 2050. The water use efficiency gains under the Green Growth Scenario relieve water shortage situation in these regions and other regions as well. Globally, IWSR is 0.619 under the BAU scenario and 0.726 under the Green Growth Scenario, which are below the global average IWSR value of 2000 (Rosegrant et al. 2013).

Irrigation water supply availability could be increased with additional investments in structures such as hydropower dams to capture additional runoff and store water for irrigation purposes. Concomitantly, as demand for energy continues to rise and environmental costs associated with burning fossil fuels mount, faster expansion of all sizes of hydropower and scaling-up existing hydropower facilities could be a viable option. Consideration of hydropower, as outlined in the next section, culminating its further development, could provide structures and power for additional irrigation water and support its distribution for growing food crops and addressing global food security issues.

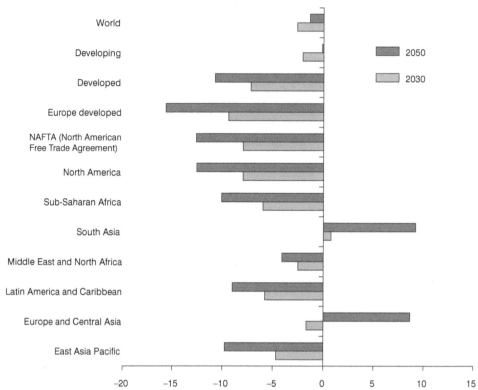

Figure 1.4 Percentage change in irrigation water consumption under the Green Growth Scenario compared to BAU, 2030 and 2050 (source: IFPRI IMPACT Model, 2012 Simulations).

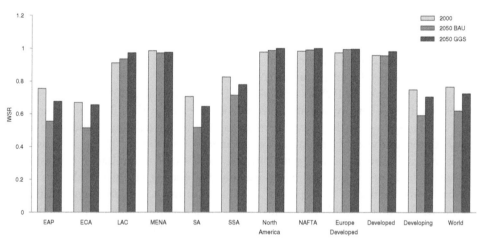

Figure 1.5 Irrigation water supply reliability under BAU and the Green Growth Scenario in 2000 and 2050 (source: IFPRI IMPACT projections 2012).
Notes: IWSR—ratio of annual irrigation water supply to demand; BAU—Business-As-Usual; GGS—Green Growth Scenario; EAP—East Asia and Pacific; ECA—Europe and Central Asia; LAC—Latin America and Caribbean; MENA—Middle East and North Africa; SA—South Asia; SSA—Sub-Saharan Africa; NAFTA—North American Free Trade Agreement.

Scenario results for the year 2050—food prices and food security

Demand and supply play a major role in influencing the prices of major commodities in the Green Growth Scenario relative to BAU. With higher yield growth and larger crop production, prices for most crops, including rice, wheat and maize, decline relative to the BAU scenario despite higher income growth, as consumer demand for cereals declines relative to demand for meat (see consumption results below). Prices for beef, lamb, and poultry increase, reflecting the impact of higher income on these commodity markets. Figure 1.6 presents the projected percent change in crop prices between the Green Growth Scenario and BAU scenario, and Figure 1.7 shows the projected change in meat prices.

In the Green Growth Scenario, higher income growth and lower crop prices result in higher food consumption of both cereals and meat (see Figures 1.8 and 1.9). The impact

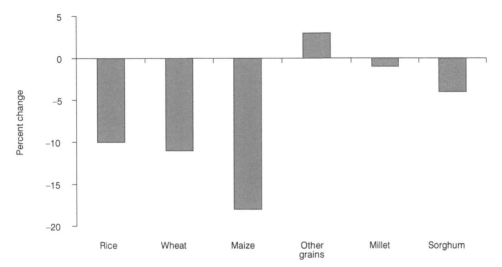

Figure 1.6 Percentage change in world prices of cereals between BAU and the Green Growth Scenario, 2050 (source: IFPRI IMPACT Model, 2012 Simulations)

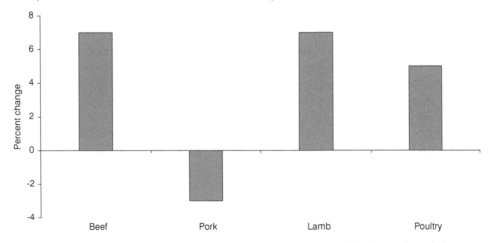

Figure 1.7 Percentage change in world prices of meat between BAU and the Green Growth Scenario, 2050 (source: IFPRI IMPACT Model, 2012 Simulations)

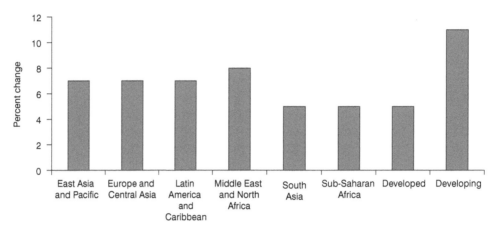

Figure 1.8 Percentage change in per capita cereal consumption between BAU and the Green Growth Scenario, 2050 (source: IFPRI IMPACT Model, 2012 Simulations)

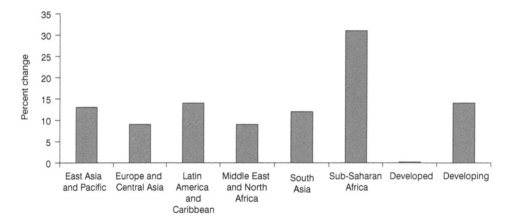

Figure 1.9 Percentage change in per capita meat consumption between BAU and Green Growth Scenario, 2050 (source: IFPRI IMPACT Model, 2012 Simulations)

of high income growth is found to be stronger in the consumption of meat relative to cereals. Middle East and North Africa has the highest percentage increase in cereal food consumption, followed by Latin America and the Caribbean, and Europe and Central Asia regions. On the other hand, the greatest percentage increase in meat consumption is in Sub-Saharan Africa, followed by Latin America and the Caribbean. This result shows that income growth leads to a subsequent change in dietary patterns in most regions, whereby consumers increase consumption of meat and move away from cereals.

Potential impacts go beyond influences on the food market, at least to the extent that overall food production increases and prices decline. In this case, the wider economy and society in general benefit as lower equilibrium food prices translate into lower costs and higher profitability in the rest of the economy, thereby freeing up resources spent on food and increasing demand for investment.

To measure changes in food security, an important indicator is the number of people facing the risk of hunger in the different regions of the world. In general, higher yield

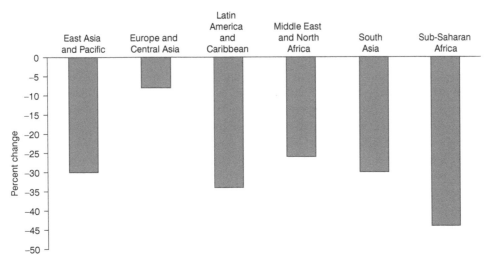

Figure 1.10 Percentage change in population at risk of hunger between BAU and the Green Growth Scenario, 2050 (source: IFPRI IMPACT Model, 2012 Simulations)

growth expands the food supply and pushes down prices, leading to an increase in food consumption and fewer people at the risk of hunger. Also, an increase in income results in consumers buying more food. Figure 1.10 illustrates the projected change in the population at the risk of hunger presented as the percent change between BAU and the Green Growth Scenario for 2050. In the Green Growth Scenario, the share of population at risk of hunger declines significantly for all the regions. Sub-Saharan Africa declines the most, with a 44 percent reduction in the population facing the risk of hunger, followed by Latin America and the Caribbean, South Asia, and East Asia and Pacific.

Potential for hydropower

Hydropower emerged as a major source of energy during the twentieth century. More than 45,000 large dams have been constructed around the world to generate electricity, irrigate crops to produce food, supply water to industry, and control floods. Many smaller dams also have been built to meet similar objectives.

Some argue that given today's rising demand for energy and food, along with environmental concerns associated with burning fossil fuels, there should be significantly more rapid expansion of hydropower, not only for clean energy, but as a source of development and income growth in developing countries and a valuable supply of irrigation water for food production. Such an approach can be controversial, as it directly and indirectly affects the lives of many people. As a result, it is necessary to recognize and address social elements while weighing global energy needs and environmental costs, which together can help policymakers work toward solutions that account for many competing interests. Also, while hydropower does not compete with food production for water as much as other energy sources such as biofuels, the offtake of water for power generation should be managed so as to minimize downstream costs and maximize downstream benefits from changes in stream flows.

Hydropower as an energy source

Hydropower is a major source of electricity, accounting for about one-fifth of the global electricity supply (World Bank 2009). Leading hydropower-producing countries include China, Canada, Brazil, United States, Russia, and Norway (International Energy Agency 2010). Hydropower accounts for more than 50 percent of national electricity in 65 countries and more than 80 percent in 32 countries. It supplies almost all electricity in 13 countries. In 2000, hydropower represented more than 90 percent of total renewable energy generated (International Hydropower Association 2000).

Hydropower has a number of attractive engineering characteristics (International Hydropower Association 2000). First, the technology for its generation is considered proven and well advanced, with more than a century of experience. Modern power plants provide one of the most energy efficient conversions at more than 90 percent. Importantly, because potential energy is stored as water in a reservoir, hydropower provides an option for energy storage (i.e., prior to generation), which optimizes electricity generation. A secondary benefit is flexible timing for energy generation. Moreover, power generation can be adjusted to meet demand instantaneously, with turbines spinning with zero load while synchronized with the electrical system before additional power is added. A fast response to peak demand enables hydropower to supplement less flexible (i.e., intermittent or less predictable) electrical power sources such as wind and solar energy.[2] In some cases, hydropower can be variable on a seasonal basis or annually due to drought or wet spells.

The initial investment for hydropower depends on the plant capacity. As summarized in Table 1.1, estimates from various sources indicate that the initial cost for large plants (10 MW or larger) averages about US$2 million–US$3 million per MW. Costs for small plants (1–10 MW) are approximately US$4 million–US$5 million per MW, although several estimates are above and below this range. Plants considered "mini" (0.1–1.0 MW) and "micro" (<0.1 MW) have substantially higher unit costs, ranging from an estimated US$5 million to US$500 million per MW. Micro and mini projects are typically used for communities and/or businesses. For example, a 5 kW micro hydropower station was built in south central China in 1992, originally to serve a village of 24 families (Dou 2011). These investment costs compare with approximately US$2 million per MW for a fossil fuel plant.

From an operational standpoint, compared with other large-scale power plants, hydropower has the lowest operating costs and longest plant life. Moreover, plant life can be extended at a relatively low cost through routine maintenance and periodic replacement of turbine parts and rewinding of generators. With proper maintenance, the life of a typical plant in service for 40 to 50 years can be effectively doubled.

Greenhouse gas emissions

Proponents of hydropower also point to significantly lower greenhouse gas (GHG) emissions compared with other energy sources.[3] According to research summarized in a report by the International Hydropower Association (2000), the GHG emissions factor is 30 to 60 times less for hydropower plants compared with fossil fuel generation. Also, development of half the world's economically feasible hydropower could reportedly reduce GHG emissions by about 13 percent, with an even more beneficial reduction in sulphur dioxide and nitrous oxide emissions (particulate emission for hydropower is

Table 1.1 Average costs of initial investment per unit of installed capacity and average costs per unit of electricity power generated

Energy source	Output (MW)	Investment cost (US$ Mil./MW)	Levelized capital cost (US$/MWh)	Operational and maintenance cost (US$/MWh)	Reference
Fossil fuel	–	2	–	–	Kosnik (2010)
Coal	–	–	65.3	28.2–42.3	U.S. Energy Information Administration (2010)
Natural gas	–	–	17.5–45.8	–	U.S. Energy Information Administration (2010)
Hydropower			74.5	10.1	U.S. Energy Information Administration (2010)
Micro	<0.1	500 (using median cost at 1 MW equivalent)	–	–	Kosnik (2010)
Mini	0.1–1.0	12 (using median cost at 1 MW)	–	–	Kosnik (2010)
	<0.5	10	–	–	Paish (2002)
	<1	5	–	1.5% to 2% of investment	Energy Technology Systems Analysis Programme (2010)
Small	Not specified	0.6–6	–	–	Domingo et al. (undated)
	<2	2.5–3.0	–	–	Paish (2002)
	1–10	4.5	–	1.5% to 2% of investment	Energy Technology Systems Analysis Programme (2010)
	<10	2–4	–	10–40	IEA (2010)
	1–30	5 (using median amount)	–	–	Kosnik (2010)
Large	>10	4	–	1.5% to 2% of investment	Energy Technology Systems Analysis Programme (2010)
	10–100	2–3	–	5–20	IEA (2010)
	100–300	2–3	–	5–20	IEA (2010)
	>300	<2	–	5–20	IEA (2010)
Large dam in Congo	196	1.2	–	–	Ministerial Conference on Water for Agriculture and Energy in Africa (2008)
Aswan high dam in Egypt	2,810	At least 0.4	–	–	Ministerial Conference on Water for Agriculture and Energy in Africa (2008)

Table 1.2 Greenhouse gas emission per unit of electricity generated

Energy source	*Emissions Pounds (CO_2/kWh)*	Emissions (kt eq. CO_2/TWh)	*Reference*
Coal	–	974	Ministerial Conference on Water for Agriculture and Energy in Africa (2008)
	–	941–1,022	Gagnon (2003)
	2.02–2.12	–	U.S. Energy Information Administration (2012)
Oil	1.57–1.70	–	U.S. Energy Information Administration (2012)
	–	841–1,177	Gagnon (2003)
Natural gas	1.12	–	U.S. Energy Information Administration (2012)
	–	551	Ministerial Conference on Water for Agriculture and Energy in Africa (2008)
	–	422–499	Gagnon (2003)
Hydropower	–	15 (with reservoir) 1 (run of river)	Ministerial Conference on Water for Agriculture and Energy in Africa (2008)
	–	10–33 (with reservoir) 3–4 (run of river)	Gagnon (2003)
	–	Negligible	Energy Technology Systems Analysis Programme (2010); Kosnik (2010)
	–	More than typical estimates	Middleton and Lawrence (undated) of the International Rivers Network

essentially zero). In cases of tropical reservoirs, the emissions factor is higher—a result of decomposition of flooded biomass—but still five times below the value for coal.

Hydro energy by itself does not emit GHG, and it has been a commonly-held belief that dam reservoirs do not emit any GHGs (Ministerial Conference on Water for Agriculture and Energy in Africa 2008). Although GHG emissions are emitted, mostly from the initially filling of reservoirs, total emissions relative to energy produced is quite small relative to conventional fuel sources (see Table 1.2).

In contrast, others such as the International Rivers Network claim that emissions from hydropower are underestimated, especially for tropical reservoirs due to an initial massive pulse released from the decay of vegetation submerged by the reservoir and subsequent flow of carbon into the reservoir in the form of sediment (Middleton and Lawrence undated). It is not clear if this accumulation of subsequent material (and eventual decay and carbon release) should be attributed to the dam.

Challenges for hydropower

Barriers to developing hydropower in developing countries include the technical and financial capacity to design, build, and manage hydropower projects and facilities. Without assistance, most communities or even larger government entities cannot undertake such projects, especially large-scale ones.

More broadly from a social standpoint, criticism of dams for hydropower and other purposes is well established. As World Wildlife Fund International (2007) points out, dam installations can force large-scale resettlement of human communities (totaling in the millions of persons) while flooding biodiversity hotspots and fertile lands. Dams can also seriously disrupt river systems and permanently alter or destroy their ecology by changing the volume, quality, and timing of water flows downstream, and by blocking the movement of wildlife, nutrients, and sediments. From a construction and financial perspective, compared with thermal plants, hydropower plants take more time to design, obtain approval, build, and recover investment.

Hydropower development can impact fish and fisheries. During construction and filling, the river habitat is lost, which is important for maintenance of fish resources. For hydropower facilities that use large reservoir storage, altering the natural river cycle adversely affects habitat availability and stability during periods of spawning and incubation. Also, the operation of long-term storage influences the river downstream from the reservoir and can adversely affect river productivity. Importantly, the report by the World Commission on Dams (2000) found that large dams have more negative than positive impacts on rivers, watersheds, and ecosystems, and in many cases their installation leads to irreversible loss of species and ecosystems. However, management of storage dams for hydropower generation can also have positive downstream benefits such as smoothing highly variable seasonal flows over the year, allowing additional irrigation cropping, providing flood control benefits during the wet season, and preventing salt water intrusion into rivers during the dry season (Ringler 2001).

Sedimentation is another problem with dams. Organic and chemical materials can be transported by a river and trapped in a reservoir rather than flushed out by the river system, potentially building up to undesirable levels. To reduce these effects, besides addressing the underlying problem of excess chemical application or runoff, conservation and agricultural practices in the catchment area can be altered to help reduce erosion prior to water entering the reservoir. Other water quality issues can also result from installing a dam. Water emerging from a dam outlet is often colder and, compared with water in the reservoir, contains altered levels of dissolved gases, mineral, and chemical content.

Social issues can be enormous. The report by the World Commission on Dams (2000) said that failure to consider and address resettlement issues and develop programs for the displaced communities led to the impoverishment and suffering of millions. The report identified a major discrepancy regarding equity in the distribution of benefits, and also concluded that it is not known if benefits of building large dams exceed costs when all issues are included. Even proponents of hydropower recognize that "Difficult ethical issues, such as ... ensuring rights of people and communities affected by a project are respected are also likely to arise" (International Hydropower Association 2000, p. 7). Relocating people and involuntary resettlement affect entire communities, local culture, and have significant implications for religion and burial sites.

Aside from the social issues, the report by the World Commission on Dams (2000) concluded that in general large dams tend to fall short of physical and economic targets. In contrast, it noted that large hydropower dams generally meet their financial targets, although performance can be highly variable.

Potential hydropower expansion

The World Bank (2009) has estimated an absolute level of feasible hydropower capacity in developing countries at more than 1,900 gigawatts, with 70 percent of the total yet to be tapped.[4] The untapped amount is not quite double the currently installed hydropower worldwide. Regionally, the unexploited potential is greatest in Africa (93 percent of potential), East Asia and the Pacific (82 percent), Middle East and North Africa (79 percent), South Asia (75 percent), and Latin American and the Caribbean (62 percent). Additional energy could be created by rehabilitating existing infrastructure. The World Bank report (2009) points out that developing Africa's hydropower to the same extent as Canada would result in an eight-fold increase in electricity supply. In the United States, the Department of Energy has identified nearly 6,000 sites with undeveloped hydropower potential representing about 40 percent of existing hydropower capacity (U.S. Department of Interior 2005).

The World Bank (2009) reports that its lending for hydropower has increased in recent years due in part to hydropower's role in a range of issues, including energy security, poverty alleviation, and sustainable development. The need for better water resource management—from the perspectives of energy, irrigation needs, human and industry consumption, and flood management—has elevated the importance of hydro infrastructure and contributed to a growing awareness that hydrology and economic growth are closely linked. In 2009, the World Bank reported that 67 hydropower projects had been approved since financial year 2003, with US$3.7 billion in contributions to support a total of US$8.5 billion and nearly 9,700 MW of capacity. New lending on an annual basis increased from US$250 million during 2002–2004, to US$500 million during 2005–2007, and to more than US$1 billion in 2008. Major projects were approved in Africa (Senegal, Democratic Republic of Congo, Sierra Leone, and Uganda) and Asia (People's Democratic Republic of Laos and India).

While funding by the World Bank remains significant, new capital from a number of countries, including China, Brazil, Thailand, and India, is also funding dam construction (Imhof and Lanza 2010; Eberhard et al. 2010). Chinese banks and companies are reportedly involved in constructing more than 200 large dams[5] in nearly 50 countries. Within China, hydropower capacity is scheduled to increase by 50 percent by 2015 in order to meet energy demands as well as reduce carbon dioxide emissions (Chinadaily.com 2010). Currently, coal accounts for more than 80 percent of China's electrical output.

Multipurpose dams and adding hydropower to existing dams

The development of multipurpose dams has potential for coincident expansion of hydropower and irrigation water supply. In Africa, water resources for hydropower and agricultural purposes remain comparatively underdeveloped, despite economically viable potential for both power generation and irrigated area. According to Rosegrant,

Ringler, and de Jong (2010), only 3.5 percent of Africa's agricultural land is equipped for irrigation, some 7 million hectares concentrated in a handful of countries. At least 1.4 million hectares could be economically developed using existing or planned dams associated with hydropower. An additional 5.4 million hectares would be viable for small-scale irrigation. Countries with the greatest potential for dam-associated large-scale investments include Ethiopia, Nigeria, Sudan, and Zimbabwe.

Similarly, adding hydropower to existing dams has the potential to increase electrical generation without incurring the cost of building new dams. For example, a potential project in Coimbra, Portugal, would integrate a small hydropower plant into an existing multipurpose dam-bridge to generate electricity for city buses and trolleys. Currently, the dam-bridge stretches across the Mondego River, creating a 1 meter drop in a "run-of-river" project (designed to not affect the natural river flow more than for daily storage). Storage is primarily for industrial and municipality supply, and flow is adjusted through spillway gates for ecological purposes and irrigation. A technical study has shown that installing "low head" turbines would provide more than enough energy for the city's transport system while addressing concerns for fish passing through the river (de Almeida et al. 2011).

Water resources are also underdeveloped in Indonesia, although the country has many irrigation weirs and dams. In many of these areas, many farmers and communities have little or no access to the electricity needed for economic development. Efforts are underway to apply hydropower to existing structures where it is technically and economically feasible (Andritz Hydro 2010).

Potential for small-scale hydropower

Given the challenges associated with large hydropower investments, including proper accounting of environmental and social costs, a number of researchers have identified a growing potential for small-scale hydropower. Kosnik (2010) points out that, while *average* investment costs for small-scale hydropower exceeds the competitive cost for fossil fuels (US$5 million per MW compared with US$2 million per MW, see Table 1.1), hundreds of small-scale hydropower sites could be developed for US$2 million or less. The implication is that while average costs remain high, considerable economic potential exists immediately where sites are favorable in the United States and elsewhere in the world.

Moreover, according to Paish (2002), small-scale hydro offers a number of advantages including a more concentrated energy resource than wind or solar power, predictable energy availability (usually continuous and available on demand), limited maintenance, long-lasting technology, and minimal environmental impact. Similarly, a hydropower resource assessment for Africa perceives substantial potential for small hydropower in the region due to low investment requirements, low environmental impacts, and viable technologies (Ministerial Conference on Water for Agriculture and Energy in Africa 2008). Importantly, small-scale hydropower is typically designed to run "in-river" rather than creating storage. This is considered more environmentally friendly because it does not interfere significantly with the flow of the river (Energy Technology Systems Analysis Programme 2010).

In Europe, small-scale hydropower is appealing because large-scale opportunities have already been exploited or are not considered since these are environmentally unacceptable (Paish 2002). Paish (2002) also indicates that small-scale hydropower appears to have similar potential in less developed countries, concluding that micro-hydro is one of the most cost-effective energy technologies to be considered for rural electrification. In China, small hydro is seen as an environmentally sound solution to improving economic growth in the country. This may be particularly true given the social and environmental impacts of large-scale dams in China over the years.

Others have pointed out that despite the fact that small-scale hydropower technology is well-developed and enormous potential exists in Africa, a relatively small number of units are in operation on the continent, suggesting that barriers other than technology persist (Klunne undated; Ministerial Conference on Water for Agriculture and Energy in Africa 2008). These barriers include lack of access to appropriate technologies; lack of infrastructure for manufacturing, installation, and operation; lack of local capacity to design and develop small hydropower schemes, including feasibility studies; and regulatory burden for small-scale operators. More generally, Klunne (undated) concludes that small-scale hydro projects need to be embedded in a national program for capacity building to foster a new industry. The Ministerial Conference paper recommends that regulations for approving projects need to be changed to accommodate small players.

Historically, the World Bank has been instrumental in developing hydropower. Its strategy is to support a range of hydropower investments, including small run-of-river projects, rehabilitations, and multipurpose projects. Another component of the lending strategy, which reflects the trend in government and private funding, focuses on energy and water planning at the county and regional level, forming partnerships with private financiers and others to leverage World Bank financing.

Reducing financial risk to investing in hydropower would speed its development. One way to do so is to promote smaller-scale projects that carry relatively less risk. For small-scale hydropower, there appears to be significant technical potential as well as economic potential, although a number of challenges outlined above must be addressed before the economic potential can be realized. More generally, hydropower is a potential solution as part of the policy mix that can help address water scarcity and food security. Additional strategies are discussed in the next section.

Strategies for addressing water scarcity and food security

Policy reforms and investments in agricultural productivity and water conservation could significantly offset impacts of water scarcity on the environment and risks to farmers. Policy reforms to bolster the water sector and improve water use efficiency are urgently required to obtain the outcomes in the Green Growth Scenario. This includes support for improved demand management, such as incentives and flexibility; the strengthening of public–private partnerships, particularly on irrigation infrastructure and water user associations; the establishment of economic incentives for efficient water use; and the increase of investment in water and irrigation. Investments in agricultural R&D need to focus on crop productivity improvements with respect to water, not just land.

Three strategies that can be used to address the challenge posed by this increasing water scarcity: (1) increasing the supply of water for irrigation, hydropower, domestic, and industrial purposes through investment in infrastructure; (2) conserving water and improving the efficiency of water use in existing systems through water management and policy reform; and (3) improving crop productivity per unit of water and land through integrated water management and agricultural research and policy efforts, including crop breeding and water management for rainfed and irrigated agriculture. The remainder of this chapter examines the potential of these three avenues for reducing water scarcity and improving food security.

Investment in infrastructure and water supply

Although the financial, environmental, and social costs are high for new water supply projects, the selective expansion of water supply capacities (including storage and withdrawal capacities) is still necessary in some regions, especially in developing countries. Storage and water distribution systems (such as water lift projects and canals) are particularly needed for Sub-Saharan Africa, some countries in South and Southeast Asia (e.g., India, Bangladesh, and Vietnam), and some countries in Latin America (Rosegrant, Cai, and Cline 2003). You et al. (2011) show that there is substantial physical and economic potential for expansion of both large-scale irrigation and small-scale irrigation in Africa.

For hydropower projects that address both energy and water needs, a significant conclusion from researchers and participants alike is that the decision-making process for large-scale projects should be broadened sufficiently to include all relevant environmental and social aspects, both positive and negative. The full social, economic, and environmental costs of development must be considered, but so must the costs of failure to develop new water sources. Project design that ensures comprehensive accounting of full costs and benefits includes not only irrigation benefits, but also health, household water use, and catchment improvement benefits. Importantly, analysis of new dam investments should include the downstream beneficiaries, not just those who will be displaced by construction.

A more inclusive approach, and continuous communications between developers and people affected, would help increase confidence in the legitimacy of the processes for decision-making and development. The reasonable approach outlined in the paper by the Ministerial Conference on Water for Agriculture and Energy in Africa (2008) is to exploit the resources while using an integrated approach to management of water resources in order to determine the optimal allocation of scarce water for competing sectors (hydropower, agriculture, industry, urban, etc.).

Thus, hard infrastructure investment has a role to play in the future in some regions, but it will be a reduced role compared to past trends, when large dam-building and expansion of irrigated area drove rapid increases in irrigated area and crop yields, particularly in developing countries. Instead, some of the increasing demand for water must be met from the carefully selected, economically efficient development of new water, both through impoundment of surface water and sustainable exploitation of groundwater resources, and through expansion in the development of non-traditional water sources. For example, small-scale hydropower for power generation and irrigation water appears to have significant potential because of its lower costs and other attributes described above. Future construction of large and small irrigation and water supply projects will require balanced development approaches that are acceptable to diverse constituencies.

Water management and policy reform

Improvements in basin-scale irrigation water use efficiency are needed to compensate for reduced irrigation growth resulting from: (1) transferring irrigation water use to household and industrial uses; (2) reducing groundwater overdraft worldwide; (3) increasing committed environmental flows; (4) higher prices for agricultural water use; and (5) slowing irrigated area development. The Green Growth Scenario showed the importance of water use efficiency gains and crop water productivity gains in reducing water scarcity and increasing food production. However, in some individual water scarce basins, there is little space for water use efficiency improvement when water is already near maximum usage. In these basins, compensation for the agricultural impacts of growing water scarcity will require alternative interventions, including more rapid crop yield growth from agricultural investments, diversification into less water-intensive crops, or broader economic diversification that reduces the role of agriculture over time (Rosegrant, Cai, and Cline, 2002).

The institutional, technical, and financial feasibility of big improvements in river basin efficiency in specific river basins requires site-specific research and analysis. Improvement of basin efficiency depends on both technological improvements in irrigation systems, domestic and industrial water use and recycling systems, and institutional settings related to water allocation, water rights, and water quality (Cai, Ringler, and Rosegrant 2001).

Improvements in the irrigation sector can be made at the technical, managerial, and institutional levels. Technical improvements can include advanced irrigation systems such as drip irrigation, sprinklers, conjunctive use of surface and groundwater, and precision agriculture, including computer monitoring of crop water demand. Managerial improvements can include the adoption of demand-based irrigation scheduling systems and improved maintenance of equipment. Institutional improvements may involve the establishment of effective water user associations, the establishment of water rights, the introduction of water pricing, and improvements in the legal environment for water allocation.

Key to inducing higher water efficiency gains is the introduction of market or market-type incentives into water use decisions. Incentive prices for water can have a major impact on water withdrawals and consumptive use in irrigation and urban water uses, thus freeing up water for the environment and other uses (Rosegrant and Binswanger 1994). Rosegrant, Cai, and Cline (2002) show in an IMPACT scenario that water markets result in reduced water consumption, improved efficiency, and higher crop production.

An important step in incentivizing water allocation is the establishment of water rights. Although some system of water rights is found to operate in virtually any setting where water is scarce, systems that are not firmly grounded in formal or statutory law are likely to be more vulnerable to expropriation. A system of marketable rights to water induces users to consider the full opportunity cost of water, including its value in alternative uses, thus providing incentives to economize on the use of water and gain additional income through the sale of saved water. The promise of efficient water use and (re)allocation of water resources without harming the welfare of irrigators and other rural users make the establishment of water rights one of the highest priorities for water reform. The establishment of base water rights would increase the political feasibility of water pricing by formalizing existing water rights, rather than being perceived as an expropriation of these rights (Rosegrant and Binswanger 1994; Rosegrant, Cai, and Cline 2002; Rosegrant, Ringler, and Zhu 2009).

Reform of water policies to increase water prices is politically difficult and can have negative impacts on poor consumers and farmers if badly designed and implemented. But in the domestic and industrial sectors, it is feasible to improve both efficiency and equity by increasing water prices to provide incentives for conservation, to cover the costs of delivery, and to generate adequate revenues to finance growth in supplies and coverage of clean piped water.

To minimize impacts on the poor, broad subsidies should be replaced with targeted subsidies for the poor. Other policies, such as block pricing, can help to ensure water availability to low-income users without a direct subsidy. This type of tariff structure is designed with a very low per-unit price for water up to a specified volume, after which users pay a higher price for different volumetric blocks up to the highest level of consumption. This system requires high-income households that use more water to cross-subsidize low-income users so they are able to purchase enough water to meet their essential needs (Rosegrant, Cai, and Cline 2002).

Over time, high irrigation water prices are likely to severely reduce farm income (Löfgren 1996; Rosegrant et al. 2000; Perry 2001; Ringler 2005). Moreover, with irrigation in much of the developing world consisting of large systems serving many small farmers, measurement and monitoring of deliveries to a large numbers of end users as would be required for volumetric charges is too costly. Economic analysis needs to focus on detailed assessment of the value and operational design of markets in tradable water rights in developing countries.

Alternative market-like pricing systems should also be explored (Rosegrant, Ringler, and Rodgers 2005). Water brokerage systems could be a model for developing practical water pricing/market systems based on water rights, which would help to introduce incentives for efficient water use, and recover operating costs, while protecting farm incomes, and minimizing transaction costs. A water brokerage scheme can achieve efficient outcomes and appears to be politically and administratively feasible. In this scheme, a base water right would be established at major turnouts to water user groups or privately run irrigation subunits (rights could be assigned directly to individual irrigators where administratively feasible). The user group would subsequently be responsible for internal water allocation. The base water right would be set based on historical allocation, but could be somewhat lower than the historical allocation in water-scarce basins to encourage water conservation. A fixed base charge would be applied to this quantity, sufficient to cover operating and maintenance costs and longer term asset replacement (depreciation) costs. For demand greater than the base water right, an efficiency price equal to the value of water in alternative uses would be charged to the user; for demand below the base right, the same price would be paid to the water user (Rosegrant, Ringler, and Rodgers 2005; Rosegrant, Ringler, and Zhu 2009).

The establishment of base water rights would increase the political feasibility of water pricing by formalizing existing water rights, rather than being perceived as an expropriation of these rights. With efficiency prices paid on only the marginal demand above or below the base right, non-punitive incentives are introduced. The reliance on water user associations to manage water "below the turnout" improves local accountability, transparency, and the flexibility of water allocation. Information costs would be reduced, because local irrigators with expert knowledge of the value of water would bear the costs and generate the necessary information on the value and opportunity costs of water below major turnouts (Rosegrant, Ringler, and Rodgers 2005; Rosegrant, Ringler, and Zhu 2009).

Crop productivity in irrigated and rainfed agriculture

The need to expand crop production in a world with finite resources, particularly water, quickly leads to a focus on increasing productivity. One potential area for improved water use is through the development of biotechnology for agriculture. Crop biotechnology can be an important substitute (and/or complement) for direct water management and provide solutions to reducing water scarcity in agriculture.

Given the increasing demands on global agriculture, it is likely that biotechnology, in addition to conventional breeding, will be needed to increase genetic diversity to achieve some of the necessary breakthroughs. Ultimately, a combination of yield improvement output per unit of water—plus improved farming systems to better use water or help retain moisture (e.g., minimum tillage)—needs to contribute greatly to increased global crop output.

The use of biotechnology has the potential to address crop production and water scarcity in several ways. These include: (1) improving the productivity and water use efficiency in irrigated (and rainfed) crops, and (2) developing drought tolerance in rainfed crops.

Morison et al. (2008) point out that advances in genetics and the molecular sciences (and technology, generally) can now help scientists exploit a new understanding of water stress and plant response, resulting in a more targeted selection program where biotechnology is combined with conventional breeding to increase biomass water use efficiency. Variations in complex traits that increase water use efficiency are the basis for crop breeding programs, and identifying this variation can lead to the association of the trait with particular regions of the plant's genome. The combination of particular deoxyribonucleic acid sequences or markers and the associated trait can be used for marker-assisted selection, which improves the efficiency of breeding programs where complex traits are involved. The process avoids costly physiological tests on all the material, and pre-selection of progeny also saves time and expense.

Many scientists, while acknowledging the challenges, expect that progress for yield gains without requiring commensurate increases in water use is far from over (Richards et al. 1993; Richards et al. 2002; Ortiz et al. 2007). Breeding can influence biomass/unit of water through transpiration rates and efficiency of biomass per unit of transpiration. They conclude that, because it is a difficult challenge to breed for these three factors, the use of biotechnology and marker-assisted selection is a necessity for significant progress in the longer term.

Christensen and Feldmann (2007) point out that a significant accomplishment in biotechnology has been the sequencing of several plant genomes. This has greatly increased the number of genes being evaluated for conferring stress tolerance, leading to over 50 genes reported to confer drought tolerance. They consider this identification as boding well for the prospects of developing genetically engineered drought-tolerant crops, but recommend that research focus explicitly on crop productivity in the field rather than on desiccation (drying) recovery or other efforts only in the laboratory. The challenge for crop improvement is driven by the tradeoff between drought tolerance and plant productivity. Therefore, identifying the genes that affect plant water relations is useful, but only if it results in the kind of drought tolerance that is valuable for agriculture productivity (Christensen and Feldmann 2007).

Rainfed agriculture is a potential key to sustainable development of water and food. Rainfed agriculture still produces about 60 percent of total cereals. Improved water management and crop productivity in rainfed areas would relieve considerable pressure on irrigated agriculture and on water resources. However, this will require increased investment in research and technology transfer for rainfed areas.

The rate of investment in crop breeding targeted to rainfed environments is crucial to future crop yield growth. Strong progress has been made in breeding for enhanced crop yields in rainfed areas, even in the more marginal rainfed environments. The continued application of conventional breeding and the recent developments in non-conventional breeding offers considerable potential for improving cereal yield growth in rainfed environments. Cereal yield growth in rainfed areas could be further improved by extending research both downstream to farmers and upstream to the use of tools derived from biotechnology to assist conventional breeding, and, if concerns over risks can be solved, from the use of transgenic breeding.

Higher priority for agricultural extension services and access to markets, credit, and input supplies should be given in rainfed areas, since successful development of rainfed areas is likely to be more complex than in high-potential irrigated areas because of their relative lack of access to infrastructure and markets, and their more difficult and variable agroclimatic environments. Progress may also be slower than in the early green revolution because new approaches will need to be developed for specific environments and tried on a small-scale before being disseminated more widely. Investment in rainfed areas, policy reform, and transfer of technology such as water harvesting will therefore require stronger partnerships between agricultural researchers and other agents of change, including local organizations, farmers, community leaders, non-governmental organizations, national policymakers, and donors.

Conclusions

A large part of the world is facing severe water scarcity. With a continued worsening of water supply and demand trends, water scarcity will pose significant challenges to food production, health, nutrition, and the environment. But the water crisis has solutions, such as increasing the supply of water for irrigation, domestic, and industrial purposes through highly selective investments in infrastructure.

Even greater impacts could be generated through water conservation and water use efficiency improvements through water management and policy reform, and improved crop productivity per unit of water and land through integrated efforts in water management and agricultural research and policy. An emphasis on including crop breeding and water management in rainfed agriculture would be particularly helpful. Also, given its potential to simultaneously address water food security, energy, and environmental issues, a renewed consideration of hydropower would appear to be worthwhile.

The optimal mix of water policy and management reform and investments, as well as the feasible institutional arrangements and policy instruments to be utilized, must be tailored to specific countries and basins. The policy approach will vary across underlying conditions and regions, including levels of development, agroclimatic conditions, relative water scarcity, level of agricultural intensification, and degree of competition for water. While specific solutions are not necessarily easy to determine and to implement, requiring time, political commitment, and money, the policy changes and investments could measurably improve global food security and reduce the number of persons at risk of hunger.

Notes

1 For details of IMPACT, please refer to Rosegrant, M.W. and the IMPACT Development Team. 2012. International Model for Policy Analysis of Agricultural Commodities and Trade (IMPACT): Model Description. International Food Policy Research Institute (IFPRI), Washington, DC. http://www.ifpri.org/book-751/ourwork/program/impact-model
2 An additional engineering feature of hydropower includes the "black start" or ability to start generation without an outside source of power. This capability allows multi-sourced systems with hydropower to restore service more rapidly than those dependent only on thermal sources.
3 GHGs include carbon dioxide from combustion and methane from processing coal and natural gas.
4 One gigawatt (GW) equals 1,000 megawatts (MW). According to Hadjerioua et al. (2012), on an annual basis, 1 MW of hydropower produces enough electricity to power nearly 400 U.S. homes. Each gigawatt could power up to 400,000 homes.
5 At least 15 meters high, or between 5 and 15 meters with reservoir capacity of at least 3 million cubic meters.

References

Andritz Hydro. (2010) "Hydroelectric Power Generation at Multi-purpose Irrigation Dams. Paper for 61st IECM and 6th ARC of ICID conference Improvement of Irrigation and Drainage Efficiency through Participatory Irrigation Development". October 14–16, 2010. Yogjakarta, Indonesia. http://atl.g.andritz.com/c/com2011/00/01/97/19757/1/1/0/-794227257/hm-en-icid2010-jakarta-hydroelectric-power-generation.pdf

Cai, X., Ringler, C. and Rosegrant, M. W. (2001) "Does Efficient Water Management Matter? Physical and Economic Efficiency of Water Use in the River Basin", EPTD Discussion Paper 72. Washington, D.C.: International Food Policy Research Institute.

Chinadaily.com. (2010) "China to Lift Installed Hydropower Capacity by 50%." October 25. http://www.chinadaily.com.cn/bizchina/2010-08/25/content_11202593.htm

Christensen, C. and Feldmann, K. (2007) "Biotechnology Approaches to Engineering Drought Tolerant Crops", in M. Jenks, P. Hasegawa, and S. Jain (eds) *Advances in Molecular Breeding Toward Drought and Salt Tolerant Crops*, Springer, Dordrecht, The Netherlands. pp. 333–357.

de Almeida, A.C. et al. (2011) "Small-hydropower Integration in a Multi-Purpose Dam-Bridge for Sustainable Urban Mobility", *Renewable and Sustainable Energy Reviews*, vol 15, no 9, pp. 5092–5103.

Domingo, N., Ferraris, F. and del Mundo, R. (Undated) "Overview of Mini and Small Hydropower in Southeast Asia", University of the Philippines Solar Laboratory, Quezon City, Philippines. http://www.ec-asean-greenippnetwork.net/documents/tobedownloaded/knowledgemaps/KM_overview_small_hydro_SEA.pdf

Dou, C. (2011) "Self-support Development of Micro-hydro Power for Village Community. Case Study for Growing Inclusive Markets", United Nations Development Programme, New York.

Eberhard, A.A., Foster, V., Briceno-Garmendia, V. and Shkaratan, M. (2010) "Power: Catching Up. In Africa's Infrastructure – A Time for Transformation", edited by V. Foster and C. Briceño-Garmendia. Washington, DC: World Bank. www.infrastructureafrica.org/system/files/WB147_AIATT_CH08.pdf

Energy Technology Systems Analysis Programme (2010) "Hydropower", Technology Brief E12. www.etsap.org

Gagnon, L. (2003) "Comparing Power Generation Options – Greenhouse Gas Emissions. Factsheet", Hydro-Quebec. http://www.hydroquebec.com/sustainable-development/documentation/pdf/options_energetiques/pop_01_06.pdf

Hadjerioua, B., Wei, Y. and Kao, S. (2012) "An Assessment of Energy Potential at Non-Powered Dams in the United States", U.S. Department of Energy. http://www1.eere.energy.gov/water/pdfs/npd_report.pdf

Imhof, A. and Lanza, G. (2010) "Greenwashing Hydropower", *World Watch Magazine*, vol 23, no 1. http://www.worldwatch.org/node/6344

IEA (International Energy Agency) (2010) "Renewable Energy Essentials: Hydropower". www.iea.org.

International Hydropower Association. (2000) "Hydropower and the World's Energy Future", http://www.ieahydro.org/reports/Hydrofut.pdf

Klunne, W. (Undated) "Sustainable Implementation of Microhydro to Eradicate Poverty in Africa". http://www.worldenergy.org/documents/congresspapers/330.pdf

Kosnik, L. (2010) "The Potential for Small Scale Hydropower Development in the US", *Energy Policy*, vol 38, pp. 5512–5519.

Löfgren, H. (1996) "Cost of Managing with Less: Cutting Water Subsidies and Supplies in Egypt's Agriculture". TMD Discussion Paper No. 7, International Food Policy Research Institute, Washington, DC.

Malcom, S., Marshall, E., Aillery, M., Heisey, P., Livingston, M. and Day-Rubenstein, K. (2012) "Agricultural Adaptation to a Changing Climate Economic and Environmental Implications Vary by U.S. Region", Economic Research Report no 136, Economic Research Service, U.S. Department of Agriculture, Washington, DC. http://www.ers.usda.gov/media/848748/err136.pdf

Middleton, C. and Lawrence, S. (Undated) "ADB Support for Large Hydro Undermines Good Energy Rhetoric". NGO Forum on ADB. http://www.forum-adb.org/BACKUP/Articles/200708-Bankwatch_3.htm

Millennium Ecosystem Assessment. (2005) "Ecosystems and Human Well-Being: General Synthesis", Island Press. Washington, DC.

Milly, P.C.D., Betancourt, J., Falkenmark, M., Hirsch, R.M., Kundzewicz, Z.W., Lettenmaier, D.P. and Stouffer, R.J. (2008) "Stationarity Is Dead: Whither Water Management?", *Science*, vol 319 no 5863, pp. 573–574.

Milly, P.C.D., Dunne, K.A. and Vecchia, A.V. (2005) "Global Pattern of Trends in Streamflow and Water Availability in a Changing Climate", *Nature*, vol 438, no 17, pp. 347–350.

Ministerial Conference on Water for Agriculture and Energy in Africa: The Challenges of Climate Change. (2008) "Hydropower Resource Assessment of Africa", December 15–17. Sirte Libyan Arab Jamahiriya.

Morison, J., Baker, N.R., Mullineaux, P.M. and Davies, W.J. (2008) "Improving Water Use in Crop Production", *Philosophical Transactions of the Royal Society*, vol 363, pp. 639–658. http://rstb.royalsocietypublishing.org/content/363/1491/639.full.pdf+html.

Ortiz, R. M., Iwanaga, M., Reynolds, H., Wu, A. and Crouch, J. (2007) "Overview on Crop Genetic Engineering for Drought-Prone Environments." *Journal of SAT Agricultural Research*. http://www.icrisat.org/journal/SpecialProject/sp3.pdf.

Paish, O. (2002) "Micro-hydropower: Status and Prospects", *J Power and Energy*, vol 216. Part A, pp. 31–40.

Perry, C. (2001) "Water at Any Price? Issues and Options in Charging for Irrigation Water", *Irrigation and Drain*, vol 50, pp. 1–7.

Richards, R.A., Rebetzke, G.J,. Condon, A.G., and van Herwaarden, A.F. (2002) "'Breeding Opportunities for Increasing the Efficiency of Water Use and Crop Yield in Temperate Cereals', *Crop Science*, vol 42, pp. 111–121.

Richards, R.A., López-Castañeda, C., Gomez-Macpherson, H., and Condon, A.G. (1993) "Improving the Efficiency of Water Use by Plant Breeding and Molecular Biology', *Irrigation Science*, vol 14, pp. 93–104.

Ringler, C. (2005) "The Role of Economic Incentives for the Optimal Allocation and Use of Water Resources – Case Study of the Dong Nai River Basin in Vietnam", in P. Michael Schmitz (ed) *Water and Sustainable Development*. Frankfurt: Peter Lang GmbH. pp. 61–92.

Ringler, C. (2001) "Optimal Water Allocation in the Mekong River Basin", ZEF Discussion Paper on Development Policy No. 38. Bonn: ZEF

Ringler, C., Zhu, T., Gruber, S., Treguer, R., Laurent, A., Addams, L., Cenacchi, N. and Sulser, T. (2015) "Role of Water Security for Economic Development – Concepts and global Scenarios. In C. Pahl-Wostl, J. Gupta and A. Bhaduri (eds) *The Handbook of Water Security*. Aldershot: Edward Elgar

Rosegrant, M., Ringler, C., and Zhu, T. (2014) "Water Markets as an adaptive Response to Climate Change". In K. William Easter and Qiuqiong Huang (eds) *Water Markets for the 21st Century: What We Have Learned*. Nerw York: Springer.

Rosegrant, M., Ringler, C., Zhu, T., Tokgoz, S., and Bhandary, P. (2013) "Water and Food in the Bioeconomy: Challenges and Opportunities for Development". *Agricultural Economics* vol. 44 (Supplement) pp. 139–150.

Rosegrant, M., Ringler, W.C. and de Jong, I. (2010) "Irrigation: Tapping Potential". In V. Foster and C. Briceño-Garmendia (eds), *Africa's Infrastructure: A Time for Transformation*, Washington, D.C.: The International Bank for Reconstruction and Development/The World Bank. http://www.infrastructureafrica.org/aicd/system/files/WB147_AIATT_Consolidated_rx8.pdf

Rosegrant, M.W., Ringler, C. and Zhu, T. (2009) "Water for Agriculture: Maintaining Food Security under Growing Scarcity", *Annual Review of Environmental Resources*, vol 34, pp. 205–222. http://arjournals.annualreviews.org/eprint/T6e4KXUcGtcSNwJxd6pE/full/10.1146/annurev.environ.030308.090351

Rosegrant, M.W., Ringler, C. and Rodgers, C. (2005) "The Water Brokerage Mechanism – Efficient Solution for the Irrigation Sector", Conference Proceedings, XII World Water Congress "Water for Sustainable Development – Towards Innovative Solutions," New Delhi, India, November 22–25, 2005.

Rosegrant, M.W. , Cai, X. and Cline, S.A. (2003) "Will the World Run Dry? Global Water and Food Security", *Environment*, vol 45, no 7, pp. 24–36.

Rosegrant, M.W. , Cai, X. and Cline, S.A. (2002) "World Water and Food for 2025: Dealing with Scarcity", International Food Policy Research Institute, Washington, DC.

Rosegrant, M.W., Ringler, C., McKinney, D.C., Cai, X., Keller, A. and Donoso, G. (2000) "Integrated Economic-Hydrologic Water Modelling at the Basin Scale: The Maipo River Basin", *Agricultural Economics*, vol 24, no 1, pp. 33–46.

Rosegrant, M.W. and Binswanger, H.P. (1994) "Markets in Tradable Water Rights: Potential for Efficiency Gains in Developing Country Water Resource Allocation", *World Development*, vol 22, no 11, pp. 1613–1625.

UN (United Nations) (2011) *World Population Prospects: The 2010 Revision*. New York: United Nations.

U.S. Department of Interior. (2005) "Reclamation – Managing Water in the West – Hydroelectric Power". http://www.usbr.gov/power/edu/pamphlet.pdf

U.S. Energy Information Administration (2010) "Levelized Cost of New Generation Resources in the Annual Energy Outlook 2011". http://205.254.135.7/oiaf/aeo/electricity_generation.html

U.S. Energy Information Administration, Office of Integrated Analysis and Forecasting, Voluntary Reporting of Greenhouse Gases Program (2012). Table of Fuel and Energy Source: Codes and Emission Coefficients available at: http://www.eia.doe.gov/oiaf/1605/coefficients.html

Wada, Y., van Beek, L.P.H. and Bierkens, M.F.P. (2011) "Modelling Global Water Stress of the Recent Past: On the Relative Importance of Trends in Water Demand and Climate Variability", *Hydrology and Earth System Sciences*, vol 15, no 12, pp. 3785–3808.

World Bank. (2010) "World Development Report Chapter 3. Managing Land and Water to Feed Nine Billion People and Protect Natural Systems", in World Development Report 2010:

Development and Climate Change. World Bank, Washington, DC. http://siteresources. worldbank.org/INTWDR2010/Resources/5287678-1226014527953/Chapter-3.pdf

World Bank. (2009) "Directions in Hydropower", Washington, DC. http://siteresources. worldbank.org/INTWAT/Resources/Directions_in_Hydropower_FINAL.pdf

World Commission on Dams. (2000) "Dams and Development-A New Framework for Decision-Making, UK and USA". http://www.sdnbd.org/sdi/issues/water/WCD%20report.htm

WWF International. (2007) "Climate Solutions – WWF's Vision for 2050", Gland, Switzerland. http://www.worldwildlife.org/climate/Publications/WWFBinaryitem4911.pdf.

You, L., Ringler, C., Wood-Sichra, U., Robertson, R.D., Wood, S., Zhu, T., Nelson, G., Guo, Z. and Sun, Y. (2011). "What is the Irrigation Potential for Africa? A Combined Biophysical and Socioeconomic Approach", *Food Policy*, vol 36, pp. 770–782.

2

WATER SCARCITY AND THE DEMAND FOR WATER MARKETS

Richard Howitt

Water markets are the most efficient and equitable mechanisms for reallocating water to its highest and best use, and thus minimizing scarcity value. By definition, market trades can only occur when there are significant differences in the scarcity value of the resource between alternative sites and uses of water, and where there is the ability to transfer water physically from one location to the other. Another major driver of water markets is an increase in scarcity value of water that makes past efficient allocations now redundant and undervalued compared with its alternative value elsewhere. In this chapter, we analyze factors driving the change in water scarcity in terms of shifts in the physical and technical supply of water, and changes in demand for water driven by population growth, industrialization, and the expansion of irrigated agriculture. Not all the news is bad, despite the small industry that regularly publishes straight-line extrapolations of population and industrial growth to project an upcoming water Armageddon. We will spend some time examining the evidence of the existence of a water Kuznets curve which has the familiar inverted U-shaped curve and where the income elasticity of demand for water is shown to change in sign and magnitude as economic development increases aggregate income. The existence of a Kuznets curve for water holds out the possibility of long-term stability between demand and supply of water. However, the rapid rate of industrialization in many countries and high population growth in some parts of the world will still mean that the macro water economy will be facing significant growth in the medium run with consequent increasing scarcity.

Changing water supply scarcity

Many commentators have said correctly that the quantity of water in the world is finite, and thus water supply is just a question of the correct reallocation. Unfortunately, three trends seriously reduce the available physical supply of water in many locations.

The first effect is global climate change. While several countries in the northern hemisphere are blithely ignoring the drivers of climate change, other countries in the southern hemisphere, in particular Australia and Chile, are already seeing significant changes in mean precipitation levels. The most striking example is in Perth, Western Australia, where in hindsight one can see that there was a discrete shift in the streamflow distribution in 1975 that has reduced aggregate supplies by 45 percent. The Western Australian case has several lessons. First, a major discrete shift in precipitation and run-off can occur suddenly. Second, we can only tell whether a discrete shift has taken place some years after it has occurred, and third, despite the difficulty of establishing causal effect, this shift in Western Australia streamflow coincides with the sharp reduction of the ice sheet and ozone layer in the Antarctic. Thus we cannot be too sanguine when assuming that global climate change will only change the distribution of precipitation and not the annual average quantity.

Chile is also experiencing lower than average precipitation levels that may be a forerunner of the southern hemisphere shifts that have occurred in Western Australia. What is not in question, is that the increasing ambient temperatures experienced over the last two decades, which are predicted to grow with climate change, will significantly reduce the natural ability to capture and distribute water in areas where snowpack and glacial run-off forms an important component of the supplies. Many of the major irrigation systems use snow and ice storage from nearby mountain ranges as an essential part of the water catchment and storage system. Such situations are dominant in the western United States, Chile, Central Asia and China. Not only is the storage capacity of mountain systems reduced in a nonlinear manner in which the greatest proportional losses occur at a lower temperature changes, but the earlier melt and increased run-off will mean that the existing dams will have to be operated at a lower storage capacity to provide the same level of flood protection. This combination of higher peaking flows and lower effective reservoir capacity will seriously change the effect of water supply. For example, in California the effective water supply is expected to be reduced by about 27 percent by 2100 if the flood risk is maintained at approximately the same level.

The second effect is that the current levels of water use, in irrigated agriculture in particular, are in many places sustained by overdrafting groundwater supplies. In a recent study Gleeson et al. (2012) used a worldwide hydrologic model to estimate that almost a quarter of the world's population, 1.7 billion people, live in regions where groundwater is being overdrafted and used in an unsustainable manner. Gleeson et al. show the regions in which areas that are important for farming also have increasing stress on groundwater aquifers.

In addition to the stress on the aquifers, many of the regions where overdrafting occurs are in the Middle East and North Africa, which are the most water stressed parts of the world with less than 1 percent of the global renewable fresh water resources. At 226 billion cubic meters this region has only 673 cubic meters of water per person per year, which is the lowest among developing regions. In contrast, Latin America and the Caribbean with 32 percent of fresh water resources has 22,800 cubic meters per person, Europe and Central Asia with 12 percent of water resources has 12,500 cubic meters per person, and East Asia and the Pacific with 21 percent of the water resources has 4400 cubic meters per person.

A third source of reduced supplies for both urban and agricultural use is due to the salinization and contamination of groundwater aquifers. Salinization is the oldest problem facing irrigated agriculture and has been responsible for removing several established irrigation systems. Like groundwater overdraft, salinization is hard to quantify on a broad scale but Pitman and Lauchli (2002) estimate that at least 20 percent of the 227 million hectares of irrigated land in the world suffers from reduced yields due to secondary salinization. Ghassemi et al. (1995) estimate the cost in 1995 dollar terms of at least US$12 billion. In addition to the costs of salinization, contamination by heavy metals and pesticides may be a more serious long-term threat to groundwater. The contaminants move slowly through the aquifer and are essentially irreversible once established. Other pollutants such as nitrogen and salinity can be treated if the groundwater is pumped and used on crops which are tolerant to salts and heavy metals and enable some removal of these contaminants by the crop harvest. There are significant technological advances from plant breeders who are on the cusp of developing salt tolerant perennial wheat and rye varieties. If such plants are developed for commercial production, they will have a significant potential for stabilizing salinity contaminated areas. However, such is the toxic level of heavy metals that one cannot imagine any crop in which they would be tolerated.

Demand driven water scarcity

Over the past 20 years there have been many studies warning of the future collapse of local and global water systems, for example Gleick (2010) and Postel (1999). While these warnings should be heeded, they are in some cases overblown in the long run due to the approach in which the authors generally use mechanistic projections of population growth and industrial development and multiply them by the current water requirements per unit of production.

In fact, the demand for water both in terms of irrigation and industrial uses changes as the per capita income changes. The concept of a water Kuznets curve offers some hope for eventual stabilization of the world's water system.

The major drivers of increased demand for water and its consequent value and scarcity value are fundamentally population growth and the change in diet that results from increased gross national product (GNP) per capita, industrialization, and increased environmental water demands. We will address these three sources of demand growth before reviewing empirical studies of the water Kuznets curve.

Population growth slows down with increased economic development, but growth is projected by the United Nations (Food and Agriculture Organization of the United Nations (FAO) 2012) to continue from the current level of 7 billion to at least 10 billion before it stabilizes. This growth in demand for food is considerably exacerbated by increased income due to shifts in the preferred diet towards animal protein. Animal protein requires substantially more water per unit calorific value than food generated directly from plant protein. For example, using average values, it takes 15 m^3 of water to produce 1 kg of beef and 6 m^3 per kg for poultry production. In contrast cereals take 1.5 m^3 to produce a kilogram of grain, and citrus fruit one cubic meter per kilogram.

In addition, most of the growth in population will take place in countries in Asia and Africa which are in the tropical climate belt, and already rely substantially on irrigation for the increase in food production that was realized over the last 50 years. Irrigated lands account for as much as 80 percent of the food production in Pakistan, some 70 percent in China, and over 50 percent in India and Indonesia. Africa, where only 10 percent of food is produced by irrigated production, has been estimated by the FAO to have only developed 30 percent of its potential irrigated area. A World Bank/United Nations Development Programme (UNDP) study estimates that irrigation could be extended to an additional 110 million hectares in developing countries producing enough grain to feed 1.5 to 2 billion people. While only 16 percent of the world's croplands are currently irrigated, these lands yield some 36 percent of the global food production. While it is possible to move virtual water throughout the world in terms of trade in food products, many countries are concerned about food security and insist on growing a substantial portion of their own supplies, usually at significantly greater cost than their availability on the international market. The recent panic in the rice market and the resulting embargoes on exports gave strong support to this risk-averse and cautionary approach to supplying stable food supplies from resources controlled by your own country.

Industrialization

While two thirds or more of the world's water supplies are used for irrigated production, industrial use of water has grown rapidly over the past 40 years and now comprises 20 percent of the total water use. This quantity includes water use for power generation, industrial processes, and thermal power cooling. A United Nations study predicts that the annual volume used the industry will rise from 750 cubic km in 1995 to 1172 cubic km in 2025. By this time industrial water use will comprise 24 percent of world's water use. In low income countries industry accounts for about 5 percent of the water withdrawals compared to 86 percent in some high income countries such as Germany. What is not clear is whether developing countries can establish their industrial base with the lower water use per unit production that has been seen in developed countries' industrial water use.

In addition, municipal water use, which used to be a small proportion of the total, is now growing rapidly in some areas with expanding urban centers, often located in arid environments such as the Middle East.

Environmental water demands

As GNP per capita in a national economy initially increases, there is a greater shift of water demands to the industrial sector. This shift in demand continues as the economy grows, with a greater service sector and a smaller emphasis on primary production and heavy industry. A new component in the demand for water in mature economies is the demand for water to provide environmental goods. This can be viewed as a growing amenity demand by workers in the service economy. In California, market purchases of environmental water have grown from 80,000 acre feet in 1988 to approximately 400,000 acre feet in 2009. At 25 percent of the flows purchased, the environmental sector now purchases more water than the agricultural sector, which buys 22 percent of flows (Hanak et al. 2011). The environmental demand for water has many of the characteristics of public goods, namely non-exclusion, a very low, or zero, marginal cost of consumption, and the ability to supply collective goods. As a result, much of the environmental demand is provided by the public sector. However, since this is a newly emerging demand, and in many countries property rights for environmental flow water are indistinct or sometimes nonexistent, additional environmental supplies are often provided through market mechanisms. From an aggregate viewpoint, commodity crops have a very low income elasticity of demand compared to most environmental goods that have a high income elasticity of demand.

Environmental demands for water also have a quality component. For example, supplies used to augment river flows for anadromous fish habitat have to be supplied at the right time of year, and at the right temperature, as well as being free of most pollutants. These environmental demands have given rise to new terms such as temperature pollution, in which the flows in the water course are the same, but the increase in temperature due to return flows from irrigation and industrial use make the streams unusable for most species of fish.

The water Kuznets curve

An implicit assumption in many of the pessimistic studies of future water scarcity is that the demand for water grows in a constant proportion to population and industrial development. This approach is a modern version of the Malthusian principal in which nonlinear population growth conflicts with finite resource capacities. The potential for demand modification and technological change in the use of water as income in the form of the GNP per capita is assumed to be unchanged. In economic parlance, it assumes that the income elasticity of demand for water is one or constant.

The demand for water can be thought of as going through several stages during economic development for the given country. In the first stage, the move is from subsistence diet to an improved diet and more water intensive products such as fruit and meat. This stage is usually accompanied by improvements in irrigation technology and distribution systems. The next stage is one of growing industrialization and urbanization and an increase in the use of water by manufacturing industries. The third stage of economic development results in a greater proportional service sector which has a very low water input per unit production, and with greater income comes the ability to both appreciate and afford the environmental goods that are associated with water and ignored in the first two stages of development. Thus one would expect the relationship between growth in GNP and water use to be constantly changing from very rapid rates initially to a reversal of the rate of change and a gradual reduction in per capita consumption.

There is a significant body of literature which is taken from Simon Kuznets' income inequality curve and applied to the level of environmental pollution. This environmental Kuznets curve is still a subject of considerable research and discussion. Projections of water scarcity have to question whether there is a Kuznets curve for water, and, if so, at what income level the income elasticity of water demand switches from positive to negative.

I found three empirical studies that estimate a water Kuznets curve. One by Yang and Jia (2005) estimates a Kuznets curve for industrial water use based on Organisation for Economic Co-operation and Development (OECD) data. A second study by Duarte et al. (2012) estimates a curve for total water use per capita without distinguishing the economic sector that it is used in, and a third study by Bhattarai (2004) estimates a Kuznets curve for irrigated agriculture.

Yang and Jia (2005) test the hypothesis that greater industrial production leads to a proportional increase in water use. Despite the trend for increased industrial use of water in many countries, it also has been observed that industrial water use decreases with economic development and income. The question whether there is an inverted U-shaped Kuznets curve for industrial water is tested using a simple quadratic specification of the data set of 20 OECD countries from 1995–2002. If a water Kuznets curve is established, a critical policy parameter is the turning point at which the gross domestic product (GDP) elasticity changes from positive to negative, and what conditions precipitate this decline. At the turning point, the industrial demand for water will still be increasing at a decreasing rate. Yang and Jia defined industrial water use to include water used for cooling, transportation and as a solvent, and as an ingredient of its products. The definition does not take into account the discharge after use of one activity that can be reused. Water used for hydropower generation is excluded from the definition. The industrial sector is defined broadly to include mining and manufacturing secondary industry and thermal power generation.

Yang and Jia (2005) plot of the industrial water use against GDP per capita for the OECD countries, transformed to log form. The range in which industrial water use peaks varies between US$10,000 to US$25,000 GDP. Yang and Jia estimate a simple quadratic model for each country and show that the quadratic model generally fits well and that the individual country turning points are within the above GDP range. A second regression using a larger panel data set of OECD and non-OECD countries shows the same phenomenon of industrial water use per capita, which is increasing at a decreasing rate. The authors conclude that the data show clear evidence of the existence of a water Kuznets curve for industrial water use.

Duarte et al. (2012) published a working paper using panel data and a smooth transition regression approach to test the existence of a water Kuznets curve. They regressed the effect of per capita use on per capita income in 65 countries over the period 1962–2008. The panel smooth transition regression (PSTR) approach used allows smooth parameter changes and the effect of alternative explanatory variables. The authors use precipitation as an indicator of water availability, and importantly the effect of institutions is measured by political freedom variables. They conclude definitively that there is a water Kuznets curve, but are also careful to warn that the elasticity estimates show great variability within the sample, reflecting a wide diversity in water use of these countries. The situation is not surprising given the wide variation in climate, of water endowment, and income. However, the size of the database also adds to the power of the conclusions. In summary, Duarte et al. find that at the very lowest levels of economic activity in income, elasticity of water is positive and increases sharply. However, other rapidly developing countries such as China show decreasing elasticities, some of which go into the negative orthant. One advantage of the PSTR approach taken in this paper is that it is able to examine individual heterogeneity within the sample, and time variability of income elasticity.

Duarte et al. show that their estimated water use elasticities range from positive to negative and decrease with income. Unlike the industrial Kuznets curve estimated by Yang and Jia (2005), the turning point estimated by Duarte et al. (2012) is a much lower income level, namely US$1000. However, the sample includes extremely poor countries where annual per capita income is less than this value. A notable result is the very rapid decrease in the elasticity over the US$1000–10,000 GDP range. Duarte et al. report that the average elasticity of developing countries is -0.6, the overall average is

-0.24, with some very low-income countries such as Zaire having a positive elasticity of 0.07. One optimistic note raised by Duarte et al. is on the ability of rapidly developing countries to shorten the adoption and technological lag by using innovations and systems from developed countries. This transfer of water technology can be seen in the rapid adoption of drip irrigation in many countries, so much so, that we have simultaneous rapid adoption of water saving technology in both developed and developing countries. Like the introduction of cell phones, water efficiency technology may leapfrog previous stages of development that developing countries used to evolve the current technology. This is probably one of the most beneficial ways that technology can be transferred. It is notable that countries such as Mexico and Malaysia showed similar negative water use elasticity values as those in the United States and Australia.

Again, it's worth emphasizing that although there is a negative effect on the elasticity of water use per capita, with growing populations and income and industrial growth, water use will continue to increase though at a decreasing rate. Duarte et al. also show that the control variables for precipitation and political institutions are significant and of the expected positive signs. From the results one can conclude that the introduction of more efficient institutions such as water markets both change the price and the quantity of water traded but also help to decrease the elasticity of income response.

A third empirical study of a water Kuznets curve focuses exclusively on irrigation development and uses data on the total area of land in irrigation and the percentage of crop land in irrigation as a measure of the intensity of use. In this study published by the International Water Management Institute (IWMI) Bhattarai (2004) uses a sample of irrigated land area from 66 tropical countries in Asia Africa or Latin America from 1972 to 1991. He also estimates a separate equation for Asia using data from 13 countries. Like Duarte et al. (2012) Bhattarai (2004) tests for the effect of other factors such as the macro-economic policy, agricultural productivity of institutions, and government structures, in short, a policy such as water markets. Like the other two studies Bhattarai tests for a changing income elasticity which is a function of per capita income and other institutional factors. The results for both the 66 country Global Tropical model and the 13 country Asian model strongly support the water Kuznets curve hypothesis. The specification is defined as a simple quadratic form, and there are fixed effects variables for each country to account for structural and climatic differences between them.

The results show estimates for both models that have very strong R^2 values of 0.95 and 0.96 respectively. In addition, the test of significance on the coefficients for both the linear quadratic terms are highly significant with t values at the 1 percent level. A time trend is also significant at 5 percent and positive for both models and shows that both regions are, on average, still predominately in the growth phase in terms of irrigation expansion.

Both models use income per capita as an explanatory variable. In addition, Bhattarai calculates the turning point of the Kuznets curve. His results of $2800 income for the global model and $5500 income for the Asian model lie between the extremes of the previous two studies by Yang and Jia (2005) and Duarte et al. (2012).

Figure 2.1 shows the water business curve estimated by Bhattarai from 13 Asian countries. The difference in income between the sample mean income and the turning point income (TPI) when the income elasticity of water switches from positive to negative explains the positive time trend in the regression results. The Kuznets curve shown in Figure 2.1 can be used in a very qualitative way to measure the expected increase

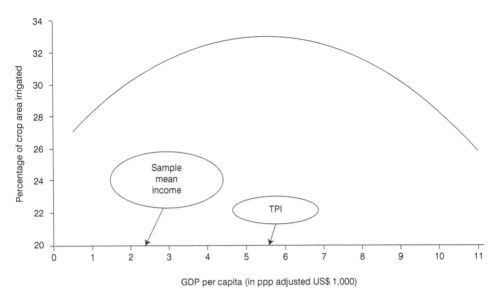

Figure 2.1 Estimated water Kuznets curve for 13 Asian countries (source: Bhattarai 2004.)

in irrigated area that is needed to reach the turning point income level. Figure 2.1 shows that for the Asian model we need an approximate doubling of GDP per capita, which the function implies increasing proportional irrigated area from 31 percent of agricultural area to 33 percent. This expansion of irrigated area is a 7 percent increase from the existing mean for the 13 countries analyzed. Essentially what the Kuznets curve is telling us is that if we can substantially increase the GNP per capita and expand irrigated area by 7 percent, the per capita water use should start falling from then on, assuming continued growth in real GNP. If the trends in population increase also stabilize with increased GDP per capita, as is suggested by several studies, the Kuznets curve holds out the possibility of achieving a steady state water use regime with higher per capita production.

The simplicity of this empirical curve analysis drives home the point that even rapidly growing Asian countries are still facing a significant amount of time before the income growth will be such that relative water scarcity starts to decrease rather than increase at an increasing rate. Bhattarai also calculates how the income elasticity changes with per capita income and obtains results which are similar to those found by Duarte et al. (2012) These results further reinforce the importance of not assuming constant income elasticities for water as was done in many earlier studies.

Bhattarai also tested both models for the effect of exogenous policies on the water Kuznets curve. He added the policy variables to the Kuznets specification one at a time, presumably to avoid multicollinearity problems, and tested for the effect of a wide range of policy variables from the price of electricity, agricultural value added, and cereal yields. The most interesting policy variable for this chapter is the one that measures the strength of institutions governing irrigated water management. The explanatory variable was composed from an index of civil and political rights and was highly significant for both the Global and Asian model. In both models the governments' policy variable have a positive sign showing the importance of institutions which, of course, includes water markets, on the sign and turning point of the Kuznets curve.

The water Kuznets curve is significant in terms of long-term analysis of scarcity in water supplies, since for the first time, it holds out some empirical evidence for hope in the long-term future. However, with continued growth of industrial production and population along with a period of increasing water use per capita, it is hard to see how we can achieve greater irrigated crop production without water resource and ecological costs, some of which will probably be irreversible.

One pessimistic projection by the FAO states that "by 2025, 1000–800 million people would be living in countries and regions with absolute water scarcity, and two thirds of the world population could be under stress conditions" (FAO 2012)

Such well-founded predictions of water scarcity are daunting but must be blunted by improved allocation and management institutions that importantly include water markets and hold out the promise of long-term stability to farmers, environmentalists, and industrial water users facing the true opportunity cost of water in their locality. Water markets are generally thought of as an esoteric institution that is only applicable in a developed country. However, as several areas of Spain show, water markets are a very old and intuitive concept, and it is really the level of transaction costs of information and enforcement that make the establishment of markets costly in developing countries. Australia now has efficient operating water markets on cell phones, and it is worth investigating whether it is possible to leapfrog old, developing country institutions in a similar manner to that of the Kenyan banking sector that is generally acknowledged to lead the world in cell phone banking transactions. If collective water rights could be established and enforced on the village basis, then a system of trades using cell phones could be as effective with water as it is with other commodities.

The importance of the water Kuznets curve is that like many other natural resources such as deforestation, land use change, and measures of air and water pollution that showed similar environmental Kuznets curves, it holds out hope for a steady state if we can just get through the current stage of growing per capita demand.

For long-term water planning it is necessary to be able to predict the peak irrigation use, to try and ensure that the ecological current capacity of the watershed in question is not exceeded by the peak water requirement. This idealized situation is unlikely to occur given the current population growth and industrial pressures for water use in developing countries. What an irrigation Kuznets curve does introduce into water policy is the ability to use dynamic analysis driven by the income gains from irrigation to show how total irrigated water use per capita will be reduced after the peak consumption. This concept, for the first time, allows the possibility of achieving the steady-state irrigated water economy in post-peak water use conditions. It does not, however, solve the difficulty of getting irrigation economies past the peak level in the earlier stages of development. Based on these three studies we can think of water markets as being a critical institution that is capable of flattening peak water use by innovative institutional changes.

References

Bhattarai. M. 2004. "Irrigation Kuznets curve, governments and dynamics and irrigation development Co. in global cross-country analysis from 1972 to 1991", Research Report # 78. International Water Management Institute, Colombo, Sri Lanka.

Duarte, R., V. Pinilla, and A. Serrano. 2012. "Is there an environmental Kuznets curve for water use? A Panel Smooth Transition Regression Approach" Documento de Trabajo 2012-03. Universidad de Zaragoza, Spain.

FAO (Food and Agriculture Organization of the United Nations). 2012. FAO global data sets. FAOSTAT-AGRICULTURE. Rome. Italy. http://www.fao.org

Ghassemi. J., A.J. Jakeman, and H.A. Nix. 1995. *Salinisation of Land and Water Resources*. University of New South Wales Press, Sydney.

Gleick, P.H. and M. Palaniappan. 2010. "The water limits to freshwater withdrawal and use", *Proceedings of the National Academy of Sciences*, 107, 11155–11162.

Gleeson, T., Y. Wada, M. Bierkens, and L. van Beek. 2012. "Water balance of global aquifers revealed by groundwater footprint". *Nature*, 488, 197–200.

Hanak, E., J. Lund, A. Dinar, B. Gray, R. Howitt, J. Mount, P. Moyle, and B. Thompson. 2011. *Managing California's Water: from Conflict to Reconciliation*. PPIC Press, Public Policy Institute of California, San Francisco, CA.

Pitman, M.G., and A. Lauchli. 2002. "Global Impact of Salinity and Agricultural Ecosystems", in *Salinity: Environment-Plants-Molecules*. A. Lauchli and U. Luttge (eds). Kluwer Academic Publishers, Dordrecht.

Postel, S.L. 1999. *Pillar of Sand: Can the Irrigation Miracle Last?* WW Norton, New York.

Yang, H. and S.F. Jia. 2005. "Industrial water use Kuznets curve: evidence from industrialized countries and implications for developing countries", *Water Science*, 11(4), 470–477.

3

ORDERING EXTRACTION FROM MULTIPLE AQUIFERS

James A. Roumasset and Christopher A. Wada

Introduction

The economics of groundwater (e.g. Burt 1967; Brown and Deacon 1972; Krulce et al. 1997; Koundouri 2004; Pitafi and Roumasset 2009) is typically directed to finding the optimal extraction profile for a single aquifer. But in many cases, water managers have multiple aquifers to coordinate, and the single-aquifer conditions do not provide instruction on which aquifer to use first. As a default, managers may use rules of thumb such as using the nearest available aquifer, even when multiple aquifers are connected to the same distribution system.

In response to this need, this chapter shows how renewable resource economics can be extended to the case of multiple resources. A natural starting point for this research is Herfindahl's (1967) demonstration that different grades of a non-renewable resource should be extracted in the order of "least-cost-first," where cost refers to the unit extraction cost of a particular grade. Subsequent research has shown that Herfindahl's result is a special case of a more general least-cost-first principle, namely that different resources should be extracted in the order of their full marginal cost, including the opportunity cost of foregone future benefits (Chakravorty and Krulce 1994; Chakravorty et al. 1997; Gaudet et al. 2001; Chakravorty et al. 2005). Less attention has been paid to the optimal ordering of renewable resources, however, partly because the analysis of renewable resources has traditionally focused on the steady state, not the transition thereto.

Several papers have examined problems involving multiple renewable resources, but most have relied on somewhat restrictive assumptions—e.g. constant unit extraction costs and homogenous growth functions (Shimomura 1984), constant resource growth (Zeitouni and Dinar 1997), and exogenous prices (Costello and Polasky 2008)—or have focused primarily on steady-state analysis (Horan and Shortle 1999). Roumasset and Wada (2012) confirmed that the least marginal opportunity cost rule—a generalization of Herfindahl's least-extraction-cost-first rule—extends to multiple renewable resources and demands but focused their application primarily on the 2-resource, 1-demand case. In this chapter, we review these findings, with an emphasis on the 2-resource, 2-demand case, and discuss the role that transport costs play in determining optimal use. After examining some particular cases of interest, we then discuss a potential extension involving the management of multiple water resources when water quality is heterogeneous across aquifers and water quality requirements are heterogeneous across users.

Optimal extraction of multiple nonrenewable resources

Herfindahl (1967) established that deposits of a nonrenewable resource, identical aside from (constant) unit extraction costs, should be extracted in order of least marginal extraction cost. Lewis (1982) comments that "it seems to be almost transparent that it is efficient to exploit low cost deposits first." Indeed, if the marginal benefit and marginal user cost for each deposit is equal, minimizing extraction cost maximizes the present value net benefit of the resource. Chakravorty and Krulce (1994) show that when more than one demand (marginal benefit curve) exists for the resource, however, the least-extraction-cost rule does not generally describe the optimal extraction path. In particular, deposits should be extracted in order of least-price plus conversion cost for each demand, where price is equal to the sum of marginal extraction cost and marginal user cost. Optimal ordering in this case is predetermined in the sense that the relative magnitude of the initial deposit shadow prices can be established using knowledge about the initial value of the resource, i.e. without solving the entire dynamic optimization problem. Assuming that marginal extraction costs are also constant, it is straightforward to see which deposit is "cheapest" initially for each demand sector and then to trace out the phases of extraction.[1]

The conversion costs described by Chakravorty and Krulce (1994) appear at first blush to be specific to energy, e.g. converting coal into a liquid substitute for oil. However, the concept is generalizable to other situations, wherein the "conversion cost" simply represents a cost in addition to extraction that is specific to a particular demand. In the context of landfills, Gaudet et al. (2001) specify the marginal cost as including both the cost of extracting the resource from a particular site and the cost of transporting it to a particular location for consumption. Their *full marginal cost* (FMC)— the sum of extraction cost, transportation cost, and imputed cost or shadow price of the resource—is therefore analogous to the efficiency price described in Chakravorty and Krulce (1994). The generalized optimal extraction and allocation rules of both studies are likewise the same: each city (demand) will, at any given date, use only the resource with the lowest full marginal cost. Thus if the FMC of a particular city using a particular deposit exceeds the marginal benefit of consumption, that city should not extract from that site during that period.

Chakravorty et al. (2005) further generalize the findings of Chakravorty and Krulce (1994) and Gaudet et al. (2001) by framing resource extraction stages in terms of Ricardian absolute and comparative advantage. They show that in the general m-resource, n-demand case, optimal specialization over time is governed by absolute advantage while specialization across sectors is determined by comparative advantage. A resource that is abundant, i.e. has a relatively low shadow price, and has absolute advantage in all demands, i.e. has the lowest extraction plus conversion cost, may be extracted for a demand sector, even if it does not have comparative advantage in that use. However, if more than one resource has absolute advantage in a given demand, specialization is based on comparative advantage. As in the 2×2 case, the optimal ordering of extraction can be established without solving the entire problem if information about the extraction plus conversion costs for each resource and demand are known.

The nonrenewable resource ordering problem can be solved in a variety of ways, but Gaudet et al. (2001) discuss a straightforward and intuitive algorithm for the case of constant extraction costs. First, assign to each resource an initial multiplier (shadow price) and let each multiplier grow at the exogenous rate of interest à la standard Hotelling condition (Hotelling 1931). For each demand sector, construct the available FMCs in each time period and select the resource with the lowest FMC in every period. Extraction in each period should occur until marginal benefit equals FMC for each resource and demand. Once the steady state is obtained, the cumulative usage of each resource can be calculated over time and across demands. If the cumulative usage for each resource exactly matches the initial stock, then the extraction paths are optimal. Otherwise, the guesses for the initial values of the multipliers must be adjusted and the process repeated until the candidate paths result in cumulative extraction that corresponds to the initial reserves.

Optimal extraction of multiple renewable resources

In this section, we use the example of coastal groundwater aquifers because leakage from a coastal aquifer increases with the head level, which implies that net resource growth is a function of the groundwater stock, a distinguishing characteristic of a renewable resource. Moreover, since the extraction costs are also stock-dependent—a function of the energy required to lift the resource vertically from the water table to the surface— the model remains fairly general. Constant unit extraction costs are a special case.

Even in the case of one demand, Shimomura (1984) has shown that Herfindahl's least-cost rule does not extend to renewables such as groundwater. Because the marginal user cost component of the optimal shadow price for each "deposit" of a renewable resource varies according to the size of each stock, scarcity of each resource with respect to a particular demand can increase or decrease over time, meaning that the Gaudet et al. (2001) least-FMC rule must instead be used. Following Roumasset and Wada (2012), the least-FMC rule for m-resources and n-demands can be summarized as

$$p_t^j = \min[FMC_t^{1j}, FMC_t^{2j}, ..., FMC_t^{mj}] \text{ for } j = 1, ..., n \tag{3.1}$$

Optimality requires that the marginal benefit (p) and FMC of groundwater be equated in every period for every demand sector. If the marginal benefit for sector j is less than the FMC of a unit of groundwater extracted from aquifer i for consumption in j, then no water should be extracted for that purpose.

One of the FMC terms in Equation 3.1 may correspond to a backstop resource such as desalinated brackish or sea water. Inasmuch as the backstop FMC is, in most cases, relatively high in earlier periods, optimal use of the groundwater alternative is delayed until later periods when groundwater is scarcer. Corner solutions or zero extraction from one or more sources are not limited to backstop resources, however. In general zero extraction may be optimal for any resource if its FMC is too high. A special characteristic of renewable resources—and one of the reasons why determining the optimal ordering is generally more challenging than for nonrenewable resources—is that corner solutions along the dynamic path correspond to decreasing scarcity of the unused resource(s) as natural replenishment occurs. This can be more clearly understood by examining the components of the FMC. One can show that the FMC of groundwater extracted from aquifer *i* for use by demand sector *j* is comprised of the stock-dependent marginal extraction cost and the marginal user cost (Roumasset and Wada, 2012):

$$FMC_t^{ij} = c_i(h_t^i) + \frac{\dot{p}_t^j - c_i'(h_t^i) f_i(h_t^i) + \mu_t^i}{r - f_i(h_t^i)} \tag{3.2}$$

where *c* is the marginal extraction cost, *f* is the resource growth function, *h* is the head level which serves as an index for groundwater volume, *r* is the discount rate, and *μ* is the multiplier associated with the minimum head constraint, determined for example by US Environmental Protection Agency (EPA) salinity requirements for potable water. Unlike in the basic nonrenewable setup wherein the FMC is equal to the unit extraction cost *c* plus $\dot{p}_t^j / r$, Equation 3.2 varies with the head level.

Because groundwater aquifers replenish naturally via precipitation, all groundwater resources must be used in a steady state; otherwise head levels will continue to rise, which contradicts the condition that the system is in a steady state. Although unlikely to be optimal, a steady state may also obtain if the aquifer is allowed to replenish to its pre-extraction level. If demand for water is also perpetually increasing due to population and per capita income growth, the backstop resource will supplement groundwater in the long run (steady state). The ordering of resources prior to the steady state will vary, however, depending on the parameters of the particular application. We explore various permutations of the 2 × 1 case and discuss which insights carry over to the 2 × 2 case, and more generally to the m × n case.

Suppose we have two coastal aquifers (A and B), each with its own FMC (Equation 3.2), serving a single demand sector. If $FMC^A < FMC^B$ initially, Equation 3.1 tells us that aquifer A should be used exclusively, while aquifer B is allowed to replenish. Eventually, the two FMCs converge, whereupon either aquifer A reaches its steady-state head level and extraction from that aquifer is limited to net recharge, or both FMCs rise together in conjunction with simultaneous (non-constant) extraction until one of the aquifers reaches its steady-state head level. Once either head constraint becomes binding (say for aquifer A), the shadow price of A rises apace with the shadow price of B, ensuring that the least-FMC rule (Equation 3.1) is not violated. The final stage of extraction is characterized by constant steady-state withdrawal from both aquifers at rates that exactly offset natural net recharge, supplemented by a backstop resource such as desalination. In the unlikely event that $FMC^A = FMC^B$ at the outset, simultaneous extraction is optimal during the entire transition to the steady state. Lastly, if both aquifer head levels lie initially below their respective optimal steady-state levels, optimal management entails using the "backstop" resource exclusively at the outset, thus allowing the aquifers to build toward their optimum steady-state levels. Desalinated water is used in every period, and groundwater is only extracted in the steady state. Figure 3.1 illustrates the hypothetical scenario where $FMC^A < FMC^B$ at the outset.

Although the least-FMC condition (Equation 3.1) prescribes optimal resource ordering for a given set of FMCs, the FMC of a renewable resource is not generally

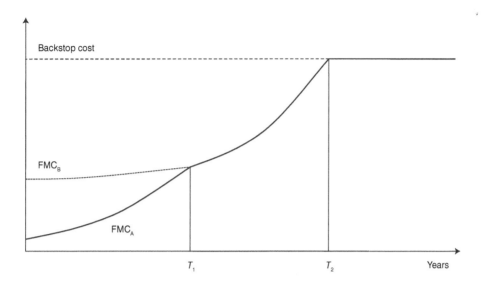

Figure 3.1 Hypothetical price path

known *ex ante*. Nevertheless, one may be able to obtain a general sense of the optimal ordering on a case-by-case basis by examining various characteristics of the resource. For example, in the case of two coastal aquifers, using the "leakier" aquifer first may be optimal. Intuitively, drawing down the leaky aquifer and allowing the other aquifer to build reduces aggregate discharge to the ocean. In the case of interior aquifers, where net recharge and extraction costs are approximately constant, Equation 3.2 varies among resources only by extraction cost and recharge. FMC is lower when extraction cost is lower and/or constant recharge is higher. Since both terms are known *ex ante*, optimal ordering can be ascertained without solving the entire problem. If, moreover, recharge is approximately equal for the two aquifers, least-FMC collapses to the Herfindahl least-cost rule.

To illustrate the 2 × 2 case, we consider a real world example on the island of Oahu, Hawaii. Pearl Harbor aquifer (PHA) is the largest groundwater aquifer in the state and is geographically located near Pearl City. To the southeast, the smaller Honolulu aquifer (HA) underlies the city of Honolulu, the most populated city in the state. Due to the difference in elevation, it is a reasonable assumption that transporting pumped water from Honolulu to Pearl City is costlier than moving water from Pearl City to Honolulu. Based on the "leakier aquifer first" logic discussed previously and the differences in transportation costs, it would likely be optimal to use PHA exclusively for Pearl City and Honolulu, while allowing HA to replenish. Once PHA reached its (minimum) steady-state level, net recharge from PHA would supply Pearl City and Honolulu, while extraction from HA would only be used in Honolulu. Assuming that demand in both districts is continuing to grow, it would be necessary to supplement extraction from HA with the backstop (desalination) to supply the Honolulu district once HA is drawn down to its (minimum) steady-state level. Eventually, demand growth in Pearl City would initiate desalination to supplement the PHA as well, such that Pearl City uses groundwater from PHA plus desalinated water produced in Pearl City and

Honolulu uses groundwater from HA plus desalinated water produced in Honolulu. In this particular example, treating the aquifers independently can result in large welfare losses—$4.7 billion or 65 percent of the present value obtained when managing the aquifers separately (Roumasset and Wada 2012).

The general least-FMC principle extends to the m-resource, n-demand case. In a given period, any number of resources may supply a particular demand simultaneously, so long as the full marginal costs of supplying water from each aquifer to that demand are equal. Because each aquifer has its own recharge, leakage, and extraction cost function, welfare maximization may entail allowing some aquifers to replenish while drawing down others over time, i.e. taking advantage of arbitrage opportunities. If demand is continuously growing, the long run steady state will always be characterized by constant water extraction, supplemented by desalination. However, increasing the number of available groundwater resources tends to delay the need for costly desalination; aquifers allowed to replenish in earlier periods come online in later periods when the scarcity value of water is higher and when implementing desalination may have otherwise been necessary to meet the growing demand.

As the number of resources and users expands, the problem becomes increasingly complicated, and preliminary ordering based on resource characteristics tends to become less feasible. Nevertheless, the Gaudet et al. (2001) approach to numerically solving the management problem can still be applied with some slight modifications. An initial guess is made for the starting value of each resource's shadow price. Each price grows according to the extended Hotelling condition, i.e. not simply at the exogenous rate of interest. In each time period, the resource with the lowest FMC is selected for each demand, and extraction occurs until the marginal benefit is equal to the FMC for each resource and demand. Eventually, all head levels reach their optimal steady-state levels (which can be calculated using the necessary conditions). If the price corresponding to the steady-state heads is equal to the backstop price, then the extraction paths are optimal. Otherwise the guesses for the shadow prices must be adjusted and the process repeated. If instead the backstop price is reached before the head levels are drawn down to their steady state optimums, the extraction paths are also suboptimal, and the guesses should be adjusted. Alternatively, cumulative extraction upon arrival at the steady state can be compared to the initial stock, after accounting for natural leakage and replenishment in accordance with the candidate head level trajectories.

The framework and derived FMC expression (Equation 3.2) are sufficiently general to be directly applicable to other renewable resources. In particular, the model is easily modified for interior, i.e. not coastal, aquifers. In the case of an interior aquifer, extraction costs may still be stock-dependent, but the resource growth may be better characterized as a recharge quantity (rather than a function) inasmuch as stock-dependent discharge to the ocean is not relevant. If water naturally flows between adjacent inland aquifers, however, the net growth function should be modified to include the head levels of all aquifers under consideration—the relative heights of the head levels largely determine the direction of groundwater flow between aquifers. If the implementation of a backstop such as desalination is deemed unfeasible, a choke price can be set to ensure that the problem is solvable.

Spatial dimensions and water markets

Incorporating transportation costs into the framework discussed in the previous section is fairly straightforward; the expressions for FMC in Eqs. 3.1 and 3.2 need only be adjusted by the cost of transporting a unit of resource i from the extraction site to the location of demand sector j. That is, the full marginal cost of extracting resource i for consumption in sector j is equal to the sum of marginal extraction cost, marginal user cost, and transportation cost. The least-FMC rule still characterizes optimal resource extraction.

The conditions described above imply that there is an efficiency value of groundwater *ex situ* given by the minimum FMC in each period. The efficiency value of delivered water is then just the *ex situ* value plus the transport cost to a particular location. A decentralized implementation of optimal groundwater allocation over space and time can then be achieved by setting the wholesale price equal to the *ex situ* value and retail prices equal to the wholesale price plus location specific transport costs.

Instead of a centralized authority announcing prices, could the same solution be obtained by markets? The problem is that the efficiency values for a particular time are different across space. If water markets are sufficiently "thick," i.e. if potential traders for water in each location are numerous, then the water authority could announce quantities for each location, and the market would provide the efficiency price. Since this condition seems unlikely, especially if location is distinguished by fine gradations, the water authority can implement a water market by announcing exchange rates, given by the ratio of the locational retail price as determined above to the wholesale price. Indeed the failure of authorities to provide such exchange rates partly accounts for the need for ad hoc review boards and the incidence of trade being much smaller than indicated by theory. On the other hand exchange rates presuppose reliable volumetric measurements of the water being traded. Until these are established, exchange rates may be premature.

Introducing marginal externality costs

In the previous sections, it was assumed that groundwater extraction does not generate any external costs, i.e. costs not directly associated with physical extraction of the resource. When externalities are present, however, not accounting for them in management decisions can result in underpricing and overharvesting of the resource. Pearce and Markandya (1989) provide a useful and intuitive framework for incorporating externalities into optimal resource extraction: when resource use creates a downstream spillover, the FMC should include marginal extraction cost, marginal user cost (MUC), and marginal externality cost (MEC), where the last term captures the downstream external cost (e.g. pollution) resulting from extraction. The efficient level of resource use is thus achieved by extracting until the marginal benefit of doing so is equal to the FMC, which includes the MEC.

Although Pearce and Markandya (1989) do not specify expressions for MUC and MEC explicitly, some of the subsequent analyses on optimal resource sequencing—particularly in the area of nonrenewable resources—have incorporated the MEC concept. Chakravorty et al. (2008), for example, include the MEC of pollution associated with the extraction and use of natural gas and coal in their conditions characterizing the optimal order of extraction. They show that when extraction costs are homogeneous across resources but the coefficients of pollutions are heterogeneous, optimal ordering is determined by the FMC, in this case the sum of MUC and MEC. With high discount and/or dissipation rates this may imply that the dirtiest resource (coal) is optimally used first. This result is entirely consistent with the least-FMC cost rule; the MEC should and does account for the dynamic aspects of pollution including its depreciation. In the polar extreme case, with zero rates of pure time preference and dissipation, the dirty resource should never be used since any amount of stock pollution will permanently reduce the level of maximum sustainable consumption (Endress et al. 2005).

More generally, resource-extraction externalities can be classified into four categories: fund externalities, stock externalities, stock-to-fund externalities, and stock-to-stock externalities. Fund externalities, such as air pollution from sulfuric and nitrous oxides, are somewhat transient such that concentrations can be proxied by emission fluxes into a particular airshed; stock externalities are dynamic, direct spillovers such as emissions-induced climate change; stock-to-fund externalities are dynamic but indirect effects such as amenity values from resources, e.g. biodiversity; and stock-to-stock externalities are indirect and dynamic spillovers such as the effect of aquifer levels on nearshore marine environments where groundwater subterraneously discharges. For direct externalities (the first two types), the MEC term is additively separable from the MUC in the full marginal cost. When the effect is instead indirect, the MEC is embedded within the MUC, meaning that the standard "Pearce equation" does not apply. In either case, however, one might expect that optimal resource ordering is determined by the least-FMC extraction rule.

The management framework for multiple groundwater resources could be extended to include various types of externalities. In the case of water pollution (e.g. nutrient leaching from irrigation or pumping-induced saltwater intrusion of a coastal aquifer) each resource would have a different coefficient of pollution, and optimization would require keeping track of the stock of pollutant or water quality over time. Given that water quality requirements typically vary across users, calculations become increasingly complicated with these added dimensions, but the general optimality principle—namely the least-FMC extraction rule—is still expected to apply for each user type. Chapter 4 in this volume discusses the example of incorporating non-potable recycled wastewater use into a groundwater management program when users are sensitive to water quality. Positive externalities (e.g. the effect of submarine groundwater discharge on nearshore aquatic plants which is dependent on the level of the water table) could similarly be incorporated into a resource optimization framework, given aquifer-specific positive spillover coefficients.

Conclusions and directions for further research

We have generalized the least-cost rule for optimal extraction from a single aquifer to the case of multiple aquifers by including extraction cost, marginal user cost, marginal externality cost, and the shadow value of the minimum head constraint in the definition of full marginal cost (FMC). In each period, water should be extracted from the least-cost source. Higher cost sources should not be used until the FMC of the source in use rises to the FMC of the higher cost source. In the two-aquifer case reviewed, the leakier aquifer is the cheaper FMC source and is optimally drawn down to its minimum head level while the other aquifer is allowed to accumulate. This simultaneous drawdown/build-up pattern contrasts starkly with the principle of drawing from both aquifers according to their (arbitrarily selected) sustainable yields. (It may also be optimal to build up both aquifers before beginning a period of drawdown.) Welfare losses of managing aquifers independently can be substantial, as in the Hawaii case. Decision makers should therefore weigh the potential gains of implementing a joint approach with the various adjustments costs that would be generated by deviating from the status quo.

The least-FMC ordering rule for groundwater extraction can be further extended to include other water resources. Surface water and recycled water, for example, can supplement groundwater extraction for a variety of end uses. Although not exactly identical in form to the full marginal cost of groundwater, the FMC for alternative water sources should still account for future effects of contemporaneous decisions (i.e. the marginal user cost), in addition to the physical costs of extraction/production and distribution.

While not explicitly modeled in the optimization framework, energy plays a key role in the operation of groundwater pumps, conveyance of surface water, and treatment of ocean, brackish and wastewater. When energy prices are rising, less energy intensive water resources become more desirable. At the same time, however, technological advancement for any of the production processes may drive the overall costs down if improvements in energy efficiency dominate rising energy prices. Consequently, a model with projected energy prices and induced innovation may tell a very different story in the short and medium term than the model developed in this chapter. In the long run, however, if the production prices of groundwater alternatives eventually stabilize, the steady state will still ultimately be characterized by constant groundwater extraction, supplemented by the backstop resource.

Note

1 For a stylized application with implications for global warming, see Chakravorty et al. (1997).

References

Burt, O.R. (1967) "Temporal Allocation of Groundwater", *Water Resources Research*, vol 3, pp. 45–56.

Brown, G. and Deacon, R. (1972) "Economic Optimization of a Single Cell Aquifer", *Water Resources Research*, vol 8, pp. 552–564.

Chakravorty, U. and Krulce, D.L. (1994) "Heterogeneous Demand and Order of Resource Extraction", *Econometrica*, vol 62, pp. 1445–1452.

Chakravorty, U., Roumasset, J. and Tse, K. (1997) "Endogenous Substitution among Energy Resources and Global Warming", *Journal of Political Economy*, vol 105, no 6, pp. 1201–1234.

Chakravorty, U., Krulce, D. and Roumasset, J.2005) "Specialization and Non-Renewable Resources: Ricardo meets Ricardo", *Journal of Economic Dynamics and Control*, vol 29, pp. 1517–1545.

Chakravorty, U., Moreaux, M. and Tidball, M. (2008) "Ordering the Extraction of Polluting Nonrenewable Resources", *The American Economic Review*, vol 98, no 3, pp. 1128–1144.

Costello, C. and Polasky, S. (2008) "Optimal harvesting of stochastic spatial resources", *Journal of Environmental Economics and Management*, vol 56, pp. 1–18.

Endress, L., Roumasset, J. and Zhou, T. (2005) "Sustainable Growth with Environmental Spillovers", *Journal of Economic Behavior & Organization*, vol 58, no 4, pp. 527–547.

Gaudet, G., Moreaux, M. and Salant, S.W. (2001) "Intertemporal Depletion of Resource Sites by Spatially Distributed Users", *The American Economic Review*, vol 91, no 4, pp. 1149–1159.

Herfindahl, O.C. (1967) "Depletion and Economic Theory", in M. Gaffney (ed) *Extractive Resources and Taxation*, University of Wisconsin Press, Wisconsin.

Horan, R.D. and Shortle, J.S. (1999) "Optimal Management of Multiple Renewable Resource Stocks: An Application to Minke Whales", *Environmental and Resource Economics*, vol 13, pp. 435–458.

Hotelling, H. (1931) "The Economics of Exhaustible Resources", *The Journal of Political Economy*, vol 39, pp. 137–175.

Koundouri, P. (2004) "Current Issues in Economics of Groundwater Resource Management", *Journal of Economic Surveys*, vol 18, pp. 703–738.

Krulce, D.L., Roumasset, J.A. and Wilson, T. (1997) "Optimal Management of a Renewable and Replaceable Resource: The Case of Coastal Groundwater", *American Journal of Agricultural Economics*, vol 79, pp. 1218–1228.

Lewis, T.R. (1982) "Sufficient Conditions for Extracting Least Cost Resource First", *Econometrica*, vol 50, pp. 1081–1083.

Pearce, D. and Markandya, A. (1989) "Marginal Opportunity Cost as a Planning Concept", in G. Schramm and J.J. Warford (eds) *Environmental Management and Economic Development*, The Johns Hopkins University Press, Baltimore.

Pitafi, B.A. and Roumasset, J.A. (2009) "Pareto-Improving Water Management over Space and Time: The Honolulu Case", *American Journal of Agricultural Economics*, vol 91, no 1, pp. 138–153.

Roumasset, J.A. and Wada, C.A. (2012) "Ordering the Extraction of Renewable Resources: The Case of Multiple Aquifers", *Resource and Energy Economics*, vol 34, pp. 112–128.

Shimomura, K. (1984) "The Optimal Order of Exploitation of Deposits of a Renewable Resource", in M.C. Kemp and N.V. Long (eds) *Essays in the Economics of Exhaustible Resources*, North Holland, Amsterdam.

Zeitouni, N. and Dinar, A. (1997) "Mitigating Negative Water Quality and Quality Externalities by Joint Management of Adjacent Aquifers", *Environmental and Resource Economics*, vol 9, pp. 1–20.

4

INTEGRATING DEMAND-MANAGEMENT WITH DEVELOPMENT OF SUPPLY-SIDE SUBSTITUTES

James A. Roumasset and Christopher A. Wada

Introduction

Freshwater scarcity has long been an important issue in many regions across the globe. The United Nations recommends a multidisciplinary approach to management, inasmuch as water scarcity "affects all social and economic sectors and threatens the sustainability of the natural resource base" (United Nations 2006). While the scope of the problem is clear, more research is needed to effectively integrate demand- and supply-side water management strategies in a systematic manner. To that end, we develop and discuss a framework for assessing and prioritizing a variety of water management instruments.

Sustaining existing water resources at current prices in the face of growing demand is not possible, especially when taking into account the effects of climate change, urbanization, and watershed degradation. Implementation of costly ground or surface water alternatives, such as desalinated seawater or recycled wastewater, can be delayed, however, with appropriate long-run planning. For example, investing in watershed conservation reduces scarcity by improving groundwater recharge. Yet, incorrect groundwater pricing can waste those potential gains. Relatedly, payments for ecosystem services such as recharge are likely to be mispriced unless system interdependence is properly accounted for.

In this chapter, we discuss the joint implementation of three management tools: (i) water pricing, (ii) watershed conservation, and (iii) wastewater recycling. We then investigate the possibility of extending the framework to include multiple groundwater resources and allowing for the effects of climate change explicitly.

Watershed conservation

We begin by examining how watershed conservation activities, integrated into a standard groundwater management framework, affect optimal outcomes for head levels (an index of the groundwater stock), extraction, and present value. Unlike alternative water sources which act to supplement groundwater extraction, watershed conservation enhances existing sources by increasing the proportion of precipitation that reaches subterraneous aquifers as recharge. There are many types of conservation activities—e.g. reforestation, removal of feral ungulates, removal of invasive plant species, construction of settlement ponds or injection wells—but the purpose of this chapter is not to develop a methodology for optimizing across a spectrum of potential conservation instruments.[1] Rather, assuming available instruments are employed in an optimal manner, we examine the interrelated nature of groundwater conservation (reduced extraction) and investment in watershed conservation capital. Both types of conservation are steered by the shadow price or scarcity value of groundwater, and both instruments should be employed until their marginal benefits equal their marginal costs.

The groundwater resource is modeled as a single-cell coastal aquifer (Krulce et al. 1997), such that the volume of stored water is directly proportional to the head level (h) or the vertical distance from mean sea level to the water table. The head level changes over time according to recharge (R), stock-dependent natural leakage to adjacent water bodies (L), and extraction (q):

$$\dot{h}_t = R(N_t) - L(h_t) - q_t \tag{4.1}$$

where $R' \geq 0$ and $R'' \leq 0$. Groundwater recharge is an increasing function of the conservation capital stock (N), but the effectiveness of additional investment declines as the stock of capital increases. The decreasing marginal productivity of N captures the idea that a dynamically optimized portfolio of conservation activities would employ the most cost-effective instruments first. Thus depending on the characteristics of the resources, it may make sense to never use one or more of the available watershed conservation instruments.

Produced conservation capital naturally degrades over time (e.g. a fence), and requires maintenance. Investment in removal and replacement of natural capital (e.g. replacing invasive weeds with native forest) also requires maintenance—invasive plant seeds can remain viable for many years. Generally, investment (I) in conservation activities at unit cost c_I can be made to both offset the natural rate of capital depreciation (δ) and to build up the existing stock:

$$\dot{N}_t = I_t - \delta N_t \tag{4.2}$$

The constant unit cost of investment could instead be modeled as an increasing and convex function of N to capture the idea that the lowest-cost areas are targeted first; in the case of fencing, this might correspond to areas with the easiest terrain to traverse. However, the concavity of the recharge function already implies optimization across available conservation instruments. Either specification is sufficient to illustrate that the marginal net benefit of conservation varies with the stock of capital.

The management problem is to choose the rates of groundwater extraction (q), an alternative backstop resource such as desalinated brackish water (b), and investment in conservation capital (I), given a non-negative discount rate (r) to maximize net present value:

$$\max_{q_t, b_t, I_t} \int_{t=0}^{\infty} e^{-rt} \left\{ \int_{x=0}^{q_t + b_t} D^{-1}(x,t)dx - c_q(h_t)q_t - c_b b_t - c_I I_t \right\} dt \tag{4.3}$$

subject to equations (4.1) and (4.2), non-negativity constraints on the control variables, and $h_t \geq h_{\min}$, where the minimum head constraint is determined, for example, by the minimum allowable salinity for potable water. The period-t benefit is measured as consumer surplus, or the area under the demand curve up to the optimal quantity. Defining the marginal benefit along the optimal trajectory or efficiency price as $p_t \equiv D^{-1}(q_t + b_t, t)$, one can derive a dynamic pricing equation that is driven by the head level and the capital stock (Roumasset and Wada 2013):

$$p_t = c_q(h_t) + \frac{\dot{p}_t - c'(h_t)[R(N_t) - L(h_t)]}{r + L'(h_t)} \qquad (4.4)$$

To incentivize extraction of the optimum quantity, the price should be set equal to the sum of the marginal extraction cost and the marginal user cost (MUC), the latter of which represents the loss in present value resulting from extracting a single unit of the resource at the time of extraction.

An equimarginality condition can also be derived to guide optimal investment in conservation capital (Roumasset and Wada 2013):

$$\lambda_t = \frac{c_I(r + \delta)}{R'(N_t)} \qquad (4.5)$$

where λ is the shadow price of groundwater or the costate variable associated with the dynamic state equation for the head level (equation 4.1). The numerator on the right hand side of equation (4.5) is the marginal opportunity cost of investment or the user cost of capital (Jorgenson 1963), which includes the forgone interest that would have accrued had the income not been invested in capital and the cost of depreciation. The denominator is the marginal product of capital in recharge and converts the user cost to dollars per unit of recharge. The entire expression can be viewed as a supply curve for recharge, inasmuch as the diminishing marginal productivity of capital implies an increasing function in N. The left hand side of equation (4.5) is the marginal benefit of recharge via conservation capital, measured as the shadow price of groundwater. Welfare is maximized where the recharge supply curve intersects the shadow price of groundwater, which is also equal to the net marginal benefit of groundwater extraction along the optimal path (Figure 4.1). In summary, the resource manager should be indifferent between conserving groundwater via consumption reduction and via investment in the watershed.

Assuming the existence of a steady state, wherein $\dot{p} = \dot{h} = \dot{N} = 0$, the optimal investment rule is to choose the maximum feasible level of investment in every period prior. The evolution of the capital stock, in turn, is determined by equation (4.2), and the optimal head and extraction paths are steered by equations (4.1) and (4.4) respectively. When demand is increasing and unbounded, the price of water eventually rises to the backstop price. Moreover, because extraction costs typically do not increase substantially as the head level is drawn down, the optimal steady state likely entails depleting the aquifer to its minimum head level to minimize leakage. In general, near constant extraction costs result in a shorter time to steady state, whereas rapidly increasing (convex) costs tend to lengthen the transition to the steady state.

Because the current value Hamiltonian corresponding to equation (4.3) is linear in investment, the dynamic paths of capital stock and investment will approach monotonically from above or below the steady state target, depending on the initial value N_0. The Hamiltonian is not linear in groundwater extraction, however, and the optimal

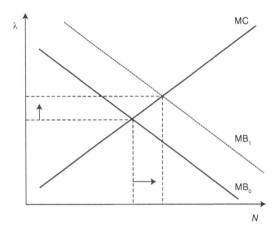

Figure 4.1 Demand and supply of recharge

trajectories of the groundwater stock, efficiency price, and extraction may take a non-monotonic approach to the steady state. In the case of constant aquifer recharge, it has been shown that a period of aquifer replenishment may optimally precede drawdown in anticipation of future scarcity (Krulce et al. 1997; Roumasset and Wada 2012).

Wastewater recycling

While watershed conservation enhances existing groundwater sources, recycled wastewater serves as a groundwater alternative, much like desalinated sea water. If treatment costs (inclusive of amortized capital) were identical, there would be little reason to prefer one alternative to the other, aside from environmental considerations—sludge left over from both treatment processes needs to be deposited somewhere. However, desalinated water is generally treated to potable standards, making it a perfect substitute for groundwater, whereas treating wastewater to potable standards for reuse may not be cost-effective. As an imperfect substitute, recycled water may optimally supply only a subset of water users, whose marginal benefit of lower quality water is still high enough to justify treatment costs. In the simplest case, and the one discussed in this section, recycled water provides the same marginal benefit as groundwater for a subset of demand sectors and does not meet quality standards for the remaining sectors. Recycled water, therefore, serves as a sector-specific supplemental resource that facilitates substitution of groundwater usage toward sectors that require high quality water.

We consider the problem of allocating three types of water—groundwater, recycled water, desalinated water—to two demand sectors. The household sector (H) can only use potable water provided by the aquifer or the desalination facility, whereas the agriculture sector (A) can draw from any of the three sources. Because recycled water is assumed to be of lower quality, a separate distribution infrastructure is required. Infrastructure investment can be accounted for by assuming a convex marginal cost of delivering recycled water, i.e. $c_R(q_t^R) > 0$, $c_R'(q_t^R) > 0$, and $c_R''(q_t^R) \geq 0$. Implicitly, treatment facilities are initially constructed near the densest area of users, and the distribution network endogenously expands over time as the proportion of users in the agriculture sector increases in accordance with rising groundwater scarcity. The groundwater hydrology is described by equation (4.1), with a slight modification; the quantity of groundwater extracted (q) must be further disaggregated into extraction for the household (q^{GH}) and agriculture (q^{GA}) sectors. The management problem is to choose groundwater extraction for each sector, desalination for the household (q^{BH}) and agriculture (q^{BA}) sectors, and water recycling for the agriculture sector (q^{RA}) to maximize net present value subject to the modified aquifer equation of motion, non-negativity constraints on the control variables, and a minimum head level constraint.[2] The marginal cost of a particular resource is the same between sectors—with the exception of recycled water whose marginal cost is implicitly infinite for the household sector—but the marginal benefit varies according to the sector-specific demand functions D_H and D_A. One can show that efficient water use in each sector is achieved by equating the marginal benefit with the lowest available marginal opportunity cost (Roumasset and Wada 2011):[3]

$$p_t^H = \min\{c_G(h_t) + \lambda_t, c_B\} \tag{4.6}$$

$$p_t^A = \min\{c_G(h_t) + \lambda_t, c_B, c_R(q_t^{RA}) + q_t^{RA} c_R'\} \tag{4.7}$$

In general, a particular resource need not be used in both sectors or at all. For example, if the scarcity value of groundwater is initially low, both demands are supplied by groundwater. As scarcity rises, it becomes cost-effective to supplement groundwater extraction with recycled water in the agriculture sector, while the household sector continues to use groundwater exclusively. In the steady state, all available resources are used in each sector.

The gains from recycling are most apparent in the sectors that are able to use recycled water. In Figure 4.2, the agriculture sector postpones the use of costly desalination from T_A in the top panel to T_A in the bottom panel. Starting from period t_s, recycled water is used to supplement groundwater, and users enjoy lower prices over a longer period of time. Households also gain from the recycling program because substituting groundwater in the agriculture sector means that scarcity is reduced for *all* groundwater users. Although the optimal marginal opportunity cost (MOC) is rising over time, total consumption may be rising or falling, since the growth in demand and the price effect work in opposite directions. If the effects are of equal magnitude or are both relatively small, optimal consumption may be nearly constant.

Systems approach to water management

Fully characterizing tradeoffs across sources and end uses requires integration of all available demand- and supply-side management strategies. In the examples discussed in this chapter, demand-side management included sector-specific pricing, and supply-side management included two types of instruments: (i) augmentation of existing groundwater resources (watershed conservation) and (ii) optimal implementation of groundwater alternatives (water recycling and desalination). With multiple demands, resources, and management instruments, the framework should more explicitly account for spatial heterogeneity.

Consider the case of two consumption districts and two aquifers, where each of the aquifers has its own watershed, each of the consumption districts has its own desalination

54

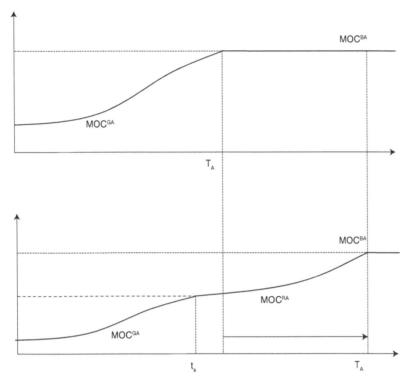

Figure 4.2 Hypothetical time paths of MOCs

and recycling substitutes, and the transport costs of water from one district to the other are given.[4] Assume moreover that if no water is transported from one district to another that each of the consumption districts is more cheaply served by the aquifer nearest to it. In this case one simply solves for each watershed-aquifer-district separately by simultaneously satisfying the optimality conditions for each system. However, for a fully internal solution with positive transport from one system to another, the two systems themselves must be simultaneously solved such that the shadow price of the recipient district is equal to the shadow price of the source district plus transport costs. In this case there is effectively one demand, the aggregation of the source demand and the recipient demand shifted up by the amount of the transport costs (Pitafi and Roumasset 2009). Accordingly, the order of extraction from the two aquifers can be solved by the principle of least-MOC-first. In the case of "leaky" coastal aquifers, this may mean that the largest and leakiest aquifer should be used first since the opportunity costs of future use is offset by the reduction of leakage (Chapter 3 in this volume). In the case of inland aquifers with nearly constant extraction costs, it will typically mean the aquifer with the lowest extraction cost is used first.

Implementation

Assuming that production of recycled water is never constrained by the availability of primary feed water, its marginal opportunity cost is captured entirely by the marginal cost of treatment, inclusive of construction costs for the treatment facilities and distribution infrastructure. Thus, implementation is relatively straightforward. In the case of watershed conservation, however, the marginal opportunity cost of groundwater depends on upstream decisions to invest in natural capital. Inasmuch as the upstream providers of groundwater recharge services generally receive only a portion of the benefits generated by their investment in natural capital, welfare maximization for society as a whole often requires some means of compensating upstream landowners, i.e. a payment for watershed services (PWS) plan.

The PWS approach falls under the broader umbrella of payments for ecosystem services (PES), which has advanced as a method for incentivizing provision of any number of ecosystem services by private landowners (Daily et al. 2009). PES, which are "voluntary transactions where a well-defined environmental service is being bought by a service buyer from a service provider if and only if the service provider secures service provision" (Engel et al. 2008), are sometimes set in accordance with the perceived opportunity cost of upstream landowners (e.g. Muñoz-Piña et al. 2008). In other cases, ecosystem services prices are viewed as exogenous, meaning private providers can be theoretically incentivized by payments according to a fixed Pigouvian subsidy (Baumol and Oates 1988). If the ecosystem service in question provides indirect benefits via another natural resource, however, the marginal benefit of the service is only revealed as part of a joint solution (Sathirathai and Barbier 2001; Barbier et al. 2002; Barbier 2007; Sanchirico and Springborn 2011; Roumasset and Wada 2013). For example, the value of groundwater recharge provided by a watershed depends on both the quantity of groundwater extraction and the amount of watershed conservation.

While most researchers appear to be in agreement that beneficiaries should at least be partially responsible for financing payments, the economic foundations underlying many PES programs remain largely informal and leave open important questions, such as how payments should be precisely determined and how conservation investment should be financed without distorting incentives. In practice, a common way to finance payments for ecosystem services is to use general revenue (Liu et al. 2008; Pagiola 2008), which has the disadvantage of increasing required tax revenues and consequently generating additional marginal excess burden (Ballard et al. 1985). A frequently employed alternative to financing from general revenue, especially for PWS programs, is a volumetric tax—e.g. Houston's (Texas) dedicated conservation fees, Salt Lake City's (Utah) Watershed-Water Rights Purchase Fund Program, Rhode Island's Watershed Land Acquisition Program, and New York City's Catskills Watershed Management Plan. Although these types of programs target beneficiaries in the spirit of PES/PWS and generate revenue, they are not welfare maximizing because user fees drive a wedge between marginal benefits and costs. One way to avoid this problem is to finance PES through intergenerational benefit taxation.

Optimality conditions (4.4) and (4.5) confirm that the marginal benefit of the upstream watershed service is dependent on the shadow price of the downstream groundwater resource. Private producers of the recharge service would only invest in watershed conservation if the payments received from doing so are at least equal to the costs, including the opportunity cost of the land. In the simplest case in which the opportunity cost of the land is negligible, e.g. in upper watershed areas not particularly suitable for urban or agricultural development, Roumasset and Wada (2013) show that the decentralized solution for ecosystem service provision emulates the social optimum

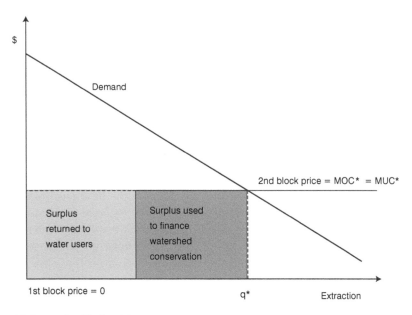

Figure 4.3 Increasing block pricing structure

if the government sets the payment for each unit of recharge equal to the shadow price of groundwater, adjusted for the discount rate and capital depreciation, and water is priced at its marginal opportunity cost. Optimal ecosystem service and water pricing generates a revenue surplus, which can be returned to water consumers in lump sum fashion with appropriate block pricing (e.g. Pitafi and Roumasset 2009). Because the water price need only be correct at the margin, the size of the first price block can be adjusted to return surplus to water users through free inframarginal units of water (Figure 4.3).

In some instances, the size of a conservation project is determined exogenously by policymakers or the conservation instrument may exhibit non-convexities that render marginal-product pricing infeasible. A large public project may incur substantial costs initially, even though most of the benefits are generated in future periods. If the revenue surplus is not sufficient to cover initial investment, bond-financing will be required, such that the present value of collections is equal to the present value of investment. A lump sum proportional benefit tax can preserve efficient incentives while maintaining a balanced intergenerational budget. Under such a program, each generation pays a fixed proportion of its watershed conservation induced benefits—lower water prices and delayed implementation of expensive alternatives such as desalination—to pay off the bond required to finance the conservation project.

Adapting to climate change

Existing climate models, although highly uncertain, predict that the frequency of storm events will increase in many regions across the globe and mean rainfall will decrease in some regions in the coming centuries. If the projections turn out to be reasonably accurate, the net effect will be an increase in runoff and a decrease in recharge to groundwater resources in many areas. Adapting to climate change, therefore, requires modifying groundwater management frameworks to allow natural recharge to decline over time. The equimarginality conditions for groundwater extraction, recycled water use, and watershed conservation remain unchanged, except that recharge is a function of both conservation capital and a time-varying component that captures the effect of climate change. The general effect of optimally employing various conservation instruments (compared to managing only extraction) is the same with or without climate change: scarcity is reduced and the resource is drawn down more slowly toward its steady state level. However, patterns of extraction and investment in conservation capital may be substantially different in transition to the steady state.

As previously discussed, a period of aquifer replenishment may optimally precede drawdown in the case of constant recharge. That is to say the head level may follow a non-monotonic path to the steady state. The incentive to build the groundwater stock may be even larger in the face of climate change, given that future water scarcity is expected to be higher owing to declining recharge. Similarly, optimal watershed conservation will initially increase in the short and medium term. But since the unit cost of supplying recharge via watershed conservation increases as rainfall declines, the groundwater scarcity effect may be eventually dominated, resulting in a non-monotonic path of optimal watershed investment.

Challenges and research needs

While many advances have been made in the theory of integrated water resource management, operationalizing optimization models over time and space remains a challenge. One particularly important challenge involves interfacing complicated physical watershed and groundwater models with an economic optimization framework. Starting with a simple single-cell aquifer, one can imagine a grid representing the watershed landscape. Given satellite or other spatially disaggregated data, one could characterize each section of the grid in terms of slope and permeability, the latter being a function of investment in natural capital and/or physical capital. The first problem is to determine the optimal allocation of investment across the grid, such that the marginal dollar invested in each cell yields the same recharge. The next step is to implement the watershed optimization model, increasing total investment in natural capital until the marginal opportunity cost equals the shadow price of water, co-determined with aquifer management. Once the optimal spatial composition is determined for each level of aggregate investment, the remainder of the problem can be solved using standard methods, given that we need only keep track of two state variables (conservation capital and groundwater).

The problem becomes more complicated when the hydrology is instead described by a higher dimensional physical model, wherein the aquifer is also divided into multiple cells, with specification of cell-to-cell gradients for lagged water movements over time. Even if aggregate pumping can be optimally distributed across available wells in every period to satisfy the total optimal rate of extraction, cell-specific head levels will depend on spatially distributed infiltration and extraction as well as cell-to-cell movements. And while complicated spatial hydrology can be modeled using numerical methods, optimization solutions become increasingly less reliable and eventually unattainable as the number of state variables—in this case head levels associated with each cell—

grows. The management problem becomes even more complex when climate change is considered. For example, Monte Carlo methods may be required to generate rainfall sequences that serve as inputs to the watershed model.

One way to address these research challenges is to improve search algorithms for numerical solutions or to find and apply existing algorithms previously unused in the field of resource economics. Another approach is to refine the simplifying assumptions underlying tractable lower-dimensional models that are currently solvable with existing algorithms. Refinements might include, for example, improving spatial aggregation methods or simplifying linkages between various model components. The two approaches are not mutually exclusive, however. The validity of proposed model refinements can and should be tested to the extent feasible using existing numerical models.

Appendix

The groundwater hydrology for the 2-sector 3-resource management problem is described by the following equation:

$$\dot{h}_t = R(N_t) - L(h_t) - (q_t^{GH} + q_t^{GA}) \tag{4.A1}$$

where the total quantity of groundwater extracted in every period is equal to the sum of household extraction (q^{GH}) and agricultural extraction (q^{GA}). The marginal benefit of a unit of water is represented by sector-specific demand functions D_H and D_A for the household and agriculture sectors respectively. The management problem is to choose groundwater extraction for each sector, desalination for the household (q^{BH}) and agriculture (q^{BA}) sectors, and water recycling for the agriculture sector (q^{RA}) to maximize net present value:

$$\underset{q_t^{GH}, q_t^{BH}, q_t^{GA}, q_t^{RA}, q_t^{BA}}{Max} \int_0^\infty e^{-rt} \left\{ \begin{array}{l} \int_0^{q_t^{GH}+q_t^{BH}} D_H^{-1}(x,t)dx + \int_0^{q_t^{GA}+q_t^{RA}+q_t^{BA}} D_A^{-1}(x,t)dx - \\ (q_t^{GH} + q_t^{GA})c_G(h_t) - (q_t^{BH} + q_t^{BA})c_B - q_t^{RA}c_R(q_t^{RA}) \end{array} \right\} dt \tag{4.A2}$$

subject to equation (4.A1), non-negativity constraints on the control variables, and a minimum head level constraint.

Notes

1 See Chapter 8 in this volume for a framework that allows for different types of conservation instruments.
2 See the appendix for a rigorous mathematical representation of the management problem.
3 See Chapter 3 in this volume for a more detailed discussion of the marginal opportunity cost and optimal ordering of water resources.
4 The framework can be further generalized by incorporating additional distribution costs within a particular district, e.g. for pumping water to users located at different elevations (Pitafi and Roumasset 2009).

References

Ballard, C.L., Shoven, J.B. and Whalley, J. (1985) "General Equilibrium Computations of the Marginal Welfare Costs of Taxes in the United States", *The American Economic Review*, vol 75, no 1, pp. 128–138.

Barbier, E.B. (2007) "Valuing Ecosystem Services as Productive Inputs", *Economic Policy*, January, pp. 177–229.

Barbier, E.B., Strand, I. and Sathirathai, S. (2002) "Do Open Access Conditions Affect the Valuation of an Externality? Estimating the Welfare Effects of Mangrove-Fishery Linkages in Thailand", *Environmental and Resource Economics*, vol 21, pp. 343–367.

Baumol, W.J. and Oates, W.E. (1988) *The Theory of Environmental Policy*, Cambridge University Press, Cambridge.

Daily, G.C., Polasky, S., Goldstein, J., Kareiva, P.M., Mooney, H.A., Pejchar, L., Ricketts, T.H., Salzman, J. and Shallenberger, R. (2009) "Ecosystem Services in Decision Making: Time to Deliver", *Frontiers in Ecology and the Environment*, vol 7, no 1, pp. 21–28.

Engel, S., Pagiola, S. and Wunder, S. (2008) "Designing Payments for Environmental Services in Theory and Practice: An Overview of the Issues", *Ecological Economics*, vol 65, pp. 663–674.

Jorgenson, D.W. (1963) "Capital Theory and Investment Behavior", *The American Economic Review*, vol 53, no 2, pp. 247–259.

Krulce, D.L., Roumasset, J.A. and Wilson, T. (1997) "Optimal Management of a Renewable and Replaceable Resource: The Case of Coastal Groundwater", *American Journal of Agricultural Economics*, vol 79, pp. 1218–1228.

Liu, J., Li, S., Ouyang, Z., Tam, C. and Chen, X. (2008) "Ecological and Socioeconomic Effects of China's Policies for Ecosystem Services", *Proceedings of the National Academy of Sciences*, vol 105, pp. 9477–9482.

Muñoz-Piña, C., Guevara, A., Torres, J.M. and Braña, J. (2008) "Paying for the Hydrological Services of Mexico's Forests: Analysis, Negotiations and Results", *Ecological Economics*, vol 65, pp. 725–736.

Pagiola, S. (2008) "Payments for Environmental Services in Costa Rica", *Ecological Economics*, vol 65, pp. 712–724.

Pitafi, B.A. and Roumasset, J.A. (2009) "Pareto-Improving Water Management over Space and Time: The Honolulu Case", *American Journal of Agricultural Economics*, vol 91, no 1, pp. 138–153.

Roumasset, J. and Wada, C.A. (2011) "Ordering Renewable Resources: Groundwater, Recycling, and Desalination", *The B.E. Journal of Economic Analysis & Policy (Contributions)*, vol 11, no 1, article 28.

Roumasset, J.A. and Wada, C.A. (2012) "Ordering the Extraction of Renewable Resources: The Case of Multiple Aquifers", *Resource and Energy Economics*, vol 34, pp. 112–128.

Roumasset, J. and Wada, C.A. (2013) "A Dynamic Approach to PES Pricing and Finance for Interlinked Ecosystem Services: Watershed Conservation and Groundwater Management", *Ecological Economics*, vol 87, pp. 24–33.

Sanchirico, J.N. and Springborn, M. (2011) "How to Get There from Here: Ecological and Economic Dynamics of Ecosystem Service Provision", *Environmental and Resource Economics*, vol 48, pp. 243–267.

Sathirathai, S. and Barbier, E.B. (2001) "Valuing Mangrove Conservation in Southern Thailand", *Contemporary Economic Policy*, vol 19, no 2, pp. 109–122.

United Nations Water (2006) "Coping with Water Scarcity: A Strategic Issue and Priority for System-Wide Action, 2006", www.unwater.org.

5

OPTIMAL CONJUNCTIVE WATER USE OVER SPACE AND TIME

Sittidaj Pongkijvorasin and James A. Roumasset

Introduction

"Water scarcity affects all social and economic sectors and threatens the sustainability of the natural resources base" (United Nations, 2006). Among the different types of water use, agriculture is the largest consumer of water worldwide, often accounting for 80 percent of water use in a country or region (Wolff and Gleick, 2002). Many economists and policy planners have been searching for a way to improve the efficiency of irrigation water allocation. Water pricing and water markets are increasingly recommended as management strategies in the face of growing scarcity. But there is no consensus, for example, on how water prices should be set. How should transportation costs, including water losses in transit, be assessed? Similar issues cloud the setting of exchange rates to facilitate water trading. This chapter[1] connects resource, environmental and hydrological economics and derives shadow prices over space and time for a model of conjunctive water-use that incorporates conveyance losses that occur during water transmission.

It is widely recognized that market-based instruments promise substantial efficiency gains in water allocation (e.g., Easter et al., 1999). But inasmuch as water is a different commodity at different points of time and space, proposed water trades must often be reviewed by committees representing a variety of stakeholders before they are approved. These administrative procedures may be an important explanator of why water markets have failed to develop to the extent originally expected (Howe, 1998). The necessity for administrative review can be reduced to the extent that the appropriate governing authorities develop efficient exchange rates for the trades in question. This requires a system for determining shadow prices over space, time, and form while fully accounting for water balance conditions. Administrative "safeguards" would then be required only for any environmental or social concerns left out of the model, reducing transaction costs accordingly.

It has been shown that the Hotelling-Herfindahl theory-of-the-mine is inadequate for the case of independent demands for different resource and a number of papers have provided a more general theory for ordering the extraction of non-renewable resources (e.g., Chakravorty et al., 2004; Gaudet et al., 2001). A theory of extraction-ordering has also been provided for the case of two groundwater aquifers (Roumasset and Wada, 2012, and Chapter 3, this volume). The problem of conjunctive irrigation management affords an opportunity to extend resource economics to the case of two renewable resources, one of which (groundwater) has a conventional stock-dependent cost function and the other (surface water) has a cost function dependent on the total amount of water sent from the headworks and conveyance efficiency. The two resources are related by the extent that surface water use helps to recharge groundwater.

Water lost during transmission, mainly from seepage and evaporation, is a common issue in water irrigation. In the presence of conveyance losses, efficient allocation requires that the quantity of received water decreases with the distance from the water source (Chakravorty and Roumasset, 1991). Chakravorty et al. (1995) develop a spatial model to determine the optimal conveyance investment. Conveyance improvement is found to have strong efficiency and equity effects. This finding reinforces the importance of conveyance loss concerns in irrigation management.

Chakravorty and Umetsu (2003) incorporated conveyance losses into a spatial model wherein the transmission costs of surface water are assumed to increase with the distance from the irrigation headworks and where groundwater extraction costs are uniform across space. In a static framework, they find that the shadow price of surface water increases with the distance, while that of groundwater is independent of location. The optimal program calls for farmers near to the headworks to use surface water and for farmers farther away to pump groundwater. The threshold distance from the headworks is endogenously determined by the condition that the shadow price of delivered surface water equals that of pumped groundwater.

In a dynamic context, changes in groundwater stock and extraction cost over time must be taken into account (e.g., Koundouri, 2004; Koundouri and Christou, 2006; O'Mara, 1984; Tsur, 1991). Tsur (1991) provides a dynamic equilibrium model of conjunctive use of ground and surface water, wherein each farmer uses the water that is cheapest to him. Smith and Roumasset (2000) model the conjunctive water management of two districts in Hawaii. They consider the case where one district has exclusive access to a surface water source and the other to groundwater source, and both have access to a third (surface water) source. The condition for an internal optimal solution is that the third source is allocated to equalize the wholesale prices in the two districts. Without the third source, the price in the surface water district would increase faster than that in the groundwater district because the demands grow and surface water supply is completely inelastic. To counter this tendency, an increasing percentage of third source water must be allocated to the surface water district in order to keep the wholesale prices in both districts the same.

Pongkijvorasin and Roumasset (2007) provide a dynamic model of conjunctive use of ground and surface water. The model in their paper accounts for the conveyance losses, groundwater recharge from conveyance and on-farm losses, and canal return-flows. They show that the results from the static model of Chakravorty and Umetsu (2003) still hold in the dynamic context; however, the boundary between surface and groundwater use may be changing over time. Beyond that, however, the model is too general to generate any unambiguous results, and they do not provide any numerical analysis to illustrate possibilities.

The current chapter extends Chakravorty and Umetsu (2003) and Pongkijvorasin and Roumasset (2007) by developing a dynamic model of conjunctive water use with the presence of conveyance losses, and groundwater recharge (from conveyance and on-farm losses). We confirm the primary spatial results found in those studies that the amount of optimal surface water use declines with the distance from the headworks and that farmers far enough from the headworks use only groundwater. We further find that when the initial head level[2] is above its steady state level, the shadow price of groundwater increases faster than that of surface water as the head level is drawn down such that the optimal area irrigated by surface water increases over time. The reverse is true if groundwater has already been depleted below its steady state level such that the shadow price of groundwater increases as head level increases and the boundary between ground and surface water moves back towards the headworks. The model is operationalized and illustrated by combining data from a coastal aquifer with evidence on conveyance losses and recharge from surface water.

The following section develops the theoretical model and illustrates the solutions. Shadow prices over space and time and the optimal patterns of conjunctive water use are derived in the third, fourth and fifth sections respectively. The fifth section provides a numerical simulation using available data from water productivity in agriculture and hydrological properties of a groundwater aquifer. The final section provides conclusions and suggestions for future research.

The Model

We assume that the surface water is supplied from an upstream point source (headworks). The water is sent via an irrigation canal to the downstream farms. There are n homogenous farms located along the canal. The distance between the headworks and farm i is denoted by y_i. In this setup, the farm nearest the headworks is designated by $i=1$, and the nth farm is furthest downstream. We denote q_i as the water sent from the headworks for farm i, and x_i as the water received at farm i.

Part of the water sent from the headworks is lost during conveyance to seepage and evaporation such that $q_i = \dfrac{x_i}{s(y_i)}$, where $s(y_i)$ is the conveyance efficiency for distance from the headworks, y_i. The conveyance efficiency is assumed to have the following properties: $0 \leq s(y) \leq 1$, $s(0)=1$, and $s'(y)<0$. The conveyance loss of water sent to farm i is then equal to $q_i - x_i = \dfrac{1-s(y_i)}{s(y_i)} x_i$. A proportion of conveyance loss, μ, percolates to the groundwater aquifer, $0 \leq \mu \leq 1$.

Let C denote the total cost of sending surface water, which depends on the total amount of water sent ($Z = \sum_i q_i = \sum_i \dfrac{x_i}{s(y_i)}$). The long run cost of surface water includes the implicit rental cost of the infrastructure (including maintenance) as well as operational expenses including volumetric measurement. The marginal cost of supplying surface water, $c(Z)$, is assumed to be increasing in the amount of water sent, i.e., $c'(Z)>0$.[3]

Following Chakravorty and Umetsu (2003), we assume that groundwater is available to all farmers in the system and that each farm faces the same head level and extraction costs. The marginal cost of groundwater extraction, $c_g(h)$, is a decreasing and convex function of head level, i.e., $c'_g(h)<0$, and $c''_g(h) \geq 0$. There is also natural leakage from the groundwater aquifer, especially for coastal aquifers (e.g., Krulce et al., 1997).

Leakage is assumed to be a positive, increasing, and convex function in the head level, i.e., $l(h) \geq 0$, $l'(h) > 0$, and $l''(h) > 0$. The amount of groundwater extraction for each farm is denoted by g_i.

The model in this chapter also takes into consideration the on-farm loss of water. For one unit of received water, k percent is effectively used for the farm production (consumptive use), while the rest is the on-farm loss. Out of the loss, β percent recharges the aquifer; the rest is the loss out of the system. Let p be the price of the farm product, $f(e)$ be the production function of delivered water,[4] and e_i be the effective or consumptive water use at farm i, which is equal to $k(x_i + g_i)$.[5] The net benefit of water use is the excess of benefit above the cost of supplying surface water and the cost of groundwater extraction, i.e., $\sum_{i=1}^{n} pf(x_i + g_i) - C(Z) - \sum_{i=1}^{n} c_g(h)g_i$.[6] Let r denote the discount rate. The social planner's problem is now to choose the amounts of surface water and groundwater use in order to maximize the stream of net benefits, i.e.,

$$\max_{x_i, g_i} \int_{t=0}^{\infty} [\sum_{i=1}^{n} pf(x_i + g_i) - C(Z) - \sum_{i=1}^{n} c_g(h)g_i] e^{-rt} dt \tag{5.1}$$

$$\text{s.t. } Z = \sum_{i=1}^{n} \frac{x_i}{s(\gamma_i)}$$

$$\dot{h} = a[R - l(h) - \sum_{i=1}^{n} g_i + \mu \sum_{i=1}^{n} \frac{1 - s(\gamma_i)}{s(\gamma_i)} x_i + \beta(1-k) \sum_{i=1}^{n} (x_i + g_i)]$$

and $\quad x_i \geq 0, g_i \geq 0$

where $\quad a$ is a conversion parameter that converts volume to height;

and $\quad R$ is the natural recharge rate of the groundwater aquifer.

Solving the problem using optimal control theory, the optimal conditions for surface water and groundwater use can be derived (see Appendix A). The optimal condition for surface water use can be written as:

$$kpf'(x_i + g_i) = \frac{c(Z)}{s(\gamma_i)} - a\lambda\mu \frac{1 - s(\gamma_i)}{s(\gamma_i)} - a\lambda\beta(1-k) \tag{5.2}$$

Equation (5.2) indicates that, when surface water is used, the water must be used such that its marginal benefit (value of marginal product of water) is equal to the full marginal cost of sending water (including conveyance losses) minus the groundwater recharge from the conveyance loss and on-farm loss. The shadow price of the received surface water ($p_x(\gamma_i)$) is equal to the value of the marginal product of the water received, $kpf'(x_i + g_i)$. If the cost of sending surface water is higher than the benefit from its conveyance loss,[7] the derivative of equation (5.2) with respect to the distance is unambiguously positive. As a result, the shadow price of surface water increases with the distance.

The optimal condition for groundwater use can be expressed as:

$$kpf'(x_i + g_i) = c_g(h) + a\lambda - a\lambda\beta(1-k) \tag{5.3}$$

Like surface water, groundwater should be used until its marginal benefit is equal to its full marginal cost—the unit extraction cost plus the marginal user cost minus the value of aquifer recharge from on-farm use. The shadow price of the groundwater (p_g) is equal to the value of the marginal product of the water, which is equal to the right-hand side of equation (5.3). From equation (5.3), the groundwater shadow price is identical across farms since no conveyance is needed.

The optimal condition for groundwater extraction over time can be written as:

$$r\lambda + al'(h)\lambda = \dot{\lambda} - c_g'(h)\sum_{i=1}^{n} g_i \tag{5.4}$$

The left-hand side is the cost of water conservation, which consists of interest on the foregone royalty and the value of an increase in leakage if water is not extracted. The right-hand side represents the marginal benefits of water conservation, including the change in royalty and the savings in extraction costs.

Optimal pricing policies over space

Considering a general case where surface water, as well as groundwater, is used by at least one farm, we can show that:

Proposition 1: Along the optimal trajectory
a there is only one location where both ground and surface water can be used jointly.
b farms located at a distance less than y_c use only surface water.
c farms located at a distance more than y_c use only groundwater.
Proof: See Appendix B.

Figure 5.1 shows the shadow prices of surface and groundwater at each location and the critical distance where farms start using groundwater. The shadow price of delivered surface water ($p_x(y_i)$) increases with the distance. The shadow price of groundwater (p_g) is identical regardless the distance. According to the principle of least opportunity cost first, farms located between the headworks and y_c use only surface water while more distant farms use groundwater.

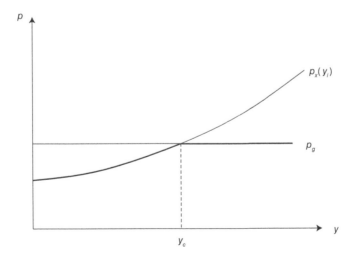

Figure 5.1 Shadow prices of surface and groundwater over space

The shadow price of water used by farms in different locations is given by the lower envelope of the ground and surface water price paths (bold line in Figure 5.1). At a particular time, the full marginal cost of groundwater acts as a backstop to the price of surface water albeit over space instead of time. This verifies that the main result of Chakravorty and Umetsu (2003)—specialization of production over space—holds in a dynamic setting as well.

Dynamics of conjunctive water use

From the first-order conditions, the optimal condition for surface water requires equality between marginal benefit and marginal cost in each period (equation 2). For groundwater, the optimal conditions involve both the equality of the marginal benefit and the marginal cost in each period (equation 3) and the condition governing scarcity rent over time (equation 4). Using phase-diagram analysis of the steady state and possible time paths, proposition 2 can be shown.

Proposition 2: If $h_0 < h\star$, then $\dot{p}_x < 0$, $\dot{p}_g < 0$ and $\dot{y}_c < 0$ until they reach the steady states. If $h_0 > h\star$, then $\dot{p}_x > 0$, $\dot{p}_g > 0$ and $\dot{y}_c > 0$ until they reach the steady states. Proof: See Appendix C.

Proposition 2 is illustrated in Figure 5.2 for the case of an initial head lower than its steady state level. Time A corresponds to the case already illustrated in Figure 5.1. At a later time, B, the optimal head level is higher as the head monotonically transitions to its steady state level. The higher head level implies that a lower shadow price of water is now acting as the "backstop" price for surface water. This has two effects. First the lower backstop price cuts off surface water usage at the closer boundary, $y_c(B)$, but also that the shadow price schedule for surface water shifts downward. That is, the shadow prices of both surface and groundwater decrease over time and the surface water irrigation boundary shrinks over time until they reach their steady state levels (see also Figure 5.3). Moreover, as illustrated in Figure 5.2, the price of groundwater decreases faster than that

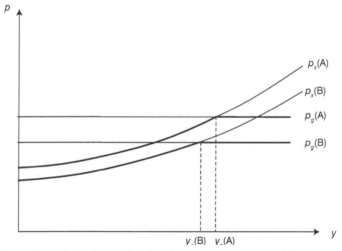

Figure 5.2 The shadow prices of ground and surface water over space and time

of surface water. Everything is reversed if the initial head level is higher than its steady state level, i.e., time B is before time A.

The order of resource use over space and time thus follows the least-opportunity-cost-first principle, previously developed in the context of the multiple, non-renewable resources model (e.g., Chakravorty et al., 2004). For the case of a falling (rising) head level, the area of optimal surface water coverage expands (shrinks). Furthermore, in the absence of set up costs, a particular farm can switch from ground to surface water (or vice versa) only once; there can be no reswitching.[8]

Numerical simulation

In this section, we provide a simple numerical illustration of the model developed above. The parameters related to irrigation and on-farm production are drawn from Chakravorty et al. (1995) and Chakravorty and Umetsu (2003). The parameters for the groundwater aquifer are extracted from Duarte (2002).

The agricultural production function is given as a quadratic function of effective water on each farm and the output price is taken to be \$0.75/lb. The revenue function as a function of effective water (m³) is given by:

$$pf(e) = -0.2224 + 1.0944e - 0.5984e^2 \tag{5.5}$$

The value of marginal product function is:

$$pf'(e) = 1.0944 - 1.1968e \tag{5.6}$$

The rising long-run marginal cost function for water supply was constructed from average cost of water supply data from 18 irrigations projects in the western United States (Chakravorty et al., 1995) as: $c'(Z) = 0.003785 + (3.785 \times 10^{-11} Z)$, where the cost is in \$ and Z is in m³. Conveyance loss occurs at the rate of 0.4 per 10 km or a conveyance efficiency of 0.6 per 10 km. The efficiency of on-farm water use depends on the technology applied in the farm. Traditional furrow irrigation yields a low efficiency rate of 0.6. This means that for one unit of water received at the farm, 0.6 will be consumptive use. In this chapter, we assume that 10 percent of conveyance loss and on-farm loss percolates and recharges the groundwater aquifer.

Groundwater parameters are based on a coastal aquifer. The aquifer's head level is determined by the width, length, and porosity of the cap-rock. We assume that 1,000,000 m³ of water contributes to the 0.00396 m change in the head level (m) (Duarte, 2002). The state equation for the aquifer can be written as: $\dot{h}_t = 3.96 \times 10^{-9}(R - l_t - q_t)$. The natural leakage is a function of the head level. Mink (1980) shows that the relationship can be expressed by: $l(h) = bh^2$, where b is a coefficient specific to an aquifer. In this chapter, we use data on Kukio region of Hawaii from Duarte (2002). The discharge function is estimated to be $l(h) = 4,800,000h^2$.

The cost of extracting groundwater is mainly the cost of energy used to lift water to the ground level. The energy cost of lifting one unit of water (1 m³) up 1 m is estimated to be \$0.00083 (Duarte, 2002). Thus, the unit cost of groundwater extraction can be expressed by: $c = 0.00083(H - h)$, where h is the head level, and H is elevation at the well site. We assume here that $H = 100$ m and $h = 1.75$ m.

Results over space

The model is run using Excel's Solver program. Groundwater and surface water are used such that the net present value over 30 years is maximized, using a discount rate of 5 percent. As described in the theoretical section, the shadow price of surface water increases with distance from the headworks while that of groundwater is invariant. The shadow price of surface water starts from less than 0.01 $/m³ at the headworks and rises exponentially, according to the conveyance efficiency (0.6), to higher than 0.7 $/m³ for a farm located 100 km away. For the first period, the shadow price of groundwater is approximately 0.1 $/m³ and the critical distance where the prices of groundwater and surface water are equal is at 56.8 km from the headworks.

Results over time

For our illustrative case, the initial head level was over-depleted such that the optimal head level increases from 1.75 m to 2.56 m in the 30-year period resulting in progressively lower shadow prices of both ground and surface water. Corresponding to the theoretical section, the shadow price of groundwater decreases at a faster rate, and the surface water boundary shrinks over time. After 30 years the boundary shrinks from 56.8 km to 56.6 km. Figure 5.3 illustrates the changing boundary over time.

Conclusions

This chapter develops a dynamic model of conjunctive water use, taking into account the conveyance losses that occur during water transportation and groundwater recharge from conveyance and on-farm use. Shadow prices of delivered surface water are equal to the marginal cost of water at the headworks plus the value of conveyance losses minus the value of water recharge to the aquifer. Charging shadow prices that increase with distance will result in optimal usage of surface water. The full marginal cost of

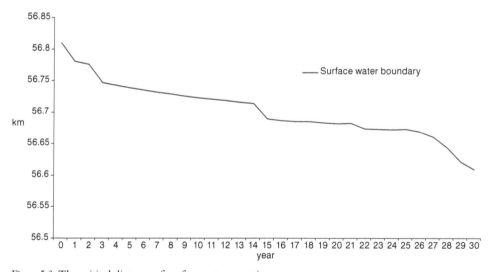

Figure 5.3 The critical distance of surface water over time

groundwater, however, is invariant with respect to location. According to the principle of least-opportunity-cost-first, there is a threshold distance from the headworks below which farmers optimally use surface water and above which they use groundwater. This implies that farmers nearer the headworks use surface water, while the more distance farms use groundwater.

For the case where the initial head level is below its steady state level, the prices of surface and groundwater decrease over time in the dynamically optimal solution resulting in an increase in the optimal use of groundwater and an increased use of surface water on farms continuing to use that source. However, the price of surface water decreases more slowly than that of groundwater resulting in shrinkage of the surface water area. Despite the identical and constant water demands by each farm, optimal water allocation changes according to changing shadow prices over time and space. For the case where the initial head level is above its steady state, the surface water irrigation area expands over time.

We have found conditions for specific patterns of optimal conjunctive use over space and time. One immediate implication is that optimal design of irrigation systems (and optimal maintenance and rehabilitation of existing systems) should take account of how the surface water command area should change over time. For example, if groundwater is being depleted over time, then the surface-water command area can be gradually expanded accordingly.

Empirical results confirm the findings from the analytical model. However, with the parameters used, the change in the surface water boundary is very small (e.g., only 200 meters, from 56.8 to 56.6 km over 30 years). If this were generally the case, the welfare effects of the ignoring spatial changes over time would also be small. There is accordingly a need to examine the robustness of this result for other parameters.

The model developed in this chapter can be readily extended to incorporate the problem of water quality. On-farm water use may induce contamination of discharged water, which flows to nearby ecosystems (e.g., canals, rivers, groundwater aquifers or lakes). Conceptually, the optimality condition—that water is used until the marginal benefit equals the marginal cost—still applies. However, in this case, the full marginal cost must include the external cost of pollution. In other words, efficiency pricing must involve charging the marginal cost for water sent from the headworks (including the conveyance losses) and the marginal externality cost of the pollution. Another interesting extension involves modeling equilibrium departures from the social optimum, e.g., extracting water according to the model of an open access resource.

Other extensions that could be incorporated into the model include uncertainty regarding surface water supply and stochastic processes affecting evaporation and recharge. The assumption of a homogenous single-cell aquifer underlying all farms can also be relaxed, e.g., modelling water gradient as a function of the slope or accounting for more detailed three-dimensional hydrological aspects. Another extension would be to allow for endogenous conveyance efficiency and on-farm technology by including canal lining and irrigation technology as controls.

In this chapter, we have focused on first-best patterns of allocation and shadow prices. Informational and political economy issues need further consideration prior to implementation of the pricing principles proposed. Straight marginal cost pricing, for example, may be burdensome and perceived to be inequitable especially when it implies that more distant farmers would have to pay a higher total charge for less water received. Some of these problems may be soluble through block pricing. Optimal policy requires charging the optimal price for only the marginal unit of water use. Alternatively, farmers can be granted water entitlements which they can supplement or sell using water markets.

In the presence of imperfect information about water productivity, mechanisms can be designed to reveal differential productivities. Schemes that approximate the proposed solution but which save on administrative costs are natural candidates for further exploration.

Appendix A: Solutions to the optimal control problem

The social planner's problem is now to choose the amounts of surface water and groundwater use in order to maximize the stream of net benefits, i.e.,

$$\max_{x_i, g_i} \int_{t=0}^{\infty} [\sum_{i=1}^{n} pf(x_i + g_i) - C(Z) - \sum_{i=1}^{n} c_g(h)g_i] e^{-rt} dt \tag{5.A1}$$

s.t. $Z = \sum_{i=1}^{n} \frac{x_i}{s(\gamma_i)}$

$$\dot{h} = a[R - l(h) - \sum_{i=1}^{n} g_i + \mu \sum_{i=1}^{n} \frac{1 - s(\gamma_i)}{s(\gamma_i)} x_i + \beta(1-k) \sum_{i=1}^{n} (x_i + g_i)]$$

and $\quad x_i \geq 0, g_i \geq 0$

where $\quad a$ is a conversion parameter that converts volume to height;

and $\quad R$ is the natural recharge rate of the groundwater aquifer.

The corresponding current-value Hamiltonian equation will be in the following form:

$$\tilde{H} = \sum_{i=1}^{n} pf(x_i + g_i) - C(Z) - \sum_{i=1}^{n} c_g(h)g_i$$

$$+ \lambda a[R - l(h) - \sum_{i=1}^{n} g_i + \mu \sum_{i=1}^{n} \frac{1 - s(\gamma_i)}{s(\gamma_i)} x_i + \beta(1-k) \sum_{i=1}^{n} (x_i + g_i)] \tag{5.A2}$$

where λ is the shadow price of groundwater in terms of the head level. The first order conditions are as follows:

$$\frac{\partial \tilde{H}}{\partial x_i} = kpf'(x_i + g_i) - \frac{c(Z)}{s(\gamma_i)} + a\lambda\mu \frac{1 - s(\gamma_i)}{s(\gamma_i)} + a\lambda\beta(1-k) \leq 0 \quad (= 0, if \ x_i > 0); \quad i = 1,...,n$$

$$\tag{5.A3}$$

$$\frac{\partial \tilde{H}}{\partial g_i} = kpf'(x_i + g_i) - c_g(h) - a\lambda + a\lambda\beta(1-k) \leq 0 \quad (= 0, if \ g_i > 0); \quad i = 1,...,n \tag{5.A4}$$

$$-\frac{\partial \tilde{H}}{\partial h} = c_g'(h) \sum_{i=1}^{n} g_i + \lambda al'(h) = \dot{\lambda} - r\lambda \tag{5.A5}$$

where conditions (5.A3) to (5.A5) hold at all times. The transversality condition is given by:

$$\lim_{t \to \infty} \lambda(t)h(t) = 0 \tag{5.A6}$$

Since $\tilde{H}$ is concave, the sufficient second-order condition for the maximization problem is satisfied (Chiang, 2000).[9] Rearranging equation (5.A4) gives the following condition:

$$kpf'(x_i + g_i) = \frac{c(Z)}{s(\gamma_i)} - a\lambda\mu\frac{1-s(\gamma_i)}{s(\gamma_i)} - a\lambda\beta(1-k) \tag{5.A7}$$

Equation (5.A7) indicates that, when surface water is used, the water must be used such that its marginal benefit (value of marginal product of water) is equal to the full marginal cost of sending water (including conveyance losses) minus the groundwater recharge from the conveyance loss and on-farm loss. The shadow price of the received surface water $(p_x(\gamma_i))$ is equal to the value of the marginal product of the water received, $kpf'(x_i + g_i)$.

The derivative of equation (5.A7) with respect to γ_i is equal to $-\dfrac{s(\gamma_i)c'(Z)s'(\gamma_i)x_i}{s^4(\gamma_i)} - \dfrac{(C(Z)-\lambda a\mu)s'(\gamma_i)}{s^2(\gamma_i)}$. If $C(Z) > \lambda a\mu$, or in other words, the cost of sending surface water is higher than the benefit from its conveyance loss,[10] the derivative of equation (5.A7) with respect to the distance is unambiguously positive. As a result, the shadow price of surface water increases with the distance.

From equation (5.A4), the optimal condition for groundwater use can be expressed by:

$$kpf'(x_i + g_i) = c_g(h) + a\lambda - a\lambda\beta(1-k) \tag{5.A8}$$

Like surface water, groundwater should be used such that the marginal benefit is equal to the total marginal cost. The full cost of extracting groundwater includes marginal extraction cost and marginal user cost minus the value of aquifer recharge from on-farm loss. The shadow price of the groundwater (p_g) is equal to the value of the marginal product of the water, which is equal to the right-hand side of equation (5.A8). From equation (5.A8), the groundwater shadow price is identical over the distance.

Equation (5.A5) explains the optimal condition for groundwater extraction over time. It can be rearranged as:

$$r\lambda + al'(h)\lambda = \dot{\lambda} - c_g'(h)\sum_{i=1}^{n} g_i \tag{5.A9}$$

The left-hand side is the cost of water conservation, which consists of interest on the foregone royalty and the value of an increase in leakage if water is not extracted. The right-hand side represents the marginal benefits of water conservation, including the change in royalty and the savings in extraction costs.

Appendix B: Proof of Proposition 1

Proof: (a) $x_i > 0$ and $g_i > 0 \Leftrightarrow \dfrac{\partial \tilde{H}}{\partial x_i} = \dfrac{\partial \tilde{H}}{\partial g_i} = 0$, which requires that

$$kpf'(x_i + g_i) - \frac{c(Z)}{s(\gamma_i)} + \lambda a\mu\frac{1-s(\gamma_i)}{s(\gamma_i)} + a\lambda\beta(1-k) = kpf'(x_i + g_i) - c_g(h) - a\lambda + a\lambda\beta(1-k) = 0$$

or

$$\underbrace{\frac{c(Z)}{s(\gamma_i)} - \lambda a\mu\frac{1-s(\gamma_i)}{s(\gamma_i)}}_{LHS} = \underbrace{c_g(h) + a\lambda}_{RHS} \tag{5.A10}$$

Inasmuch as $\dfrac{\partial RHS}{\partial y} = 0$ and $\dfrac{\partial LHS}{\partial y} > 0$, there exist a unique y (hereafter, y_c) that solves (5.A10).

(b) Denote y_x for the location of farms that use only surface water. $x_i > 0$ and $g_i = 0 \Leftrightarrow$

$$\dfrac{\partial \tilde{H}}{\partial x_i} = 0 \text{ and } \dfrac{\partial \tilde{H}}{\partial g_i} < 0 \text{ which requires that}$$

$$kpf'(x_i + g_i) - \dfrac{c(Z)}{s(y_x)} + \lambda a\mu \dfrac{1 - s(y_x)}{s(y_x)} + a\lambda\beta(1-k) > kpf'(x_i + g_i) - c_g(h) - a\lambda + a\lambda\beta(1-k)$$

or

$$\underbrace{\dfrac{c(Z)}{s(y_g)} - \lambda a\mu \dfrac{1 - s(y_x)}{s(y_x)}}_{LHS} > \underbrace{c_g(h) + a\lambda}_{RHS}$$

Inasmuch as $\dfrac{\partial RHS}{\partial y} = 0$ and $\dfrac{\partial LHS}{\partial y} > 0$, then $y_x < y_c$.

(c) Denote y_g for the location of farms that use only groundwater. $x_i = 0$ and $g_i > 0 \Leftrightarrow$

$$\dfrac{\partial \tilde{H}}{\partial x_i} < 0 \text{ and } \dfrac{\partial \tilde{H}}{\partial g_i} = 0 \text{ which requires that}$$

$$kpf'(x_i + g_i) - \dfrac{c(Z)}{s(y_g)} + \lambda a\mu \dfrac{1 - s(y_g)}{s(y_g)} + a\lambda\beta(1-k) < kpf'(x_i + g_i) - c_g(h) - a\lambda + a\lambda\beta(1-k)$$

or

$$\underbrace{\dfrac{c(Z)}{s(y_g)} - \lambda a\mu \dfrac{1 - s(y_g)}{s(y_g)}}_{LHS} < \underbrace{c_g(h) + a\lambda}_{RHS}$$

Inasmuch as $\dfrac{\partial RHS}{\partial y} = 0$ and $\dfrac{\partial LHS}{\partial y} > 0$, then $y_g > y_c$.

Appendix C: Proof of Proposition 2

We begin the analysis from the groundwater side by developing a phase diagram for head level and the groundwater extraction. Using the equation of motion for head, the $\dot{h} = 0$ locus is given by:

$$\sum_{i=1}^{n} g_i = R - l(h) + \mu \sum_{i=1}^{n} \dfrac{1 - s(y_i)}{s(y_i)} x_i + \beta(1-k) \sum_{i=1}^{n} (x_i + g_i). \tag{5.A11}$$

At the steady state, total groundwater extraction is a negative function of the head level.[11] Substituting the equation of motion of the head level and equation (5.A3) into (5.A4), the $\dot{p}_g = 0$ (analogous to $\dot{g} = 0$) locus can be rewritten as:

$$p_g = c_g(h) - \dfrac{ac'(h)[R - l(h) + \mu \sum_{i=1}^{n} \dfrac{1 - s(y_i)}{s(y_i)} x_i + \beta(1-k) \sum_{i=1}^{n} x_i]}{[r + al'(h)]} \tag{5.A12}$$

The right-hand side of equation (5.A12) has a negative derivative with respect to head level.[12] As a result, along the $\dot{p}_g = 0$ locus, groundwater extraction is positively related to the head level. Figure 5.4 shows a phase diagram explaining head level (h) and groundwater extraction in the optimal solution. Assuming that the initial head is lower

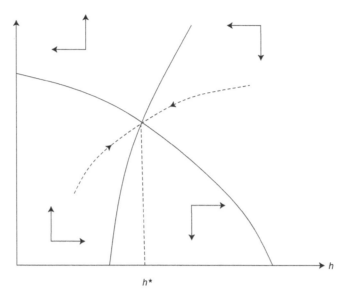

Figure 5.4 Phase diagram for the head level and groundwater extraction

than the steady state level, optimal groundwater extraction starts below its steady state level. Thereafter, head level and groundwater extraction increase over time (along the lower arm of the saddle path towards the steady state). If the initial head level is higher than its steady state, groundwater extraction and the head level decrease over time (along the upper arm of the saddle path).

Case 1: If $h_0 < h^\star$, *then* $\dot{p}_x < 0$, $\dot{p}_g < 0$ *and* $\dot{y}_c < 0$.

If $\dot{p}_g > 0$, then $\dot{g}_i < 0$. In order to satisfy $\dot{G} > 0$, $\dot{y}_c < 0$. This implies $\dot{p}_x > 0$ and $\ddot{p}_x > \ddot{p}_g$, and thus $\dot{x}_i < 0$. With $\dot{x}_i < 0$ and $\dot{y}_c < 0$, then $\dot{Z} < 0$, and $\dot{c}(Z) < 0$. This contradicts $\dot{p}_x > 0$. Thus, $\dot{p}_g \leq 0$.

If $\dot{p}_g = 0$, in order to satisfy $\dot{G} > 0$, $\dot{y}_c < 0$. This implies $\dot{p}_x > 0$, and thus $\dot{x}_i < 0$. With $\dot{x}_i < 0$ and $\dot{y}_c < 0$, then $\dot{Z} < 0$, and $\dot{c}(Z) < 0$. This contradicts $\dot{p}_x > 0$. Thus, $\dot{p}_g < 0$.

When $\dot{p}_g < 0$, if $\dot{p}_x > 0$, then $\dot{y}_c < 0$ and $\dot{x}_i < 0$. As a result, $\dot{Z} < 0$, and $\dot{c}(Z) < 0$. This contradicts $\dot{p}_x > 0$. Thus, $\dot{p}_x \leq 0$.

If $\dot{p}_x = 0$, then $\dot{y}_c < 0$. As a result, $\dot{Z} < 0$, and $\dot{c}(Z) < 0$. This contradicts $\dot{p}_x = 0$. Thus, $\dot{p}_x < 0$.

The only possible case is that $\dot{p}_x < 0$, $\dot{p}_g < 0$ *and* $\dot{y}_c < 0$. In this case, $\dot{g}_i > 0$, $\dot{G} > 0$, $\dot{x}_i > 0$, $\dot{Z} < 0$, and $\dot{c}(Z) < 0$.

Case 2: If $h_0 > h^\star$, *then* $\dot{p}_x > 0$, $\dot{p}_g > 0$ *and* $\dot{y}_c > 0$.

If $\dot{p}_g < 0$, then $\dot{g}_i > 0$. In order to satisfy $\dot{G} < 0$, $\dot{y}_c > 0$. This implies $\dot{p}_x < 0$ and $\ddot{p}_x > \ddot{p}_g$, and thus $\dot{x}_i > 0$. With $\dot{x}_i > 0$ and $\dot{y}_c > 0$, then $\dot{Z} > 0$, and $\dot{c}(Z) > 0$. This contradicts $\dot{p}_x < 0$. Thus, $\dot{p}_g \geq 0$.

If $\dot{p}_g = 0$, in order to satisfy $\dot{G} < 0$, $\dot{y}_c > 0$. This implies $\dot{p}_x < 0$, and thus $\dot{x}_i > 0$. With $\dot{x}_i > 0$ and $\dot{y}_c > 0$, then $\dot{Z} > 0$, and $\dot{c}(Z) > 0$. This contradicts $\dot{p}_x < 0$. Thus, $\dot{p}_g > 0$.

When $\dot{p}_g > 0$, if $\dot{p}_x < 0$, then $\dot{y}_c > 0$ and $\dot{x}_i > 0$. As a result, $\dot{Z} > 0$, and $\dot{c}(Z) > 0$. This contradicts $\dot{p}_x < 0$. Thus, $\dot{p}_x \geq 0$.

If $\dot{p}_x = 0$, then $\dot{y}_c > 0$. As a result, $\dot{Z} > 0$, and $\dot{c}(Z) > 0$. This contradicts $\dot{p}_x < 0$. Thus, $\dot{p}_x > 0$.

The only possible case is that $\dot{p}_x > 0$, $\dot{p}_g > 0$ *and* $\dot{y}_c > 0$. In this case, $\dot{g}_i < 0$, $\dot{G} < 0$, $\dot{x}_i < 0$, $\dot{Z} > 0$, and $\dot{c}(Z) > 0$.

Notes

1 We would like to acknowledge the support of the Department of the Interior, U.S. Geological Survey and the Water Resources Research Center, University of Hawaii, under Grant Nos. 05HQGR0146 and 06HQGR0081. Any views and conclusions are those of the authors and should not be attributed to our sponsors. We are also indebted to Kimberly Burnett, Jane Kirton, Lee Endress, Chris Wada, Alberto Garrido and participants in the ASSA Annual Meeting (Chicago, 2007), session on New Frontiers in Water Resource Economics, for helpful comments and suggestions. Any remaining errors are the authors' responsibility.

2 Hydraulic head is the altitude of an unconfined aquifer. Its conversion to groundwater stock is discussed in the numerical section.

3 Any economies of scale in dam construction are eventually exhausted and the returns to higher dams in terms of water captured are eventually diminishing. Moreover, more sent water requires greater conveyance infrastructure both for lining and extending the command area. The assumption of increasing marginal cost may also be thought of as forgone benefits from other uses, e.g., downstream non-agricultural use, recreation, hydro-electricity generation, and buffering capacity. As marginal benefits of other uses increase as water is withdrawn from those uses, the marginal cost of using surface water for irrigation increases with amount of water sent.

4 The production function is assumed to have traditional properties, i.e., $f'(q) > 0$, $f''(q) < 0$. We assume that on-farm technology is exogenously determined and fixed across time, thus abstracting, e.g., from the endogenous choice of drip irrigation.

5 We assume that surface and groundwater are perfect substitutes in farm production.

6 Other input costs can be regarded as fixed for the purposes here. Alternatively one can assume that the costs of any inputs that increase with water use are embedded in the cost functions.

7 If the cost of sending surface water is lower than the benefit from its conveyance loss, the optimal strategy is to let the water percolate and recharge the aquifer.

8 See Gaudet et al. (2001) for the role of set up costs in reswitching between non-renewable resources.

9 As $\tilde{H}_{x_i x_i} = kpf''(x_i + g_i) - \dfrac{c'(Z)}{s^2(\gamma_i)} < 0$, $\tilde{H}_{g_i g_i} = kpf''(x_i + g_i) < 0$,

$\tilde{H}_{x_i g_i} = kpf''(x_i + g_i) < 0$, and $\tilde{H}_{x_i x_i} \tilde{H}_{g_i g_i} - (\tilde{H}_{x_i g_i})^2 = -kpf''(x_i + g_i) \dfrac{c'(Z)}{s^2(\gamma_i)} > 0$, $\tilde{H}$ is concave.

10 If the cost of sending surface water is lower than the benefit from its conveyance loss, the optimal strategy is to let the water percolate and recharge the aquifer.

11 There are two effects when the head level is high. First, natural leakage increases, so the steady state extraction must be lower. Second, the cost of groundwater is lower. Surface water will be used less. As a result, there will be less recharge resulting from surface water irrigation. Both effects indicate that, at the steady state, the higher the head level is, the lower the groundwater extraction will be.

12 If h increases, then $c'(h)$ decreases, $l'(h)$ increases, x_i decreases, and $l'(h)$ increases. Consequently, the right-hand side of equation (5.A12) decreases with the head level.

References

Chakravorty, U., Hochman, E. and Zilberman, D. (1995) "A Spatial Model of Optimal Water Conveyance", *Journal of Environmental Economics and Management*, vol 29, pp. 25–41.

Chakravorty, U., Krulce, D. and Roumasset, J. (2004) "Specialization and Non-Renewable Resources: Ricardo Meets Ricardo", *Journal of Economic Dynamics and Control*, vol 29, no 9, pp. 1517–1545.

Chakravorty, U. and Roumasset, J. (1991) "Efficient Spatial Allocation of Irrigation Water", *American Journal of Agricultural Economics*, vol 73, pp.165–173.

Chakravorty, U. and Umetsu, C. (2003) "Basinwide Water Management: A Spatial Model", *Journal of Environmental Economics and Management*, vol 45, pp. 1–23.

Chiang, A.C. (2000) *Elements of Dynamic Optimization*, Lng Grove, IL: Waveland Press, IL.

Duarte, T.K. (2002) "Long-term management and discounting of groundwater resources with a case study of Kuki'o, Hawaii'", PhD dissertation, Massachusetts Institute of Technology.

Easter, K.W., Rosegrant, M.W. and Dinar, A. (1999) "Formal and Informal Markets for Water: Institutions, Performance, and Constraints", *World Bank Research Observer*, vol 14, no 1, pp. 99–116.

Gaudet, G., Moreaux, M. and Salant, S.W. (2001) "Intertemporal Depletion of Resource Sites by Spatially Distributed Users", *American Economic Review*, vol 91, no 4, pp. 1149–1159.

Howe, C.W. (1998) "Water Markets in Colorado: Past Performance and Needed Changes", in K.W. Easter, M.W. Rosegrant and A. Dinar (eds), *Markets for Water: Potential and Performance*, Kluwer Academic Publishers, Boston, MA.

Koundouri, P. (2004) "Current Issues in the Economics of Groundwater Resource Management", *Journal of Economic Surveys*, vol 18, no 5 pp. 703–740.

Koundouri, P. and Christou, C. (2006) "Dynamic Adaptation to Resource Scarcity and Backstop Availability: Theory and Application to Groundwater", *Australian Journal of Agricultural and Resource Economics*, vol 50, pp. 227–245.

Krulce, D.L., Roumasset, J.A. and Wilson, T. (1997) "Optimal Management of a Renewable and Replaceable Resource: The Case of Coastal Groundwater", *American Journal of Agricultural Economics*, vol 79, pp. 1218–1228.

Mink, J. (1980) *State of the Groundwater Resources of Southern Oahu*, Honolulu Board of Water Supply, Honolulu.

O'Mara, G.T. (1984) "Issues in the Efficient Use of Surface and Groundwater in Irrigation", World Bank Staff Paper no 707. Washington, DC: World Bank.

Pongkijvorasin, S. and Roumasset, J. (2007) "Optimal Conjunctive Use of Surface and Groundwater with Recharge and Return Flows: Dynamic and Spatial Patterns", *Review of Agricultural Economics*, vol 29, pp. 531–539.

Roumasset, J.A. and Wada, C.A. (2012) "Ordering the Extraction of Renewable Resources: The Case of Multiple Aquifers", *Resource Energy Economics*, vol 34, pp. 112–128.

Smith, R. and Roumasset, J. (2000) "Constrained Conjunctive-Use for Endogenously Separable Water Markets: Managing the Waihole-Waikane Aqueduct", *Agricultural Economics*, vol 24, pp. 61–71.

Tsur, Y. (1991) "Managing Drainage Problems in a Conjunctive Ground and Surface Water System", in A. Dinar and D. Zilberman (eds), *The Economics and Management of Water and Drainage*, Kluwer Academic Publishers, Norwell, MA.

United Nations (2006) "Coping with Water Scarcity: A Strategic Issue and Priority for System-Wide Action", www.unwater.org.

Wolff, G. and Gleick, P.H. (2002) "The Soft Path for Water", in P. Gleick (ed), *The World's Water: The Biennial Report of Freshwater Resources 2002-2003*, Island Press, Washington DC.

PART II

Private behavior and regulatory design

6

STRATEGIC BEHAVIOR AND REGULATION OVER TIME AND SPACE

C.-Y. Cynthia Lin and Lisa Pfeiffer

Introduction

The management of groundwater resources for use in agriculture is an issue that reaches far and wide; many of the world's most productive agricultural basins depend on groundwater and have experienced declines in water table levels. Increasing competition for water from cities and environmental needs, as well as concerns about future climate variability and more frequent droughts, have caused policymakers to look for ways to decrease the consumptive use of water.

There is a socially optimal rate of extraction that can be modeled, measured, and achieved through policy and a complete definition of the property rights that govern groundwater. Social optimality can incorporate environmental amenities that provide value to people, ecosystems, or environments. Complete, measured, enforceable, and enforced property rights that consider the physical properties of the resource will induce the socially optimal rate of extraction in many cases. Where externalities occur, whether they are caused by the physical movement of water, by environmental damages or benefits, or by other causes, well thought-out policy can provide the incentives to move an individual's extraction path back to the socially optimal one.

There are two main reasons why the farmers may be over-pumping relative to what would be the socially optimal water pumping. First, owing to institutional reasons farmers may not be optimizing dynamically: they might not be considering the effects of current pumping on the amount of water that would be available to pump in the future. For example, as we explain below, the prior appropriation doctrine is an institution that distorts the incentive to optimize dynamically over the life of the resource, because farmers are unable to bank any unused portions of the water allocation in a particular year for use in future years. The second reason farmers may be over-extracting the resource is owing to a common pool resource problem: because farmers are sharing the aquifer with other farmers, other farmers' pumping affects the amount of water they have available to pump.

In this chapter we discuss our research on several aspects of strategic behavior and regulation over time and space. The first aspect we discuss is the behavioral response to voluntary, incentive-based water conservation programs for irrigated agriculture. These programs are often considered win–win policies; their objective is to reduce the consumptive use of water for agriculture, and they also often contribute to an increase in the earning potential of farms through the yield-increasing effect of efficient irrigation technology (Cox, 2013). For this reason, these programs are extremely popular and politically feasible, especially where the resource is considered scarce. However, when behavioral responses of the irrigator are ignored, such policies can have unintended or even perverse consequences. In Pfeiffer and Lin (2014), we examine the effect of programs that subsidize efficient irrigation technology on water extraction. Our results show that programs that subsidize efficient irrigation technology cause farmers to respond by switching to more water intensive crops, thereby increasing, not decreasing, water extraction.

The second aspect of strategic behavior and regulation over time and space we discuss is how the prior appropriation doctrine affects dynamic optimality. Groundwater users extract water under an institutional setting that governs their property rights to the groundwater and affects constraints they face and the choices they make. A variety of property rights doctrines and institutions governing groundwater have evolved in the western United States. Many more institutions, both formal and informal, are in place in other locations around the world. Our focus is on Kansas, a state that overlies a portion of the High Plains Aquifer. Current water rights in Kansas follow the prior appropriation doctrine. The prior appropriation doctrine allots water rights based on historical use, with priority going to those who claimed their right first. Appropriation contracts are stated in terms of a maximum acre-feet of extraction per year with a "use it or lose it" clause. Farmers must use their allocation each year and are unable to bank any unused portions of the water allocation in a particular year for use in future years. However, since the groundwater is in part a nonrenewable resource, since the availability of water is stochastic, since demand for water is greater when it is less available, farmers could operate in a more dynamically efficient manner if the appropriator could use less water in some years and more in others. In Pfeiffer and Lin (2013), we develop an empirical model to test whether groundwater users faced with the prior appropriation doctrine are behaving in a manner consistent with the Hotelling model for dynamically optimal nonrenewable resource extraction.

The third aspect of behavior and regulation over time and space we discuss is strategic behavior in the face of spatial externalities. One reason farmers over-extract water is due to a common pool resource problem: they share the aquifer with other farmers, and thus other farmers' pumping affects water availability. This property gives rise to a spatial externality whereby pumping by one user affects the extraction cost and total amount that is available to other nearby users. In Pfeiffer and Lin (2012), we empirically examine whether the amount of water one farmer extracts depends on how much water his neighbor extracts. Our results provide evidence of a spatial externality that causes farmers to over-extract water.

Our research focuses exclusively on the groundwater used for agriculture in the High Plains (Ogallala) Aquifer system of the Midwestern United States. There, 99 percent of the water extracted is used for crop production; the remaining 1 percent is used for livestock, domestic, and industrial purposes. The economy of the region is based entirely on irrigated agriculture. The corn, soybeans, alfalfa, wheat, and sorghum grown there are used for local livestock production or exported from the region. The small local communities support the agricultural industry with farm implement dealers, schools, restaurants, and other services. The state governments are also greatly concerned with supporting their agricultural industry.

We first describe the High Plains Aquifer in Kansas and the data we use for our research, and then proceed to discuss the three aspects of strategic behavior and regulation over time and space in turn.

The High Plains Aquifer in Kansas

Use of the High Plains Aquifer system, located in the Midwestern plains of the United States, began in the late 1800s but greatly intensified after the "Dust Bowl" decade of the 1930s (Miller and Appel, 1997). Aided by the development of high capacity pumps and center pivot systems, irrigated acreage went from 1 million acres in 1960 to 3.1 million acres in 2005, and accounts for 99 percent of all groundwater withdrawals (Kenny and Hansen, 2004). Irrigation converted the region from the "Great American Desert" into the "Breadbasket of the World."[1]

The High Plains Aquifer underlies approximately 174,000 square miles, and eight states overlie its boundary. It is the principle source of water in the Great Plains region of the United States. Also known as the Ogallala Aquifer, the High Plains Aquifer system is now known to include several other aquifer formations. The portion of the aquifer that underlies western Kansas, however, pertains mainly to the Ogallala, and this is why the name persists.

The High Plains aquifer is underlain by rock of very low permeability that creates the base of the aquifer. The distance from this bedrock to the water table is a measure of the total water available and is known as the saturated thickness. The saturated thickness of

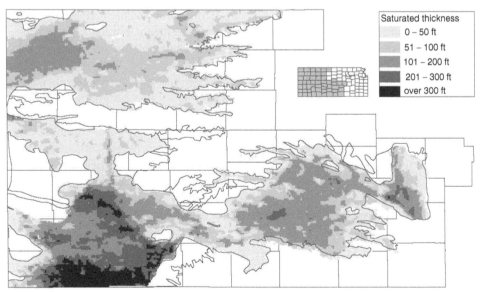

Figure 6.1 Predevelopment saturated thickness of the Kansas portion of the High Plains Aquifer (source: Kansas Geological Survey)

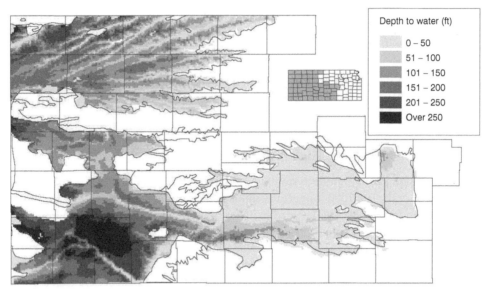

Figure 6.2 Average 2004–2006 depth to groundwater in the Kansas portion of the High Plains Aquifer (source: Kansas Geological Survey.)

the High Plains aquifer in Kansas ranges from nearly zero to over 300 feet (Buddemeier, 2000). Figure 6.1 shows the predevelopment saturated thickness of the Kansas portion of the High Plains Aquifer.

The depth to water is the difference between the altitude of the land surface and the altitude of the water table. In areas where surface and groundwater are hydrologically connected, the water table can be very near to the surface. In other areas, the water table is much deeper; the depth to water is over 400 feet below the surface in a portion of southwestern Kansas (Miller and Appel, 1997). Figure 6.2 shows the average 2004–2006 depth to groundwater in the Kansas portion of the High Plains Aquifer.

Recharge to the Kansas portion of the High Plains aquifer is primarily by percolation of precipitation and return flow from water applied as irrigation. The rates of recharge vary between 0.05 and 6 inches per year, with the greatest rates of recharge occurring where the land surface is covered by sand or other permeable material (Buddemeier, 2000).

Irrigation accounts for 99 percent of all groundwater withdrawals. The main crops grown in western Kansas, in order of decreasing water intensiveness, are alfalfa, corn, soybean, grain sorghum, and wheat. Corn production accounts for more than 50 percent of all irrigated land (Buddemeier, 2000); in 2005, the last year of our data set, corn production accounted for 40.0 percent of the irrigated land. Soil types and access to high volumes of irrigation water determine the suitability of a particular piece of land to various crops.

Data

We use a rich data set for our empirical analyses of strategic behavior and regulation over time and space. Kansas has required the reporting of groundwater pumping by water rights holders since the 1940s, although only data from 1996 to the present are considered to be complete and reliable. The data are available from the Water Information Management and Analysis System (WIMAS). Included are spatially referenced pumping data at the source (well or pump) level, and each data point has the farmer, field, irrigation technology, amount pumped, and crops grown identified. A sample of about 9,000 points of diversion for each of the 10 years from 1996 to 2005 is used for the analysis. We combine this data with spatial data sets of recharge, water bodies, and other geographic information.

The United States Geological Survey's (USGS's) High Plains Water-Level Monitoring Study maintains a network of nearly 10,000 monitoring wells. Data from these wells will be used to estimate yearly water levels. The USGS also has information on hydroconductivity, and precipitation data come from the PRISM Climate Group (2014). Relevant information from the geographic files will be captured at the points of diversion (well) level using ArcGIS. There are about 8,000 sampled points of diversion for each of the 10 years from 1996 to 2005.

The crop price data we use are a combination of spring futures contracts for September delivery for commodities with futures contracts and average price received for crops without futures contracts. Futures prices are from the Commodity Research Board (CRB), and price received is from the U.S. Department of Agriculture (USDA) Economic Research Service. Crop price ratios are constructed for the estimation and consist of the crop price divided by a weighted sum of the prices of all crops. The weights used are the average proportions of irrigated acres planted to each crop over the 1996 to 2005 time period in the counties included in the estimation.

Natural gas prices come from the U.S. Energy Information Administration, and are used for irrigators in counties with natural gas production. In other areas, the price of electricity is used as the price of energy. Over 50 percent of the acres irrigated from groundwater wells in Kansas are powered by natural gas (FRIS, 2004).

Soil characteristics come from the Web Soil Survey of the USDA Natural Resources Conservation Service.

We use this data to empirically analyze whether programs that subsidize efficient irrigation technology decrease water extraction, whether groundwater users faced with the prior appropriation doctrine are behaving in a manner consistent with the Hotelling model for dynamically optimal nonrenewable resource extraction, and whether the amount of water one farmer extracts depends on how much water his neighbor extracts. We now describe each of these empirical analyses in turn.

Behavioral response to incentive-based water conservation programs

In many places, policymakers have attempted to decrease rates of extraction through incentive-based water conservation programs. Between 1998 and 2005, the state of Kansas spent nearly US$6 million on incentive programs, such as the Irrigation Water Conservation Fund and the Environmental Quality Incentives Program, to fund the adoption of more efficient irrigation systems. Such programs paid up to 75 percent of the cost of purchasing and installing new or upgraded irrigation technology, and much of the money was used for conversions to dropped nozzle systems (NRCS, 2004). These policies were implemented under the auspices of groundwater conservation, in response to declining aquifer levels occurring in some portions of the state due to extensive groundwater pumping for irrigation (O.A.M.A. Committee, 2001).

Voluntary, incentive-based water conservation programs for irrigated agriculture are often billed as policies where everyone gains. These programs are politically feasible, farmers are able to install or upgrade their irrigation systems at a reduced cost, resulting in substantial increases in profits, less groundwater is "wasted" through runoff, evaporation, or drift, marginal lands can be profitably retired, and farmers can choose whether to participate. However, such policies can have unintended, even perverse, consequences.

In recent work (Pfeiffer and Lin, 2014), we find that policies that encourage the adoption of more efficient irrigation technology may not have the intended effect. Irrigation is said to be "productivity enhancing"; it allows the production of higher value crops on previously marginal land. Thus, a policy of subsidizing more efficient irrigation technology can induce a shift away from dry-land crops to irrigated crops. They may also induce the planting of more water-intensive crops on already irrigated land, as by definition, more efficient irrigation increases the amount of water the crop receives per unit extracted.

Similarly, land and water conservation and retirement programs may not necessarily reduce groundwater extraction, although they are billed as such. An example of a land retirement program is the Conservation Reserve Program (CRP) created by the federal government in 1985 to provide technical and financial assistance to eligible farmers and ranchers to address soil, water, and related natural resource concerns on their lands in an environmentally beneficial and cost-effective manner. These programs include payments to landowners to retire, leave fallow, or plant non-irrigated crops on their land. Such programs operate on an offer-based contract between the landowner and the coordinating government agency. The contractual relationship is subject to asymmetric information, and adverse selection may arise because the landowner has better information about the opportunity cost of supplying the environmental amenity than does the conservation agent. There is substantial evidence that farmers enroll their least productive, least intensively farmed lands in the programs while receiving payments higher than their opportunity costs, thus accruing rents. It is quite unlikely that an irrigated parcel, which requires considerable investment in a system of irrigation (which, in turn, enhances the productivity of the parcel), will be among a farmer's plots with the lowest opportunity cost and thus enrolled in the program. Instead, farmers may opt to enroll non-irrigated plots in the CRP, which does not have any effect on the amount of irrigation water extracted.

In our study, which has recently been cited in the *New York Times* (Wines, 2013), we focus on incentive-based groundwater conservation policies in Kansas and find that measures taken by the state of Kansas to subsidize a shift toward more efficient irrigation systems have not been effective in reducing groundwater extraction. The subsidized shift toward more efficient irrigation systems has in fact increased extraction through a shift in cropping patterns. Better irrigation systems allow more water-intensive crops to be produced at a higher marginal profit. The farmer has an incentive to both increase irrigated acreage and produce more water-intensive crops.

We find similar results in our analysis of the effects of land and water conservation and retirement programs on groundwater extraction. Theoretically, we know that because the programs are offer-based, farmers will enroll their least productive land. Our empirical results support this conclusion; we find essentially no effect of land conservation programs on groundwater pumping, which occurs, by definition, on irrigated, and thus, very productive land.

How the prior appropriation doctrine affects dynamic optimality

Groundwater users extract water under an institutional setting that governs their property rights to the groundwater and affects the constraints they face and the choices they make. The hydrological characteristics of an aquifer affect the way that property rights over its water should be defined. For example, if water flows easily in an aquifer, the inefficiencies associated with the exploitation of common property resources are more likely (Dasgupta and Heal, 1979; Eswaran and Lewis, 1984). Additionally, the amount of recharge that an aquifer receives helps determine the economically efficient extraction path. Because the portion of the High Plains aquifer that lies south of Nebraska receives very little recharge, it is least partially a nonrenewable resource and its social welfare maximizing extraction path can be described by a Hotelling-like model (Hotelling, 1931).

Our focus is on Kansas, a state that overlies a portion of the High Plains Aquifer. Kansas is the only state where a rich set of data on the recent history of groundwater extraction is available. Current water rights in Kansas follow the prior appropriation doctrine. Before 1945, Kansas applied the common law of absolute ownership doctrine to groundwater. Water rights were not quantified in any way (Peck, 2007). In 1945, following multiple conflicts between water users and several major water cases that reached the Kansas Supreme Court, the "Arid Region Doctrine of Appropriation" was adopted, which permitted water extraction based on the principle of "first in time, first in right" (Peck, 1995).

The earliest appropriators of water maintain the first rights to continue to use water in times of shortage or conflict. The water right comes with an abandonment clause; if the water is not used for beneficial purposes for longer than the prescribed time period, then it is subject to revocation (Peck, 2003). To obtain a new water right, an application stating the location of the proposed point of diversion, the maximum flow rate, the quantity desired, the intended use, and the intended place of use must be submitted to and approved by the Department of Water Resources (Kansas Handbook of Water Rights, n.d.). Since 1945, Kansas has issued more than 40,000 groundwater appropriation permits (Peck, 1995).[2] The permits specify an amount of water that can be extracted each year and are constant over time.

Through the 1970s, the period of intensive agricultural development in Kansas, groundwater pumping permits were granted to nearly anyone who requested them. Some permits are as old as 1945, but the majority (about 75 percent) were allocated between 1963 and 1981.

In the early 1970s, it was recognized that Kansas's groundwater resources were being depleted at a rapid rate in some locations. By 2008, in parts of southwestern Kansas, the water table had declined by over 150 feet since predevelopment.[3] This area was the first to be intensively developed, and continues to have the highest average extraction per square mile (Wilson et al., 2002).

In 1972, owing to concerns that the aquifer was over-appropriated, Kansas created five groundwater management districts (GMDs). The GMDs regulate well spacing and prohibit new water extraction within a designated radius of existing wells, which varies by GMD. The adoption of the prior appropriation doctrine, together with the development of GMDs to regulate new appropriations of water rights, arguably eliminated uncontrolled entry and the resulting over-exploitation commonly associated with common property resources. However, appropriation contracts distort the incentive to optimize dynamically over the life of the resource, because the farmer is essentially guaranteed his appropriated amount of water until the resource becomes so scarce that it is no longer economical to pump.

In ongoing work (Pfeiffer and Lin, 2013), we investigate how farmers manage groundwater over time and under an existing property rights regime. Specifically, our empirical analysis focuses on the portion of western Kansas that overlies the High Plains (Ogallala) Aquifer. An area with a relatively well-defined rights system, Kansas has used the doctrine of prior appropriation to govern the management of groundwater since 1945. Hotelling (1931) argues that the socially optimal rate of extraction of a nonrenewable resource over time is achieved in a competitive market equilibrium, provided that the social discount rate equals the market interest rate and that there are no market failures, such as externalities or incomplete property rights. The prior appropriation doctrine is an example of an incomplete property rights system, and thus may distort the incentive of a groundwater user to manage a resource dynamically, causing extraction to occur at a rate faster than is socially optimal.

We develop an empirical model to test whether groundwater users faced with the prior appropriation doctrine are behaving in a manner consistent with the Hotelling model for dynamically optimal nonrenewable resource extraction. In particular, do groundwater managers (individual farmers, in this case) consider the scarcity rent or shadow value of their resource when making extraction decisions? Or, do other behavioral assumptions, such as myopic optimization, provide a better basis for explaining their behavior? This is one of the first studies to empirically test the hypotheses of the theoretical groundwater management literature.

We find that despite the incentives given to groundwater users to pump their maximum allowable amount in each year by the prior appropriation doctrine, farmers extract water consistent with a dynamic model of resource extraction. While producers are allotted a time-invariant maximum amount that they can extract each year, they still consider the effects of recharge, their remaining stock, pumping by nearby neighbors that may affect their stock in the future, and projections of future commodities prices when making crop choice and pumping decisions. Our results therefore provide evidence that farmers recognize the nonrenewable nature of the resource that they manage, even though their property rights do not.

Strategic behavior in the face of spatial externalities

When farmers share the same aquifer with other farmers, a spatial externality arises whereby pumping by one user affects the extraction cost and total amount that is available to other nearby users. The spatial externality has been disaggregated into different types of effects, including a pumping cost externality and a stock or strategic externality (Negri, 1989; Provencher and Burt, 1993). The pumping cost externality arises because withdrawal by one user lowers the water table and increases the pumping cost for all users. The strategic externality arises because the property rights on the water in an aquifer are generally undefined. What a farmer does not withdraw today will be withdrawn by other farmers, which undermines their incentive to forgo current for future pumping (Negri, 1989). Theoretically, these externalities are potentially important causes of welfare loss (Dasgupta and Heal, 1979; Eswaran and Lewis, 1984; Negri, 1989; Provencher and Burt, 1993; Brozovic, Sunding and Zilberman, 2002; Rubio and Casino, 2003; Msangi, 2004; Saak and Peterson, 2007), but empirically we have little evidence to determine whether farmers react to these externalities or have an idea of their magnitude.

In Pfeiffer and Lin (2012), we investigate the behavior of farmers who share a common pool resource. We develop a spatial dynamic physical-economic model to characterize agricultural groundwater users' pumping behavior. We compare a social planner's optimal decisions with those of a group of profit maximizing individuals who have full property rights to the land, but whose groundwater is an incomplete common good because they cannot fully capture the groundwater beneath their land.

We then use data from western Kansas to econometrically determine if the pumping behavior of neighbors affects the groundwater extraction decision. The estimations are spatially explicit, taking advantage of detailed spatial data on groundwater pumping from the portion of western Kansas that overlies the High Plains Aquifer system.

Measuring interactions between neighbors is challenging because of simultaneity (individuals affect their neighbors and their neighbors simultaneously affect them) and spatial correlation in observable and unobservable characteristics (Manski, 1993; Glaeser, Sacerdote and Scheinkman, 1996; Brock and Durlauf, 2001; Moffitt, 2001; Conley and Topa, 2002; Robalino and Pfaff, 2012). The interaction of neighbors has been studied in oil extraction (Libecap and Wiggins, 1984; Lin, 2009). It has also been investigated in land use change using physical attributes of neighboring parcels as instruments to identify the effect of the behavior of neighbors on an individual (Irwin and Bockstael, 2002; Robalino and Pfaff, 2012).

We use an instrumental variables approach to purge neighbors' decisions of the endogenous component. Groundwater users in Kansas extract water under the doctrine of prior appropriation, meaning that they are allotted a maximum amount to extract each year. This annual amount was determined when the user originally applied for the permit. The permit amount for one's neighbors is a strong determinant of the actual pumping by one's neighbors, but is uncorrelated with one's own actual pumping, except through the effect of neighbors' pumping on one's own pumping. We therefore use the permit amount of one's neighbors as an instrument for neighbors' water pumping.

To take into account the way in which water moves through an aquifer, we weight our instrument by a function of the distance between each neighbor and the difference in lift height between neighbors that takes into account the way in which water moves through an aquifer. These weights adjust the amount pumped by the effect that it should have. If the distance between two wells is greater, the effect should be smaller. If the height gradient is larger, the effect should be greater.

This is the first study to empirically measure economic relationships between groundwater users. If externalities in groundwater use are significant, it lends insight into the causes of resource over-exploitation. If they are not significant or are very small in magnitude, a simpler model of groundwater user behavior, where each user essentially owns his own stock, is sufficient. Both outcomes would give guidance to policymakers, although it is important to note that the results are highly dependent on the hydrological conditions of the aquifer.

Details of our analysis are reported in Pfeiffer and Lin (2012). According to our results, we find evidence of a behavioral response to this movement in the agricultural region of western Kansas overlying the High Plains Aquifer. Spatial externalities resulting from the inability to completely capture the groundwater to which property rights are assigned cause some degree of over-extraction in theoretical models. Using an instrumental variable and spatial weight matrices to overcome estimation difficulties resulting from simultaneity and spatial correlation, we find that on average, the spatial externality causes over-extraction that accounts for about 2.5 percent of total pumping. Kansas farmers would apply 2.5 percent less water in the absence of spatial externalities (if, as an unrealistic example, each farmer had an impenetrable tank of water that held his or her portion of the aquifer).

Strengthening the evidence of the behavioral response to the spatial externalities caused by the movement of groundwater is the empirical result that when a farmer owns multiple wells, he does not respond to pumping at his own wells in the same manner as he responds to pumping at neighboring wells owned by others. In fact, the response to pumping at his own wells is to marginally decrease pumping, thus trading off the decrease in water levels between spatial areas and internalizing the externality that exists between his own wells.

Conclusion

In this chapter we discuss our research on several aspects of strategic behavior and regulation over time and space. The first aspect we discuss is the behavioral response to voluntary, incentive-based water conservation programs for irrigated agriculture. The second aspect of strategic behavior and regulation over time and space we discuss is how the prior appropriation doctrine affects dynamic optimality. The third aspect of strategic behavior and regulation over time and space we discuss is strategic behavior in the face of spatial externalities.

Strategic behavior and regulation over time and space can affect the optimality of the farmers' water extraction rate. Complete, measured, enforceable, and enforced property rights that consider the physical properties of the resource have the possibility of inducing the socially optimal rate of extraction in many cases. Where externalities occur, whether they are caused by the physical movement of water, by environmental damages or benefits, or by other causes, well thought-out policy can provide the incentives to move an individual's extraction path back to the socially optimal one. However, in practice, not all regulations over time and space induce the socially optimal rate of extraction. Incentive-based groundwater conservation programs are a prime example of a well-intentioned policy that may have perverse consequences, for they may actually increase rather than decrease groundwater extraction. Similarly, property rights regimes such as prior appropriation may adversely impact the dynamic optimality of water extraction. When designing policies and regulation, policymakers need to be wary of any potential perverse consequences of their policies, and also be aware of the implications of their policies over time and space.

Notes

1 However, although irrigation was a large breakthrough, the non-irrigated regions to the east of the 100th meridian are larger and more productive (e.g., Iowa, Illinois, Indiana, Ohio, southern Missouri).
2 In the 2007 census, there were 65,531 farms in all of Kansas, of which approximately 29,039 were located in regions that roughly overlie the aquifer (USDA, 2011).
3 "Predevelopment" is defined as the water level in about 1960, when the first measurements were made.

References

Brock, W.A. and Durlauf, S.N. (2001) "Discrete choice with social interactions", *Review of Economic Studies*, vol 68, pp. 235–260

Brozovic, N., Sunding, D.L. and Zilberman, D. (2002) "Optimal management of groundwater over space and time", *Frontiers in Water Resource Economics*, vol 29, pp. 109–135

Buddemeier, R.W. (2000) "An Atlas of the Kansas High Plains Aquifer", Kansas Geological Survey, URL: http://www.kgs.ku.edu/HighPlains/atlas/

Conley, T.G. and Topa, G. (2002) "Socio-economic distance and spatial patterns in unemployment", *Journal of Applied Econometrics*, vol 17, pp. 303–327

Cox, C. (2013) "Programs to reduce ag's water use must be strengthened, not cut", AgMag BLOG, Environmental Working Group, 28 May 2013, http://www.ewg.org/agmag/2013/05/programs-reduce-ag-s-water-use-must-be-strengthened-not-cut

Dasgupta, P. and Heal, G.M. (1979) *Economic Theory and Exhaustible Resources*, Cambridge University Press, Cambridge

Eswaran, M. and Lewis, T. (1984) "Appropriability and the extraction of a common property resource", *Economica*, vol 51, no 204, pp. 393–400

FRIS. (2004) "2003 Farm and Ranch Irrigation Survey", vol 3, special studies part 1, Washington, DC: National Agricultural Statistics Service.

Glaeser, E.L., Sacerdote, B. and Scheinkman, J.A. (1996) "Crime and social interactions", *The Quarterly Journal of Economics*, vol 111, no 2, pp. 507–548

Hotelling, H. (1931) "The economics of exhaustible resources", *Journal of Political Economy*, vol 39, pp. 137–175

Irwin, E.G. and Bockstael, N.E. (2002) "Interacting agents, spatial externalities and the evolution of residential land use patterns", *Journal of Economic Geography*, vol 2, no 1, pp. 31–54

Kansas Handbook of Water Rights. (n.d.) http://www.ksda.gov/appropriation/content/240

Kenny, J.F. and Hansen, C.V. (2004) "Water Use in Kansas, 1990-2000", Technical Report Fact Sheet 2004-3133, Lawrence, KS: Kansas Department of Agriculture-Division of Water Resources and the Kansas Water Office

Libecap, G. and Wiggins, S. (1984) "Contractual responses to the common pool: Prorationing of crude oil production", *The American Economic Review*, vol 74, no 1, pp. 87–98

Lin, C.-Y.C. (2009) "Estimating strategic interactions in petroleum exploration", *Energy Economics*, vol 31 no 4, pp. 586-594.

Manski, C. (1993) "Identification of endogenous social effects: The reflection problem", *Review of Economic Studies*, vol 60, no 3, pp. 531–542

Miller, J.A. and Appel, C.L. (1997) "Ground Water Atlas of the United States: Kansas, Missouri, and Nebraska", Number HA 730-D. Reston, VA: U.S. Geological Survey

Moffitt, R.A. (2001) "Policy interventions, low-level equilibria, and social interactions", in S.N. Durlauf and H.P. Young (eds) *Social Dynamics*, Cambridge, MA: MIT Press, pp. 45-79.

Msangi, S. (2004) "Managing groundwater in the presence of asymmetry: Three essays", Ph.D. thesis, University of California at Davis.

Negri, D.H. (1989) "Common property aquifer as a differential game", *Water Resources Research*, vol 25, no 1, pp. 9-15

NRCS. (2004) "Farm bill 2002: Environmental quality incentives program fact sheet", U.S. Department of Agriculture, Washington, DC, http://www.nrcs.usda.gov/programs/farmbill/2002/products.html

O.A.M.A. Committee (2001) "Discussion and recommendations for long-term management of the Ogallala Aquifer in Kansas", Technical report. Topeka, KS: O.A.M.A. Committee

Peck, J.C. (1995) "The Kansas Water Appropriation Act: A fifty-year perspective", *Kansas Law Review*, vol 43, pp. 735-756.

Peck, J.C. (2003) "Property rights in groundwater: Some lessons from the Kansas experience", *Kansas Journal of Law and Public Policy*, vol XII, pp. 493-520

Peck, J.C. (2007) "Groundwater management in the High Plains Aquifer in the USA: Legal problems and innovations", in M. Geordano and K.G. Villholth (eds) *The Agricultural Groundwater Revolution: Opportunities and Threats to Development*, Wallingford: CAB International, pp. 296-319

Pfeiffer, L. and Lin, C.-Y.C. (2012) "Groundwater pumping and spatial externalities in agriculture", *Journal of Environmental Economics and Management*, vol 64, no 1, pp. 16-30

Pfeiffer, L. and Lin, C.-Y.C. (2013) "Property rights and groundwater management in the High Plains Aquifer", Working paper, Davis, CA: University of California at Davis.

Pfeiffer, L. and Lin, C.-Y.C. (2014) "Does efficient irrigation technology lead to reduced groundwater extraction?: Empirical evidence", *Journal of Environmental Economics and Management*, vol 67, no 2, pp. 189-208

PRISM Climate Group. (2014) "Recent years (Jan 1981 – Jul 2013)", Northwest Alliance for Computation Science and Engineering, URL: http://www.prism.oregonstate.edu/recent/

Provencher, B. and Burt, O. (1993) "The externalities associated with the common property exploitation of groundwater", *Journal of Environmental Economics and Management*, vol 24, no 2, pp. 139-158

Robalino, J.A. and Pfaff, A. (2012) "Contagious development: Neighbors' interactions in deforestation", *Journal of Development Economics*, vol 97, no 2, pp. 427-436

Rubio, S.J. and Casino, B. (2003) "Strategic behavior and efficiency in the common property extraction of groundwater", *Environmental and Resource Economics*, vol 26, no 1, pp. 73-87

Saak, A.E. and Peterson, J.M. (2007) "Groundwater use under incomplete information", *Journal of Environmental Economics and Management*, vol 54, pp. 214-228

U.S. Department of Agriculture (USDA). (2011) "Kansas Farm Facts 2011", National Agricultural Statistics Service, http://www.nass.usda.gov/Statistics_by_State/Kansas/ Publications/Annual_Statistical_Bulletin/ff2011.pdf

Wilson, B.B., Young, D.P. and Buddemeier, R.W. (2002) "Kansas Geological Survey Open File Report 2002-25D", Technical report. Lawrence, KS: Kansas Department of Agriculture-Division of Water Resources and the Kansas Water Office

Wines, M. (2013) "Wells dry, fertile plains turn to dust", *New York Times*, 19 May 2013.

7

WATER ALLOCATION UNDER DISTRIBUTION LOSSES

A perspective

Ujjayant Chakravorty and Yazhen Gong

Introduction

Increasing water scarcity, arising from a growing population and periodical droughts due to climate change, calls for better management of existing water projects and more efficient water use at the farm level. Although a lot of effort has been made in the past century to generate water for irrigation and other uses, poor management of existing water projects and farmers' inefficient use of irrigation water are common in many parts of the world. For example, losses in water conveyance may range from 25–50 percent of the water carried, even in developed economies. For better management and more efficient use of water at the farm level, some possible solutions, such as institutional reforms through creation of water user associations and water markets, have been pursued in developed and developing countries.

Although some recent studies have suggested that dams have a small impact on agricultural productivity (see Duflo and Pande, 2007), most water for irrigation today is still supplied by surface water systems. In some parts of the world such as South Asia and China, groundwater irrigation is growing rapidly, but from a smaller base. Most surface water systems suffer from large distribution losses, estimated to be 50–75 percent of the water supplied. However, it is true that a portion of this water lost from seepage and percolation is often retrieved elsewhere in the system. But this retrieval process may need more energy and other capital inputs. Because conveyance loss is inevitable for surface water distribution and it is characterized by increasing returns to scale, developing a good understanding of issues related to irrigation water allocation over space and organization structure of water generation/distribution is important for better management of irrigation water in the future.

The theoretical and empirical research that characterizes the spatial dimension and organization structure of water generation/distribution is quite limited. In this chapter, we focus on irrigation water only and aim to summarize this literature and present some stylized implications for better management of irrigation water.

Water management: issues and challenges

An alarming water shortage, arising from an increasing gap in demand and supply, has become a major challenge faced by human society in this century. Rapid world population growth, economic development and expansion of irrigated land have contributed to a rapidly increasing demand for water (Volosmarty et al., 2000; Pereira et al., 2002); clean freshwater has become less and less accessible to human society due to water pollution. During the first half of this century, the world population is projected to increase from 6.1 billion in 2000 to 8.9 billion in 2050, with an annual growth of world population by 57 million (United Nations, 2004). Such rapid population growth implies that about 60 percent more food needs to be produced in the first half century and requires more water to be used to produce enough food. Urbanization and industrialization (particularly in developing countries) will also lead to an increasing demand for industrial and residential use of water. On the supply side, less than 1 percent of the world's freshwater (or about 0.007 percent of all water on earth) is accessible for direct human uses. Most freshwater occurs in the form of permanent ice or snow, locked up in Antarctica and Greenland, or in deep groundwater aquifers (UNEP, 2002). The heavy pollution arising from intensive use of chemical inputs in the agricultural sector and industrial water discharges in many parts of the world have resulted in less accessible clean water for human use. It is estimated that about one third of the world's population will live under the condition of water stress in this century (Volosmarty et al., 2000; Wallace, 2000), with around 450–460 million people under severe water stress (Volosmarty et al., 2000).

The agriculture sector, which is currently the largest user of water and is facing increasing competition from other sectors, is the sector that is most sensitive to water shortage (FAO, 2012). Currently, about 70 percent of freshwater withdrawn is used for irrigation (United Nations, 2004; FAO, 2012). However, fast-paced industrialization and urbanization in many developing countries, changes in human diets and increasing living standards have all intensified competition from industry and residential sectors for industrial and residential water uses. Pollution, arising from intensive chemical use in the agriculture sector, has caused serious environmental problems. In addition, periodic droughts associated with climate change give rise to supply shocks for agriculture use. Furthermore, a dramatic expansion of ground water irrigation (particularly in South Asia and China), which contributed to the increased agricultural productivity in the second half of the last century, has led to aquifer depletion and has become one major constraint for agricultural production. All of the above will ultimately limit the agriculture sector's capability for securing food supply for a growing world population.

To cope with water shortage for agricultural use, some potential solutions may be sought through enhancement in demand management and supply management (Pereira et al., 2002; FAO, 2012). Demand management can generally involve increasing water productivity, reducing water loss and re-allocating water across different sectors and within sectors. Planting less water-intensive demand crop varieties or cropping patterns, adopting water-saving technologies and improving farming practices are typical examples of options that can help reduce water loss at the farm level. Options for supply enhancement may be through increasing water storage capacity (e.g. reservoirs), increasing irrigation conveyance and distribution systems (e.g. canals), developing new sources of water supply, adopting on-farm water conservation practices to reduce runoff, increasing infiltration and storage of water in soils, and controlling pollution. However, as the cost of building infrastructure for water storage and getting access to new sources of water supply has become increasingly high, even prohibitively high, enhancing the management of existing water supply infrastructure (e.g. distribution/conveyance systems) may be a better solution.

There is a clear message from demand and supply options: enhancing agricultural water management on demand and supply sides both involve a spatial dimension. The spatial dimension spans from a basin-wide scale (e.g. through conveyance and distribution systems) to farm-level (e.g. adoption of water conservation practices and water-saving technology at the farm level) and different scales are often inter-linked. For example, at the basin-wide scale, some water "lost" between the source and the end users may be returned to the hydrological system, either through percolation into the aquifers or as return flows into the river systems. Therefore, effective policies to reduce water use have to be designed at the basin-wide scale that can essentially link hydrological, economic, institutional and agronomic relations among water users located at the farm level. However, such an endeavor is technically challenging. This may partly explain why empirical evidence for the spatial dimension of water management in the existing economics literature is still very limited. However, to improve the management of agricultural water, economic models need to tackle this challenge and more empirical studies need to be conducted for better policies in the future.

Losses in water distribution

Distribution losses in water have important implications for water pricing, allocation and conservation. In this section, we discuss four major issues related to the economics of distribution losses in irrigation.

The first issue is the interaction of conveyance losses with the design of alternative water pricing regimes. Due to seepage, percolation and evaporation, conveyance loss increases with the distance between the source (say the dam or the point of water diversion) and the end point where the water is received (e.g. farms) (Chakravorty and Roumasset 1991). Even if the same volume of water is delivered from the source to farmers at different locations, the loss in distribution results in an unequal water distribution between farmers located at the head and those at the tail of the canal. Ideally, an efficient pricing scheme should take into account this heterogeneity between farmers located along the canal.

An optimal water allocation regime needs to be designed by determining an optimal amount of the water stock at the source and the amount of water delivered to farms along the canal. The correct amount of water stock at the source is determined by maximizing social welfare, i.e. the sum of consumer and producer surplus, while the optimal allocation of water along the canal needs to make the value of the marginal product of water at the source equal across all farms along the distribution system (e.g. canals).

Due to increasing conveyance costs with increasing distance, an optimal pricing scheme must adhere to the following rule: the shadow price paid by individual farmers per unit of water received at the farm level should increase with the distance from the source, as the on-farm value of the marginal product of water rises with the distance from the source. However, many pricing schemes in the real world are designed by simply assuming an equal allocation of irrigation water along the canal. One typical example of this is the spatially uniform pricing scheme.

In reality, different water charge regimes are often designed by making trade-offs among efficiency, equity and environmental effectiveness. For example, a water charge regime based on marginal pricing, i.e. farmers pay the shadow price per unit of water reaching their farms, may lead to the optimal allocation of water, but it gives rise to inequity among farmers, and those located farther from the source often get lower rents. In contrast, a water charge regime may be designed to make the rents among farmers located at the different parts of the canal equal thus implementing another notion of spatial equity. However, equity is achieved at the sacrifice of efficiency loss and over-expansion of the irrigation system beyond a point that can be supported by the system capacity. The over-expansion could be largely driven by rent-seeking incentives pursued by farmers at the boundary of the irrigation system.

The second issue is that investments in water distribution are often underprovided, which may affect the performance of water allocation and pricing schemes. Because of the public good nature of the conveyance system (e.g. canals), these investments are usually provided by the public sector. Individual farmers will not invest in canals if they are unable to recoup the full benefits from the asset. However, due to government budget constraints, underprovision of water conveyance is pervasive in the real world. Most canals that transport water are often unlined ditches.

Consider the situation in which a central planner (e.g. a water utility) supplies water to individual users and makes a public investment in the distribution system, while water users, uniformly distributed along the distribution system, make their private investment in water conservation. Chakravorty et al. (1995) characterize the optimal conveyance, water allocation and pricing scheme that may affect efficiency, equity and water users' conservation behaviors. They showed that: (1) the optimal investments in distribution, which decrease with the distance from the water source, lead to a lower level of overall water use and increased level of water conservation practiced by water users; (2) since the opportunity cost of delivering water to individual water users at different locations increases with the distance from the source due to these conveyance losses, the optimal allocation of water requires downstream water users to pay a higher price per unit of received water and, under some conditions, to pay higher total water charges. Although the above optimal pricing scheme may incentivize downstream water users to invest in water conservation and reduce their water use, it leads to a skewed income distribution in favor of upstream water users. Moreover, while the investment in conveyance may help save water, it is likely to shift economic rents from upstream to downstream water users.

The underprovision of distribution canals (lack of maintenance, for example) can affect the performance of water allocation and pricing schemes. A typical example is the effect on the performance of a water market along the canal. Ideally, with the introduction of a water market, the price of traded water at each location equals its shadow price and would result in optimal water allocation. However, when conveyance is provided below an optimal level, the water market may lead to the same outcome as that from a uniform pricing scheme, i.e. water users have little incentive for water conservation but increase water use. Moreover, the introduction of the water market may provide incentives for downstream water users to invest privately in water conservation, but may also result in a skewed distribution of water rents in favor of upstream water users.

In order to save water and increase the efficiency of the end use of water (e.g. irrigation water use at the farm level), it is important to upgrade existing projects through investment in conveyance and pricing policies. However, for such policies to succeed, it is important to design accompanying compensation packages to ensure equity between upstream and downstream water users and sustain downstream water users' incentives for water conservation.

The third issue is the spatial externality of the conveyance loss on efficiency, equity and water conservation. This spatial externality arises from water losses in distribution; the lost water, which replenishes the groundwater table, can be returned to farms in the form of groundwater pumped out from the aquifer underneath. In essence, when the spatial externality of conveyance loss is considered, groundwater emerges as an endogenous backstop analogous to a Hotelling (1931) type nonrenewable resource model (Chakravorty and Umetsu, 2003).

To analyze water allocation, technological changes in conveyance and on-farm use and distribution of water rents in the entire hydrological system, Chakravorty and Umetsu (2003) developed a basin-wide spatial model for conjunctive use of water by considering the spatial externality of conveyance loss. In their model, water is supplied by a central planner to farms uniformly located along the distribution system, while farmers have the choice of using surface or groundwater. They show that the optimal solution leads to specialization of upstream and downstream water use, i.e. upstream water users draw surface water while downstream water users use groundwater. The price of surface water, which increases with distance due to conveyance costs, is bounded above by the price of groundwater. That is, when the price of using surface water goes beyond that of using groundwater, farmers switch to using groundwater. Graphically, their model is demonstrated in Figure 7.1.

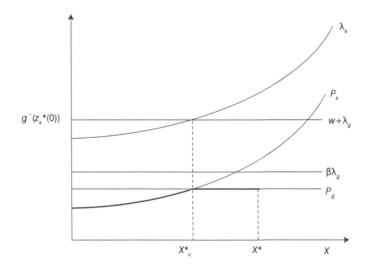

Figure 7.1 Surface and groundwater price paths

The figure shows the surface, p_s, and groundwater p_g price paths over space, which is measured along the x-axis. The origin represents the canal head or the source location of the irrigation water. The shadow price of surface water λ_s at the source is equal to the marginal cost of water generation, $g'(z_s(0))$ and increases exponentially with distance as shown. The "real" price of surface water p_s at any location is $\lambda_s - \lambda_g \beta(1-h)$ which lies at a constant distance below λ_s. Similarly, the real price of groundwater p_g is less than the shadow price of groundwater λ_g by a constant. On the left of the boundary X_c^*, $p_s < p_g$; hence, only surface water is used for production. However, with increasing distance, transporting surface water becomes more expensive and groundwater becomes cheaper beyond X_c^*, and is used exclusively until the project boundary X^*. X_c^* is the optimal length of the canal. Intuitively, since production with surface water entails conveyance losses, it is best done close to the water source in the upstream reaches of the canal. Similarly, as production with groundwater is independent of location, it is optimally located downstream of the canal. Thus, groundwater emerges as an "endogenous" backstop technology.

In essence, the spatial feature of the conjunctive model demonstrates the importance of incorporating return flows in the allocation of water resources and in determining pricing and conveyance investment when irrigation efficiencies are low, which is common in many developing countries. In this case, ignoring the spatial externality of conveyance loss might lead to overinvestment in distribution facilities. From a basin-wide perspective, when the transaction costs of upgrading existing distribution systems are high, it might be more cost-effective to allow for high distribution losses and encourage greater use of groundwater downstream. However, one ought to be careful because pumping water from the groundwater aquifer requires energy. To the extent that energy such as diesel or electricity is subsidized in developing countries, this may lead to over-pumping, and higher conveyance losses in a managed system. There is also the issue of uncontrolled pumping that goes beyond the recharge into the aquifer.

The basin-wide model also suggests some options for studying dynamic interactions between upstream and downstream water users. Since the groundwater available for downstream use depends on canal maintenance and surface water allocation in the canal system as well as on-farm investment undertaken by upstream water users, downstream and upstream water users may pursue strategic behavior. One can then study alternative institutional arrangements, such as the creation of water users' associations downstream or the introduction of water markets, and their impact on water use efficiency, equity and resource use.

Finally, one can extend this framework to study the effect of market power in generation, distribution and the end use of water. Creation of a water market has been considered an effective institutional arrangement to achieve efficient water allocation and an optimal level of social welfare in many parts of the world. Nonetheless, water markets are often characterized by market power, which may occur at the supply, distribution and end use stage of water use. Specifically, the organizational structure of market power may have the following four microstructures: (1) a decentralized distribution system, where a large number of users withdraw water from a distribution system and make their private investment, taking other users' investment as a given; (2) the water user association, which is a price-taker in the output market but has market power (monopsony) in water generation; (3) the canal operator, who owns the canal, buys water from the generator and operates as a monopoly in distributing water to users along the canal; (4) the producer cartel, which supplies water at each location through a trading or rationing scheme and operates as a monopoly in the end use market. The existence of market power leads to different levels of social welfare, water prices and project size at the equilibrium.

Chakravorty et al. (2009) develop a model to specifically compare differences in welfare and resource allocation across a social planning regime, a competition in distribution (CID) regime, and a market power regime involving the four microstructures discussed above. In their model, water is generated as a point source and distributed to users (farms) uniformly along the distribution system (i.e. canal). The water may be generated and distributed independently or by a vertically integrated firm and the number of users is proportional to the length of the canal. Given the existence of different forms of market power in the form of water user associations and producer cartels, effective regulatory policies need to be developed in order to achieve second best outcomes. For example, given the increasing returns to scale in water distribution under the canal operator regime, regulatory policies may need to focus on prices charged by the canal operator to water users located along the canal. In the absence of regulation, there may be overinvestment in distribution, less than optimal water losses and reduced welfare.

Figure 7.2 shows the effect of market power in different micro-markets in the irrigation system, generation, distribution and the market for output. It shows that compared to the social planning regime, a competitive distribution regime and a producer cartel results in less water to be provided and a higher price to be charged in the aggregate.

In Figure 7.2, $C^*(Y)$ is the optimal solution for a social planner's total cost of producing of a given output Y. The consumers' inverse demand function for aggregate output Y is denoted as $D^{-1}(Y)$ with $D^{-1}(Y)<0$. The equilibrium aggregate output

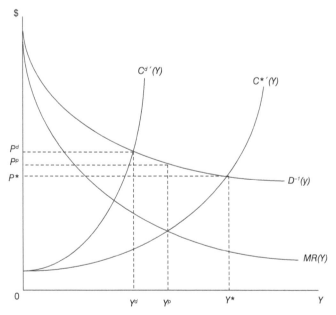

Figure 7.2 Equilibrium price and quantity under social planner, competitive and producer cartel

Y^* and price P^* solve the social welfare function $\max \int_0^Y D^{-1}(\theta)\,d\theta - C^*(Y)$ which yields $D^{-1}(Y^*) - C^{*\prime}(Y) = 0$, i.e. price is equal to marginal cost. Similarly, Y^p and P^p are aggregate output and price that solve the optimization problem in the producer cartel regime; Y^d and P^d are the same variables for the CID regime. Across these three regimes, the orders of the aggregate output and price are $Y^* > Y^p > Y^d$ and $P^* < P^p < P^d$. In essence, the above information shows that institutions characterized by market power may have unintended consequences when the use and distribution of water is accounted for.

Despite their poor performance in terms of welfare measures, institutions with market power may be desired for the sake of equity and environmental conservation. For example, the regime of the canal operator with monopoly power in distribution performs poorly in terms of delivering social welfare, and it may be desirable to ensure delivery over a relatively large service area for a larger population to get access to the water; it may also induce water users' private conservation, which is beneficial for resource conservation. Therefore, to find appropriate institutions that fit a given situation or reach a certain policy goal, it is critical to consider the elasticity of water generation, distribution or end use, geographic characteristics of the location and the socio-economic goals of the regulator.

Concluding remarks

It is important to recognize the role of distribution in water allocation and pricing. Although new technologies such as the availability of PVC pipes has reduced the cost of distribution between farms, especially if farm sizes are small, as in developing countries, most surface water is still carried by means of canals. Government investment in canals is often relegated to the primary canal that may bring water from the river or a diversion to a village, but beyond that the transport of water is often left to individuals, with the resulting market failure in conveyance provision. In the absence of state capacity to make these investments, stakeholders may provide these public goods. But whether an independent canal operator makes the investment and charges farmers monopoly prices, or a farmers' collective runs the canal, there are implications for water pricing, allocation and output, as well as the rents that accrue to each agent.

The next step may be to take these insights to the field and see how surface water irrigation systems are being organized. Although evaluations of large scale surface water and dam projects have shown that their benefits are quite small relative to the expectations *ex ante* of project commissioning, they still account for a large share of the irrigated acreage both in the developing and the developed worlds, and hence are important in meeting food security goals for an increasing population.

References

Chakravorty, U. and Roumasset, J. 1991. Efficient spatial allocation of irrigation water. *American Journal of Agricultural Economics*, 73(1), 165–173.

Chakravorty, U., Hochman, E. and Zilberman, D. 1995. A spatial model of water conveyance. *Journal of Environmental Management and Economics*, 29, 25–41.

Chakravorty, U. and Umetsu, C. 2003. Basinwide water management: a spatial model. *Journal of Environmental Management and Economics*, 45, 1–23.

Chakravorty, U., Hochman,E., Umetsu, C. and Zilberman, D. 2009. Water allocation under distributional losses: comparing alternative institutions. *Journal of Economic Dynamics & Control*, 33, 463–476.

Duflo, E. and Pande, R. 2007. Dams. *Quarterly Journal of Economics*, 122(2), 601–646.

FAO. 2012. Coping with water scarcity: an action framework for agriculture and food security. FAO Water Reports No. 38. Rome, Italy.

Hotelling, H.1931. The economics of exhaustible resources, *Journal of Political Economy*, 39, 137–175.

Pereira, L., Oweis, T. and Zairi, A. 2002. Irrigation management under water scarcity. *Agricultural Water Management*, 57, 175–206.

UNEP, 2002. *Global Environment Outlook 3: Past, Present and Future Perspectives*. London: Earthscan.

United Nations, 2004. *World Population to 2300*. New York: UN.

Volosmarty, C., Green, P., Salisbury, J. and Lammers, R. 2000. Global water resources: vulnerability from climate change and population growth. *Science*, 289, 284–288.

Wallace, J. 2000. Increasing agricultural water use efficiency to meet future food production, *Agriculture, Ecosystems & Environment*, 82(1–3), 105–119.

8

THE GOOD, BAD, AND UGLY OF WATERSHED MANAGEMENT

Kimberly Burnett, James A. Roumasset,
and Christopher A. Wada

Introduction

Payment for Ecosystem Service (PES) programs, which incentivize private landowners to protect ecosystem services when making land use decisions, have grown rapidly in the past few decades (Daily et al. 2009; Nelson et al. 2009; Sanchirico and Springborn 2011). Forested watersheds in particular provide a wide range of services, including those related to water resources, species habitat, biodiversity, subsistence activities, hunting, aesthetic values, commercial harvest, and ecotourism, among others, and those services can have substantial value. For example, Kaiser et al. (1999) estimate that the 99,000-acre Koolau Mountain Range on the island of Oahu (Hawaii) provides forest benefits valued between US$7.4 and US$14 billion. Though the motivation for watershed conservation activities is clear, the abundance of potential conservation instruments—removal of any number of invasive species, reforestation or afforestation, construction of capital such as fencing or settlement ponds, etc.—creates a considerable challenge for resource managers: what type of conservation should be employed and in what order?

The dearth of quantitative information about the costs and benefits of alternative conservation instruments, combined with the lack of a specific measurable objective, often leads to ad hoc decisions regarding investment of limited resources for conservation in practice. The objective of this chapter is to develop a framework for making efficient decisions about employing available watershed conservation instruments. By building the management problem around the concept of present value, we provide a quantifiable objective, and hence a means to rank alternatives. Taking Oahu as a case study, we also find that more scientific research is needed to quantify the effects of various conservation activities on ecosystem services of interest.

General conservation framework

To maintain clarity in the discussion that follows, we focus on freshwater services, while keeping in mind that the framework could be extended to multiple ecosystem services. Groundwater recharge depends on a variety of factors including precipitation, evapotranspiration, soil type, and land cover. Although many of those factors are out of our control, land cover can be altered in a variety of ways. Particularly high transpiration rates have been documented for some invasive plant species. Such plants tend to capture a relatively high proportion of precipitation before infiltration to groundwater resources can occur. Other plants have been shown to extract existing water directly from the groundwater resource itself via deep taproots. Feral ungulates can similarly inhibit groundwater recharge by altering landscapes; digging, trampling, and consumption of existing vegetation creates stretches of bare land that are more conducive to runoff during rain events. Although each of these examples has a negative impact on water resources, the degree of impact varies, and managing each problem entails different costs.

As an alternative to or in addition to (bad) invasive species removal, watershed conservation can entail investment in (good) plant species that tend to augment the watershed's ability to recharge underlying aquifers. An advantage of investing in natural capital is the potential for growth in the future without further investment. However, quantifying the net recharge benefit of reforestation activities (relative to the existing or counterfactual land cover) is often very challenging, and limited data is available for Hawaii. Investment in (good) produced capital such as settlement ponds, injection wells, and fencing for feral animals can also be made to increase the percentage of rainfall converted to groundwater recharge, although unlike natural capital, produced capital typically requires higher levels of maintenance over time.

To illustrate the tradeoffs between different conservation instruments, we consider two types of investment: fencing for feral pigs (*Sus scrofa*) and removal of invasive strawberry guava (*Psidium cattleianum*). In both cases, we assume that protection against new or recurring invasions is lumped into removal costs. More generally, one can imagine separate protection or maintenance costs incurred in periods following initial removal. The cost of removing one acre of (bad) strawberry guava (c_S) is assumed to increase as the current stock declines due to the loss of economies of scale and increasing unit search costs. However, the stock can only be reduced by removing more plants than natural growth (F). In the analytical model and the application that follows, we assume that the number of invasive plants removed is proportional to the number of acres cleared (x^S). One could imagine a model that tracks individual plant units with a non-uniform distribution over space, but implementation would require spatial details that are unavailable in our case. The dynamic behavior of the strawberry guava stock in terms of acres (S) is described by the following equation:[1]

$$\dot{S}_t = F(S_t) - x_t^S \tag{8.1}$$

By not explicitly modeling space, we are implicitly assuming that high priority areas, i.e. those that would generate the largest recharge gains, are targeted first. Consequently, the recharge function (R) is concave with respect to guava stock—as management shifts toward lower priority areas over time, gains in recharge are smaller, and removal costs may be higher for the reasons explained above.

Feral animal control also entails reducing or slowing a continuously growing stock but typically requires investment in (good) capital such as fencing, which may have large upfront costs for the initial fence construction and removal of pigs from the fenced-off area. The stock of produced capital, in this case fencing, could be modeled explicitly, including the initial installation/removal costs and yearly maintenance expenditures. For the purpose of illustrating the management principle of equimarginal benefits, however, we will assume that the cost of fence installation (c_Z) includes the present value of maintenance, reckoned at the time an acre is cleared. The population of pigs in terms of acres (Z) increases over time in accordance with its natural growth function G and is reduced by investments in fencing (x^Z), which includes initial removal:

$$\dot{Z}_t = G(Z_t, \sum_{\tau=0}^{t} x_\tau^Z) - x_t^Z \tag{8.2}$$

The growth function is dependent on not only the stock of feral animals but also on cumulative removal up to the current period (implicitly the acres of fencing installed). Like for strawberry guava, we assume that the location of future fence installations is predetermined, based on existing knowledge of the watershed. Typically, one starts at a high elevation, builds additional fencing to enclose previously fenced areas, then removes pigs between the fences. We also assume a proportional relationship between the number of acres fenced and the number of pigs removed. In a more general formulation, G would depend directly on growth-retarding capital such as fencing. Then, fencing installation and existing fencing stock would enter the management problem as additional control and state variables respectively.

Each type of capital—good, bad, natural, and/or produced—feeds into the groundwater recharge function. The stock of groundwater, indexed by the head level (h) converted to volume, changes over time according to recharge (R), stock-dependent natural leakage to adjacent water bodies (L), and extraction (q):

$$\dot{h}_t = R(S_t, Z_t) - L(h_t) - q_t \tag{8.3}$$

where $R_S < 0$ and $R_z < 0$.

The objective of the management problem is to maximize the present value of net benefits from water use, taking into account both groundwater extraction costs and watershed-conservation costs, subject to Eqs. 8.1–8.3:

$$\underset{q_t, b_t, x_t^S, x_t^Z}{Max} \int_0^{\infty} e^{-rt} \left\{ \int_0^{q_t+b_t} D^{-1}(\theta_t) d\theta_t - c_q(h_t) q_t - c_b b_t - c_S(S_t) x_t^S - c_Z x_t^Z \right\} dt \tag{8.4}$$

Gross benefits of water consumption are measured as consumer surplus, or the area under the inverse demand curve for water. The unit cost of groundwater extraction (c_q) is stock-dependent because the distance water must be lifted to the surface varies with the aquifer head level. In addition to groundwater extraction, we allow for the production of desalinated seawater (b) at unit cost (c_b) as a backstop resource. Desalinated water need not be used in every period, and, in most cases, is likely to serve as a supplemental resource only in the very long run.

The dynamic optimization problem can be posited in an optimal control framework, and it is straightforward to derive the following efficiency price equation for optimal groundwater extraction from the necessary conditions (Roumasset and Wada 2013):

$$p_t = c_q(h_t) + \frac{\dot{p}_t - c_q'(h_t)[R(S_t, Z_t) - L(h_t)]}{r + L'(h_t)} \tag{8.5}$$

Equation 8.5 says that water should be extracted until the marginal benefit, measured as the efficiency price (p), is equal to the marginal opportunity cost (MOC) of groundwater. The MOC includes the physical extraction cost, as well as the marginal user cost (MUC), which accounts for the forgone use of the marginal unit of groundwater when the price is higher in the future, as well as the effect of today's extraction on leakage and marginal extraction cost in the future. Along the optimal path, the MUC is equal to the shadow price of groundwater, which is by definition, the increase in net present value resulting from an additional unit of groundwater stock.

Each of the invasive species stocks has an optimality condition for efficient removal. Strawberry guava should be eliminated until the marginal benefit of the last unit removed, measured as the shadow price of groundwater (λ), is equal to the MOC of control:

$$\lambda_t = \frac{c'_S(S_t)F(S_t) + c_s(S_t)[F'(S_t) - r]}{R_s} \tag{8.6}$$

The MOC of strawberry guava removal includes the effect on both the stock-dependent marginal extraction cost and the stock-dependent growth. An analogous condition can be derived for pig control:

$$\lambda_t = \frac{c_Z[G'(Z_t) - r]}{R_Z} \tag{8.7}$$

The assumption of a constant unit cost of fence installation eliminates costs related to changes in the marginal cost of control. Consequently the MOC of pig removal is driven primarily by the effect on the growth rate. Inspection of the two optimality conditions for watershed conservation reveals that optimal management is steered by the system shadow price of water (λ). The shadow price, in turn, is determined endogenously by the joint maximization problem (Equation 8.4), which means that present value is only maximized if the watershed and aquifer are managed simultaneously.

For an internal solution, i.e. one in which removal of both invasives is optimal, the above conditions imply an equimarginality condition for the two removal instruments. Control measures should be implemented until both MOCs are equal to each other and simultaneously equal to the shadow price or scarcity value of recharge. In certain periods, however, it may not be optimal to allocate any resources toward one or more watershed conservation instruments. For example, if the MOC of pig removal is higher than the shadow price of recharge for any level of control in a given period but the MOC of strawberry guava is lower than the shadow price over some range of control, only strawberry guava removal is optimal and a corner solution obtains.

Generally, watershed conservation activities—enhancing good and removing bad or ugly watershed capital—can be arranged according to their effect on recharge. The supply cost of protective capital such as fencing is the implicit rental cost or user cost of capital. When protective and removal actions are treated separately, removal can also be thought of as capital, where depreciation is the maintenance required to keep new invasive species out of the controlled area. Interdependence between different instruments for increasing recharge—e.g. a fence complements reforestation and serves as a substitute for invasive tree removal—means that the integrated management problem must be solved in its entirety, i.e. not piecewise. For graphical purposes, however, we can assume that individual supply curves for recharge are drawn *mutatis mutandi*, i.e. with other inputs at their optimal quantities. Upon adding the supply curves horizontally, the resulting aggregate supply curve for recharge intersects the marginal benefit curve for water at the shadow price of recharge (Figure 8.1).

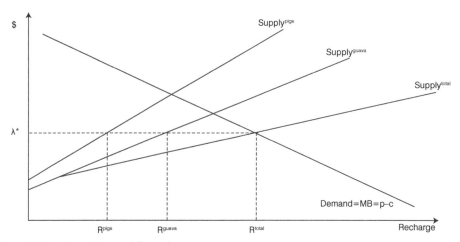

Figure 8.1 Supply and demand for recharge

The relative slopes of the supply curves will depend on the cost and effectiveness of each respective conservation instrument. If the land management authority can rank locations according to the cost per recharge saved, e.g. because of the location of plants in more or less critical watershed areas, then a least-cost approach implies an upward sloping supply curve of recharge via strawberry guava removal. The same applies to pigs, except that current conservation practices seem to suggest that a top-down approach is most cost-effective, presumably because the more critical watershed areas tend to be at higher elevations.

Application: Koolau Watershed

Watershed management can generally entail investment in a wide variety of conservation activities, possibly including installation of both natural and produced capital. In the application that follows, the problem is to remove and protect against *bad* strawberry guava and remove and protect against *ugly* pigs. Protection against new strawberry guava (e.g. making sure the herbicide reaches the roots and conducting occasional surveillance for new growth) is lumped into removal costs. The cost of removing feral pigs is included in the construction cost of *good* (preventative) fences. For reasons of tractability, investment in good natural capital (e.g. reforestation) is not considered here.

Nonnative feral pigs (*Sus scrofa*) reproduce rapidly, starting from as early as 6–12 months in age. Sows usually produce litters of five to seven piglets and having two litters per year is common (Pavlov et al. 1992). In Hawaii, survivorship of piglets is particularly high, owing to lack of predators and abundant food sources. A yearly production of ten piglets, with some mortality, accords with the notion that uncontrolled pig populations in Hawaiian rainforests are currently capable of doubling every four months (Katahira et al. 1993). Although reliable data is currently unavailable regarding both the existing population and the growth rate of feral pigs in Hawaii, we make some simple assumptions to illustrate how the optimization framework could inform resource management decisions. Suppose that the initial stock of pigs in the Koolau watershed on the island of Oahu is 5,000 and that the carrying capacity of the approximately 100,000-acre watershed is 100,000 pigs. Because Katahira et al.'s (1993) estimate of the pig population doubling every four months was made over 20 years ago, the rate of growth has likely changed substantially as the stock of pigs has continued to expand. We assume that feral pig growth can be roughly approximated by a logistic growth function, and that the intrinsic growth rate is 25 percent, the upper range of an estimate for *Sus scrofa* annual intrinsic growth in Texas (Texas A&M Institute of Renewable Natural Resources, 2011). The natural growth of pigs is described by the following function:

$$G(Z_t) = 0.25Z_t(1 - Z_t / 100000) \tag{8.8}$$

While there are many possible control methods, fencing is often considered the standard conservation instrument for removing and then excluding feral pigs from a watershed area. Kaiser et al. (1999) estimate that fencing would cost roughly US$40,000 per mile in the Koolau watershed, not including monitoring and depreciation costs. Cost data from recently completed projects on Oahu suggest that costs have since risen substantially and are likely more in the range of US$92,000–US$159,000 per mile (M. Burt, Oahu Army Natural Resources Program, personal communication). Because these costs only include clearing, scoping, construction, and gear preparation, the total costs (including helicopter time and materials) may be much larger. For the purposes of our illustration, we assume that fence construction costs US$200,000 per mile. Given that one acre is equivalent to 43,560 square feet, installing a 165 foot by 264 foot (1-acre) rectangle with a perimeter of 858 feet (0.1625 miles) costs US$32,500.[2] If additional rectangles are placed adjacent to the existing fencing, however, every rectangle except the one at the starting point in each row would only require two additional sides. We therefore assume that the average cost of fencing one acre is roughly US$16,250. Pig removal ($x^Z$) in every period will be determined by investment in fencing (y^Z) and the stock of pigs (Z) in the following fashion:

$$x_t^Z = y_t^Z \left[\frac{Z_t}{100,000 - \sum_{\tau=0}^{t} y_\tau^Z} \right] \tag{8.9}$$

In other words, the number of pigs removed in period t is equal to the number of new acres fenced, multiplied by the existing pig population, divided by the number of remaining unfenced acres in the watershed. The implied direct correspondence between fenced area and pigs and the assumption of uniform pig density over space are consequences of limited information in this particular application.

Feral pigs negatively impact native flora and fauna in Hawaii's forested watersheds and cause soil compaction. Although much more work needs to be done to fully understand the relationship between pigs and runoff, results from a recent study in the Manoa watershed on the island of Oahu suggest that exclusion by fencing may reduce pig-induced runoff (Dunkell et al. 2011). Of the seven study sites observed over a one-year period, runoff was roughly 10 percent less in fenced areas at two plots, 20 percent less at one plot, no different at three plots, and 10 percent higher at one plot. On average, fencing appears to have reduced runoff by approximately 4 percent. Given that the carrying capacity of the 100,000-acre Koolau watershed is assumed to be 100,000 pigs, we suppose that removing one pig effectively reduces runoff within a one-acre area. Further, assuming that pig-induced runoff would have instead recharged the aquifer and that the recharge function is multiplicatively separable, i.e. $R(S,Z) = R^{max} \cdot R(S) \cdot R(Z)$, the last term in the recharge function can be written as follows:

$$R(Z_t) = 1 - 0.04 \left(\frac{Z_t}{100,000} \right) \tag{8.10}$$

When the population of pigs is zero, recharge is equal to the maximum level of recharge, adjusted only for the effect of strawberry guava, $R(S)$. If the population of pigs reaches the carrying capacity, R^{max} is reduced by 4 percent in every period, in addition to the effect of strawberry guava. For pig stocks between those levels, total maximum recharge is reduced, but not by the full 4 percent.

Psidium cattleianum is a small tree (2–6 m) that tends to form dense monotypic stands. The red-fruited variety, commonly known as strawberry guava, is native to Brazil. In Hawaii, strawberry guava is a highly invasive species that is very difficult to eradicate, due to lack of natural predators and imperfect biological control agents. Growth rates of strawberry guava in native Hawaiian forests can be very high. At 3,000 feet on Hawaii Island, average annual increases of over 12 percent in stem density and 9 percent in total basal area have been measured (Geometrician Associates LLC, 2011). Statewide, it is estimated that roughly 38 percent of forested areas have "likely dense infestation" of strawberry guava (Geometrician Associates LLC, 2011). Although tracking individual trees would provide the best estimate of the effect on watershed services, doing so would be infeasible given currently available data. We assume that the unit of measure for strawberry guava is one acre—which implies that S_0 is equal to 38,000—and growth is described by a logistic function with an intrinsic growth rate of 10 percent:

$$F(S_t) = 0.1S_t(1 - S_t / 100,000) \tag{8.11}$$

Controlling strawberry guava requires cutting of stems and application of herbicide to prevent resprouting. Because many areas in Hawaii's upper watersheds are not easily accessible by road, costs can increase dramatically if equipment needs to be transported over rugged terrain. For example, if the target area is more than 1.5 miles away from a road, working crews typically camp onsite, and a helicopter is used to transport equipment and camping gear. In East Hawaii conservation areas, it is estimated that controlling dense infestations cost US\$10,500/acre near roads, US\$12,200/acre at moderate distance from roads, and US\$23,315/acre in remote areas (Geometrician Associates LLC, 2011). Assuming that the easiest-to-access infestations are removed first—i.e. US\$10,500 is the cost of removing the first acre when the entire area is infested and US\$23,315 is the cost for the last remaining infested acre—we fit a quadratic function to the three data points to describe the stock/area-dependent cost function:

$$c_S(S_t) = 23315.2 - 0.230275S_t + 1.02122 \times 10^{-6} S_t^2 \tag{8.12}$$

Strawberry guava reduces groundwater recharge because of very high evapotranspiration rates. Compared with forests dominated by native ohia (*Metrosideros polymorpha*), areas densely infested by strawberry guava have a lower proportion of net rainfall reaching the forest floor—110 versus 123 percent of rainfall (Takahashi et al. 2011). In other words, replacing a stand of strawberry guava with ohia would increase the amount of water available for groundwater recharge by up to 12 percent. Given that the study was conducted on a different island, however, and that invasive removal in our model does not include reforestation thereafter, we view the 12 percent estimate as an uncertain upper bound. For our application, we assume that strawberry guava removal increases recharge by 8 percent, twice that of pig removal. Recalling that the recharge function is expressed as $R(S,Z) = R^{max} \cdot R(S) \cdot R(Z)$, the last term in the recharge function can be written as follows:

$$R(S_t) = 1 - 0.08 \left(\frac{S_t}{100,000} \right) \tag{8.13}$$

When the population of strawberry guava is zero, recharge is equal to the maximum level of recharge, adjusted for the effect of feral pigs, $R(Z)$. If the entire area becomes invaded by strawberry guava, R^{max} is reduced by 8 percent in every period, in addition to the effect of pigs. For strawberry guava stocks between those levels, total maximum recharge is reduced, but not by the full 8 percent.

Assuming a (real) discount rate of 2 percent and plugging in specific functions to describe costs, growth, and recharge, the MOC of pigs—equivalently the supply curve for recharge via pig removal—from Equation 8.7 can written as follows:

$$MOC_t^Z = \frac{-9.34375 \times 10^9 + 203125 Z_t}{R^{max}(1 - 8 \times 10^{-7} S_t)} \tag{8.14}$$

For $R^{max}=220$ million gallons per day, the initial point on the recharge supply curve for pig removal is –US$0.039 per thousand gallons, which means that pig removal is optimal at the outset. Because the marginal benefit of recharge, equivalently the net price of groundwater, is currently positive, the demand and supply curves for recharge will intersect to the right of the initial point. The MOC of strawberry guava removal from Equation 8.6 can similarly be written as

$$MOC_t^S = \frac{5.10612 \times 10^{-6}(-29793.7 + S_t)(1.53258 \times 10^{10} - 209323 S_t + S_t^2)}{R^{max}(1 - 4 \times 10^{-7} Z_t)} \tag{8.15}$$

Again assuming $R^{max}=220$ million gallons per day, the initial point on the recharge supply curve for strawberry guava removal is US$0.002 per thousand gallons, which means that removing strawberry guava may also be optimal at the outset. When the recharge supply curves for pig removal and strawberry guava removal are added horizontally, the resulting aggregate recharge supply curve, in conjunction with the marginal benefit curve for water, determine how much of each conservation activity is warranted in the initial and future periods. Although the contemporaneous cost of strawberry guava removal is slightly lower than that of pig removal initially, pig removal may be preferable because optimality is driven by dynamic growth and cost effects. In this particular example, the pig population grows much more rapidly and the unit cost of removal is static. Strawberry guava, on the other hand, grows relatively slowly and removal cost increase dramatically as the stock is reduced. Ultimately, the scarcity of groundwater determines exactly if and when each of the conservation instruments is optimally implemented.

The example detailed in this section is meant to be illustrative, and adjustments to any of the parameter values may have large impacts on the optimal timing and ordering of the conservation instruments. In particular, the growth rates and initial populations of both invasive species are highly uncertain. Suppose, for example, that the initial population of pigs was 50,000 instead of 5,000 and the intrinsic growth rate was 10 percent instead of 25 percent. In that case, the initial point on the recharge supply curve for pig removal is US$0.004 per thousand gallons, higher than the initial point for strawberry guava removal. At that population level, the marginal pig contributes less to future growth (moving into the concave portion of the growth curve), and therefore the growth effect in the MOC term is smaller. Moreover because the intrinsic growth rate is lower, the incentive to choose pig removal over strawberry guava removal is also lower in future periods, thus affecting the optimal transition path to the watershed's steady state.

Concluding remarks

Increasing water scarcity warrants greater attention to managing watershed resources. Multiple instruments of watershed conservation can be managed according the *equimarginal* principle, to wit: positive investments in different instruments should satisfy the condition that the marginal dollar invested in each instrument should yield the same increase in the present value of recharge benefits (holding any non-recharge benefits constant). Instruments that yield lower marginal benefits even for the first dollar invested should not be exercised. But increased investment in watershed conservation may be wasted unless the downstream groundwater is well managed. The principles of integrated water management are straightforward. In addition to the equimarginal principle just described, the marginal benefits of increasing recharge must be equal to the MUC of groundwater. A decentralized method of implementation is to charge consumers the MOC of groundwater (extraction cost plus MUC) for the marginal unit consumed and to give landowners payments for watershed services (PWS) according to the (same) marginal value of recharge. Note, however, that these charges and payments need not apply to inframarginal units, thereby rendering PWS fiscally feasible and conferring maximum benefits to consumers.

In our illustrative application, we found that pig removal may be preferable to strawberry guava initially, even though the contemporaneous unit cost of strawberry guava is slightly lower. The seemingly counterintuitive result can be explained when one considers the dynamic growth and cost effects underlying the model. While the pig population grows rapidly and the cost of fencing an acre remains constant, strawberry guava grows relatively slowly and removal costs increase dramatically as the stock is reduced over time. Consequently, allowing strawberry guava to grow initially while focusing resources on the explosive pig population is the most cost-effective approach. However, the results discussed in the application are largely dependent on the accuracy of underlying cost, growth, and recharge parameters. More research is necessary to quantify initial populations, intrinsic growth rates, and the effect that various invasive species have on reducing potential recharge.

While the focus of the developed framework is to characterize optimal investment patterns for a variety of conservation instruments over time, real-world decisions may be constrained by limited budget allocations for conservation or other issues with financing potentially large projects. For example, if all investments in watershed conservation were completely financed through taxes, the model would need to be adjusted to account for the rising marginal cost of public finance. In that case, one could imagine that the proportion of conservation instruments with large initial outlays (e.g. fence construction) would be reduced in favor of options with low upfront cost but possibly higher maintenance costs (e.g. invasive plant removal). The result would be more uniform expenditures on conservation and lower taxes over a longer period of time.

A natural extension of the basic model would include reforestation and other methods of enhancing natural and/or produced capital to increase recharge. For example, consider the installation of settlement ponds that convert runoff into recharge. These facilities should be expanded until the marginal present value of increased recharge is equal to one, the marginal cost of investment. The same principle applies to investments in natural capital such as reforestation. In the latter case, however, one should add the marginal value of other ecosystem services to the marginal recharge benefits. Implementation of such a model may require detailed watershed-hydrological modeling in order to estimate the effects of different land cover scenarios on recharge.

Acknowledgements

This research was funded in part by NSF EPSCoR Grant No. EPS-0903833.

Note

1 In the equations that follow, a dot over any variable indicates its derivative with respect to time, in this case, $\partial S/\partial t$.
2 Assuming that fencing of the watershed starts from the highest elevation, one can imagine constructing a semi-circle of fence part way down one side of the mountain. A one-acre semi-circle has radius of 166 feet, and can be roughly approximated by a rectangle with dimensions of 165 feet by 264 feet. One could then move further down the mountain and install another semi-circle that encloses the original fence and remove the pigs in between. This process could be repeated to cover the entire watershed. Additional iterations would require more fencing, however, and thus be more expensive than the previous steps. For our purposes, we suppose that fences are installed in one-acre rectangular units that would roughly follow a semi-circular pattern down the mountain.

References

Daily, G.C., Polasky, S., Goldstein, J., Kareiva, P.M., Mooney, H.A., Pejchar, L., Ricketts, T.H., Salzman, J. and Shallenberger, R. (2009) "Ecosystem services in decision making: time to deliver", *Frontiers in Ecology and the Environment*, vol 7, no 1, pp. 21–28.

Dunkell, D.O., Bruland, G.L., Evensen, C.I. and Litton, C.M. (2011) "Runoff, sediment transport, and effects of feral pigs (*Sus scrofa*) exclusion in a forested Hawaiian watershed", *Pacific Science*, vol 65, no 2, pp. 175–194.

Geometrician Associates LLC (2011) "Biocontrol of Strawberry Guava by its Natural Control Agent for Preservation of Native Forests in the Hawaiian Islands", Draft Environmental Assessment prepared for the State of Hawaii Department of Land and Natural Resources.

Kaiser, B., Krause, N. and Roumasset, J. (1999) "Environmental Valuation and the Hawaiian Economy", University of Hawaii Economic Research Organization Working Paper, http://www.uhero.hawaii.edu/assets/HawaiiEnviroEvaluation.pdf.

Katahira, L.K., Finnegan, P. and Stone, C.P. (1993) "Eradicating feral pigs in Montane Mesic habitat at Hawaii Volcanoes National Park", *Wildlife Society Bulletin*, vol 21, pp. 269–274.

Nelson, E., Mendoza, G., Regetz, J., Polasky, S., Tallis, H., Cameron, D.R., Chan, K.M.A., Daily, G.C., Goldstein, J., Kareiva, P.M., Lonsdorf, E., Naidoo, R., Ricketts, T.H. and Shaw, M.R. (2009) "Modeling multiple ecosystem services, biodiversity conservation, commodity production, and tradeoffs at landscape scales", *Frontiers in Ecology and the Environment*, vol 7, no 1, pp. 4–11.

Pavlov, P.M., Crome, F.H.J. and Moore, L.A. (1992) "Feral pigs, rainforest conservation, and exotic disease in North Queensland", *Wildlife Research*, vol 19, no 2, pp. 179–193.

Roumasset, J. and Wada, C.A. (2013) "A dynamic approach to PES pricing and finance for interlinked ecosystem services: Watershed conservation and groundwater management", *Ecological Economics*, vol 87, pp. 24–33.

Sanchirico, J.N. and Springborn, M. (2011) "How to get there from here: Ecological and economic dynamics of ecosystem service provision", *Environmental and Resource Economics*, vol 48, no 2, pp. 243–267.

Takahashi, M., Giambelluca, T.W., Mudd, R.G., DeLay, J.K., Nullet, M.A. and Asner, G.P. (2011). "Rainfall partitioning and cloud water interception in native forest and invaded forest in Hawaii Volcanoes National Park", *Hydrological Processes*, vol 25, no 3, pp. 448–464.

Texas A&M Institute of Renewable Natural Resources (2011) "Feral Hog Statewide Population Growth and Density", AgriLIFE Research & Extension Fact Sheet, http://feralhogs.tamu.edu/files/2011/05/FeralHogFactSheet.pdf

9

EXTERNALITIES AND WATER QUALITY

Renan-Ulrich Goetz and Àngels Xabadia

Introduction

The availability of potable water in sufficient quantity is closely linked to the history of mankind from its earliest beginnings. With the adoption of an agrarian way of life and the establishment of permanent settlements, the permanent access to clean water became a prerequisite for urbanization and the formation of structured societies. In ancient times Greeks and Romans were already using settling tanks, sieves and filters to improve the quality of water. Yet poor management of waste water presented a major risk of waterborne infections. These illnesses, together with water-related sicknesses like malaria or bilharziosis, were probably one of the main causes of death.

Today, with the development of the modern urban infrastructure, health-related problems have been resolved in the industrialized world. Nevertheless, they still wreak havoc in the developing countries and continue to be life threatening for millions of people. However, although water pollution is not now associated with water-related diseases in the developed world, it still presents a threat to human health and the environment, and will be the principal concern of this chapter.

Water quality within the European Union (EU) is regulated by the Water Framework Directive, WFD, 2000/60/EC (European Commission, 2000). It requires that all water in Europe should have "good status" by 2015, i.e., the ecological and chemical status for surface water and the quantitative and chemical status for groundwater should be at least good (Art.2 WFD, 2000). A more detailed definition of "good status" is given in Appendix V of the WFD. However, as the report No. 9/2012 by the European Environment Agency (EEA, 2012) indicates, 25 percent of groundwater across Europe has poor chemical status. With respect to surface water, although only 10 percent has poor chemical status, one has to bear in mind that the chemical status of 40 percent of surface water is unknown. The data provided by the EEA report suggests that the poor chemical status of surface waters is unevenly distributed within the EU. River basins, which constitute over 40 percent of arable land and have a population density of over 100 inhabitants per km^2, show a less than good state in over two-thirds of their

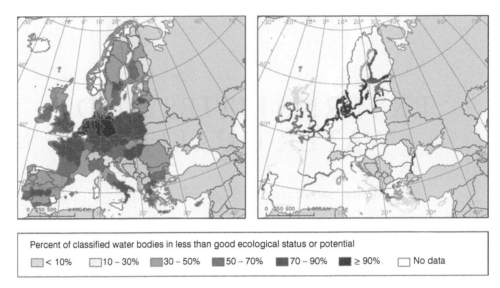

Percent of classified water bodies in less than good ecological status or potential

☐ < 10%   ☐ 10 – 30%   ▨ 30 – 50%   ▨ 50 – 70%   ■ 70 – 90%   ■ ≥ 90%   ☐ No data

Figure 9.1 Water pollution in Europe (source: European Commission, 2012)

water bodies. In total, ten member states report a poor chemical status in more than 20 percent of their rivers and lakes and for five member states this share rises to 40 percent. Emissions of phosphorus or nitrogen (nitrite, nitrate, ammonium) compounds can often be singled out as the most important pollutant pressures on water bodies and 30–50 percent emanate from diffuse sources, with agriculture being the most important.

The declining trend of certain water pollutants over the last 15 years suggests that a good chemical status of European surface water of total ammonium and total phosphorus will be reached by the years 2016 and 2028 respectively (EEA, 2012). Yet, the trend for nitrate is far less promising and indicates that, even in the year 2028, European rivers and lakes will be far from having a good chemical status. It indicates that additional and radically different measures have to be taken to meet the WFD objectives.

Water quality, however, is not only jeopardized by typical water contaminants but by new and largely unknown substances which have started appearing in the aquatic environment: for instance polychlorinated biphenyls (PCBs),[1] polycyclic aromatic hydrocarbons (PAHs), which are a byproduct of fuel burning; as well as heavy metals, phthalate (plasticizers), pesticides and medicines. Some of these substances have been shown to disrupt the hormonal balance in humans and animals, and present a true challenge for the management of water pollution. A new threat to water quality is also rising contamination by mercury, which is not biodegradable and accumulates in living organisms. Besides natural sources, human-related sources of mercury include coal combustion, chlorine alkali processing, waste incineration, and metal processing.

Water quality in Europe can also be analysed graphically. Figure 9.1 shows the proportion of classified surface water bodies in different River Basin Districts in less than good ecological status for rivers and lakes (left panel) and for coastal and transitional waters (right panel) (European Commission, 2012). An immediate conclusion that might be drawn is that water pollution is closely related to population density and the intensity of economic activities. Yet, since every pollutant is different with respect to the damage it causes, it is difficult to draw general conclusions. Moreover, even if the emissions

Table 9.1 National summary causes of impairment in assessed rivers and streams, reporting year 2010

Cause of impairment group	Miles threatened or impaired
Pathogens	138,225
Sediment	91,563
Polychlorinated biphenyls (PCBs)	76,657
Organic enrichment/oxygen depletion	68,890
Nutrients	63,511
Mercury	61,848
Habitat alterations	56,415
Metals (other than mercury)	53,076
Temperature	46,990
Flow alteration(s)	38,423

Source: http://www.epa.gov/waters/ir/index.html

Table 9.2 National summary causes of impairment in assessed lakes, reservoirs, and ponds, reporting year 2010

Cause of impairment group	Acres threatened or impaired
Mercury	7,197,986
Nutrients	2,902,104
Polychlorinated biphenyls (PCBs)	2,605,758
Turbidity	1,436,061
Organic enrichment/oxygen depletion	1,187,130
Metals (other than mercury)	901,094
pH/acidity/caustic conditions	723,037
Algal growth	619,611
Nuisance exotic species	507,737
Sediment	461,695

Source: http://www.epa.gov/waters/ir/index.html

are related to the factors mentioned above, it does not mean that water pollution will necessarily follow the evolution of these factors one by one because abatement policies and water treatment are different between countries.

Although this short overview of the current status of water quality concentrates on the situation in the EU, it is reflective of most developed countries. Reports for the Organisation of Economic Co-operation and Development (OECD) countries (OECD, 2008, 2012) or for the United States of America (US EPA, 2013) confirm the similarities within developed countries with respect to water quality and the tendencies over the last 10–20 years. As an example, the ten most important causes of water impairment in US rivers and streams and US lakes, reservoirs and ponds are presented in Tables 9.1 and 9.2 respectively.

For the evaluation of the quality of rivers, lakes, coastal zones or groundwater the WFD (Annex V) establishes three different quality aspects which are grouped into chemical, ecological and quantitative elements. Each element constitutes an array of different indicators. This qualification system, together with standardized monitoring, sampling and analysis procedures, allows for a correct evaluation of the water quality in its broad sense and a comparison between water bodies of different member states. An important aspect for standardization is the frequency of the required monitoring schemes.[2]

Water quality and economic decisions

The emission of water pollutants, their movement across the landscape and their fate in the receiving water body are complex. Below we present a conceptual model that accounts for the most important aspects of the processes involved.

We look at a single watershed. For simplicity we assume that it is not connected with other watersheds. Within the watershed there exists a competitive industry formed by heterogeneous production units which produce the good y. For the ease of the exposition, but without loss of generality, we consider the case of a single good. For the production of the good the industry employs the vector of inputs $\bar{x}$ with the components x_i, $i = 1,\dots, I$, per production unit. The inputs x_i are either productive or serve to control pollution.

Hence, the available production option for the firms is given by the production function

$$y(\bar{x}). \tag{9.1}$$

Yet, water pollutants usually accumulate at the final receptor and stay there over a certain time. Let us denote calendar time by t. Moreover, the emission of a specific quantity of pollutants at location A is likely to have different effects on the water quality than its emission at location B. Traditionally, economists think of space as a measurement of distance. However, the distance of the location is only one of many factors which determine the amount of the pollutant that reaches the water body. Georeferenced data like the slope of the land, soil texture, exposition of the land, etc., together with the characteristics of the receiving water body are fundamental for determining the amount of the pollutant that actually contributes to impair water quality. Thus, it is possible to relate emissions of a contaminant at a particular location (source) with the actual contamination of the final receptor. For instance, Sharpley et al. (2003), Richardson et al. (2008) and Duriancik et al. (2008) report on the generalization of this approach for the case of phosphorus. Xabadia et al. (2008) employed this concept to reflect differences in land quality with respect to their effect on the saturation of the subsurface with drainage water.

Let the land of a watershed be classified by a vulnerability index that ranges from 0 to 1. We associate $v = 0$ with locations where the emissions do not impair the water quality and $v = 1$ with locations with the highest impairment of water quality. Hence, there may exist several locations with the same value of v. However, for the brevity of the exposition we refer to it as location v although it would be more correct to talk about locations with vulnerability v. The use of a vulnerability index allows us to capture the fate and transport processes of the contaminant while it travels from the origin to the receptor, i.e., from the source to the sink. Likewise, as it takes into account the hydrological state of the receiving water body (physical, chemical, biological and geological conditions, for instance, as well as the trophic state, fauna, flora, sedimentation, inflow, outflow, etc.) it is not necessary to model its hydrologic components.[3]

To reduce the complexity of the model we assume that only one firm exists for each value of v. Nonetheless, each firm has several production units that can be dedicated to the production of the good. In the case of agriculture, a production unit would correspond to a particular size of arable land (acre or hectare). The total amount of production units within the watershed is equal to N. The distribution of the production units is given and equal to the density function $h(v)$ with $\int_0^1 h(v)dv = N$.

Based on the availability of a vulnerability index we can now write the production possibilities at location v as

$$y(\overline{x}(t,v), f(v))\, h(v),\qquad (9.2)$$

where $f(v)$ presents the relationship between the vulnerability index and the "productivity" of a location. Depending on the specific case this relationship may be positive, negative or nonexistent.

Likewise we can now relate the employed inputs or pollution control activities with the contamination of the water body. This function is given by

$$e(\overline{x}(t,v), v, \varepsilon),\qquad (9.3)$$

where e increases with the application of contaminating inputs and decreases with the employment of pollution control activities. The term ε reflects the stochastic nature of the vulnerability index. It is driven by stochastic environmental variables like the temperature, wind speed, and solar radiation or by the incomplete knowledge of the true underlying processes. However, in order to simplify the analysis we assume throughout the chapter that $\varepsilon = 0$.

Let the water quality impairment at time t be defined by the variable $s(t)$. It is characterized by the concentration of a pollutant in the receiving water body. The change in the concentration of the pollutant over time is described by the equation

$$\frac{ds}{dt} = \int_0^1 e(\overline{x}(t,v), v) h(v)dv - \gamma(s),\qquad (9.4)$$

where $\gamma(s)$ denotes the natural degradation of the pollutant in the water body net of the natural background load.

For the following we assume that there exists a social planner who maximizes the net benefits, i.e., the difference between the producer's benefits on the one hand and the private and external costs on the other. Given the size of the watershed in relation to the entire territory we assume that prices are influenced neither by the production within the watershed nor by the magnitude of water pollution. The decision problem of the social planner over the panning horizon T is given by

$$\max_{x_i} \int_0^T \exp(-rt)\left(\int_0^1 \left(p\, y(\overline{x}(t,v), f(v)) - \sum_{i=1}^I c_i\, x_i(t,v) \right) h(v)\ dv - D(s(t)) \right) dt,$$

subject to

$$\frac{ds}{dt} = \int_0^1 e(\overline{x}(t,v), v) h(v)dv - \gamma(s),\ s(0) = s_0,$$

$$x_i \geq 0, i = 1,...,I,$$

$$\qquad (9.5)$$

where r denotes the social discount rate, p the price of the good, c_i the price of input or pollution control x_i, s_0 the initial value of the concentration of the pollutant in the water body and $D(s(t))$ denotes the monetary damage from water pollution.[4]

We assume an interior solution with respect to x_i in order to concentrate on the solution, which is more interesting from the economic point of view.[5] To simplify the notation we suppress the arguments of all functions unless they are required for an unambiguous presentation. Based on the current-value Hamiltonian of the optimization problem stated in equation (9.5) the first order conditions yield

$$\frac{p \partial y}{\partial x_i} - c_i - \lambda \frac{\partial e}{\partial x_i} = 0, \ \forall t, v \tag{9.6}$$

$$\frac{d\lambda}{dt} = r\lambda - D'(s) + \lambda \gamma '(s), \tag{9.7}$$

$$\frac{ds}{dt} = \int_0^1 e(\bar{x}(t,v),v) \, h(v) \, dv - \gamma(s), \quad \text{with } s(0) = s_0, \tag{9.8}$$

together with the non-negativity constraint on x_i. The variable $\lambda(t) > 0$ indicates the social loss from time t to T from an increase of the pollution at time t. The variable λ has been multiplied by minus one to facilitate its interpretation as costs that have a positive sign.

Equation (9.6) requires that at every moment of time and at every location with vulnerability v, the value of the marginal product of the productive input x_i minus its marginal cost is equal to the marginal costs of the emissions caused by an additional unit of the productive input. In the case that x_i presents pollution control it has to hold that the sum of the value of the marginal loss in production and the marginal cost has to be equal to the marginal value of the reduced emissions as a result of an additional unit of the control activity.

Based on the work by Xabadia et al. (2006) it is possible to characterize the long-run solution of problem (9.5). They focus on the case of productive inputs but do not consider inputs that serve to control pollution. Let us assume for now that the length of the planning horizon tends to infinity, $T \to \infty$. At the steady state we have that $\frac{ds}{dt} = \frac{d\lambda}{dt} = 0$ and the solution of equations (9.6)–(9.8) yield constant values denoted by s^∞, x_i^∞ and λ^∞. If the initial value of water pollution, s_0, is above the steady-state equilibrium then x_i is chosen in the beginning below x_i^∞, and as time advances, the values of x_i approach the value x_i^∞, i.e., production intensity is initially reduced drastically and is increased slowly over time. Under the same circumstances, the good might be produced only at locations where the vulnerability is low and as time advances the production is also extended to locations with a higher vulnerability until the steady-state equilibrium distribution is reached.

Although the model aims to be complete, one has to keep in mind that water quality is only part of a management plan with respect to quantitative, ecological and chemical criteria. The accomplishment of these three objectives is only somewhat complementary but not entirely. Hence the focus on one particular objective implies restricting the management options. However, following the economic literature we concentrate on the chemical status of water quality.

Water quality management

The formulation of the social planner's problem in equation (9.5) assumes that the planner has full information about the choices made with respect to productive inputs, pollution control and the monetary losses associated with water pollution s. In reality these conditions are frequently not met and water management entails choices with respect to which pollutant to target, how much pollution to allow and which sources to regulate and how to regulate them. The introduction shows that water quality management has many facets and the WFD (European Commission, 2000) requires the watershed to be considered as the basis for any policy design. Moreover, water quality is often not defined in absolute terms but in relative terms taking into account the intended use of the water. For this purpose the US Environmental Protection Agency distinguishes between criteria related to human health, microbial status, recreation or aquatic life (http://water.epa.gov/scitech/swguidance/standards/criteria/current/). The WFD is less flexible and require good status for water in general. However, paragraph 31 of the preamble relaxes this objective if it is unfeasible or unreasonably expensive.

Most commonly, decision makers do not have enough information to specify the monetary loss function $D(s)$. As a result it is not possible to achieve the first-best solution as defined by the solution of equations (9.6)–(9.8). Alternatively one may resort to the physical, chemical and ecological criteria that have to be met. In fact, this approach has been proposed by the WFD and also by the US Clean Water Act (US Government, 1972). In this case the function $D(s)$ is eliminated from problem (9.5) and an upper limit on s is included

$$s(t) \leq \hat{s}. \tag{9.9}$$

The upper limit $\hat{s}$ is chosen based on physical, chemical and ecological criteria. The WFD has formulated slightly differently as it marks milestones that should be achieved at certain points in time in the future. This case is closer to our formulation and equation (9.9) should be replaced by $s(t) \leq s_t$ for different t. In particular it takes account of the fact that the demand for water quality is not constant but increases over time.

With the elimination of $D(s)$, and setting λ equal to one, the solution of equations (9.6), (9.8) and (9.9) still provides the producers' maximal net benefits but is subject to the constraint of a standard with respect to water pollution. In other words, it offers the least cost solution for meeting the standard.

Provided that $D(s)$ is eliminated from problem (9.5), one may ask whether the solution of equations (9.6), (9.8) and (9.9) with λ set equal to one, is also efficient if equation (9.9) is replaced by the social optimal trajectory of s. In other words we choose the standard such that it is always identical to the socially optimal value of the concentration of the pollutant. The solution will be least cost, but will it be efficient in the absence of the information about the monetary losses associated with impaired water quality? The answer to the question is yes, provided that the sum of emissions of the different firms actually leads to socially optimal water pollution s.[6] However, in practice, the regulator may find it difficult to guarantee this equivalence since it requires the emissions of all firms to be monitored, and especially when the emissions do not originate from a single point. In the following paragraphs we will analyse water quality policies related to emissions that originate from a single point (point source pollution) or from many different points (diffuse or nonpoint source pollution) so that is not possible to locate their origin with sufficient accuracy.

Point source pollution

In our model, emissions from a point source originate from a single point that can be identified. Yet, the hydrological conditions of the receiving water body (stream flow, internal circulation of the water, pH value, etc.) may be completely different for distinct point sources. As a result an identical amount of pollutant emitted from two different point sources is likely to lead to very distinct water quality losses.

Bearing industrial air pollution in mind, the early literature on the regulation of water pollution (Kneese and Bower, 1968) focused on the control of point sources that can be metered and monitored at acceptable costs, and, as they result from an industrial process, the emissions can be completely controlled by the firm. This assumption seems fairly reasonable for the case of air pollution since pollutants are often uniformly mixed in the air. However, for the case of water pollution it would require that the value of the vulnerability index v is identical for all point sources. For very large water bodies like the Baltic Sea or the Caspian Sea this assumption does not hold, since pollution is often concentrated, for instance, in the proximity of industrial complexes like the petrochemical industry, or the inflow of tributaries, rivers or creeks. Likewise, it does not hold for upstream pollution which impairs water quality downstream. Consequently, the condition of uniform mixing is not likely to hold given the natural length of many rivers and the associated variations of external factors along their course. These reasons suggest concentrating on the more relevant case where the point sources have different v, that is, the impact of their emissions on the water quality is different.

Emission standard/taxes

The implementation of the socially optimal outcome can be realized in principle by a command and control approach, i.e., the regulator prescribes an emissions standard. As in the Clean Water Act (US Government, 1972) the choice of the standard is also accompanied by the obligation to install the best available and economically achievable technology. Alternatively the regulator can choose to tax the emissions related to the production of the good at location v. In this case, the tax revenue on the emissions is given by $\tau e(\bar{x}(v), v)$, where τ denotes the tax. The decision problem of the social planner (9.5) can be converted into the decision problem of the firm if the term $D(s)$ is replaced by $\tau e(\bar{x}(v), v)$ and the differential equation $\dfrac{ds}{dt}$ is eliminated. The first order condition (9.6) now yields $\dfrac{p \partial y}{\partial x_i} - c_i - \tau \dfrac{\partial e}{\partial x_i} = 0$. Considering the steady-state solution of equations (9.7) and (9.8) suggests that the private (point of view of the firm) and social optimum coincide if the tax is given by

$$\tau \approx \lambda^\infty \approx \frac{D'(s^\infty)}{r + \gamma'(s^\infty)}. \tag{9.10}$$

The tax is independent of calendar time because we look at a long-run solution. Moreover, it is independent of location v because the tax base is the quantity of the pollutant that enters the water, i.e., each firm's emissions are quantifiable when they reach the receptor. Since we assumed that only one firm is present at every location v we can calculate the tax for this firm related to the production of the good. It is given by the marginal damage of the contamination weighted by the discount rate and the marginal natural degradation of the pollutant.

Following the same argumentation the optimal tax on the emissions can also be determined in the case of a finite planning horizon. The tax is calculated by solving equation (9.7) and is given by

$$\tau(t) = \lambda(t) = \exp\left(\int_t^T r + \gamma'(s(\xi))d\xi\right)\left(\int_t^T \exp\left(-\int_u^T r + \gamma'(s(\xi))d\xi\right)D'(s(u))du + \lambda(T)\right), \quad (9.11)$$

where $\lambda(T)$ is the marginal damage caused by water pollution at the end of the planning horizon. Along the trajectory of the tax, it depends on the values of the accumulated pollutant in the water, which in turn depends on the optimal values of the choice variables (Cropper and Oates, 1992) at every moment of time and every location v. This first-best tax is firm specific given the location v and varies over time because the optimal value of the choice variable $\bar{x}^*$ changes over time.[7] However, at time t, τ is a fixed amount charged for every unit of emissions.

Instead of a tax, firms often advocate for a subsidy on the reduction of the emissions. Apart from changing the burden of the regulation from firms to taxpayers, subsidies may give incentives for new firms to enter into production or for existing firms to expand production. Spulber (1985) has shown that the correction of this undesired effect would require that firms are not only taxed on the costs of the marginal damages of their emissions but also on the total costs caused by their emissions.

Emission trading

The implementation of emission taxes requires the regulator to have full information about functions y, f and e. Yet it may well be that the regulator is not fully informed, neither about the production function y nor about the function f that relates productivity with the vulnerability index. Given that firms have different productivity due to the value of f, their costs for emissions reductions in terms of forgone benefits are also different. In other words some firms are able to reduce emissions at lower costs than others. Obviously, in contrast to the regulator, the firms know their abatement costs. The regulator may induce the firms to reveal their true abatement costs by implementing a cap and trade system of emission permits between the firms. The total available amount of permits is determined by the regulator so that water pollution is less or equal to $\hat{s}$ (cap). Each firm needs to hold permits that are equivalent to their emissions, i.e., they may buy and sell permits so that they correspond to their emissions. Provided that in every moment of time $\hat{s}$ is identical with the socially optimal value of s the price of the permit ζ is given by $\zeta = \lambda(t)$ where $\lambda(t)$ is defined in equation (9.11). Although the price of the permit does not depend on the location v the permit expenditure does, since it is given by $\zeta e(\bar{x}(v),v)$. Given that every location has a different impact on the resulting water quality (emissions), the firms cannot trade permits one by one (Tietenberg, 1995; Goetz and Zilberman, 2007). Instead the trading ratio is related to the damage of the marginal emissions on water quality. For example, if discharges at location v_1 increase the concentration of pollution in the water body twice as much as discharges at location v_2, the firm at location v_1 needs to hold two permits while the firm at location v_2 only needs to hold one permit. In other words the permits are not defined in terms of discharges at the source but in terms of emissions that reach the receptor (Hoag and Hughes-Popp, 1997).

The implementation of a trading program may be challenging with respect to information requirements and administrative complexity. A possible simplification would be to divide the range of the index v into different subsets and each subset would correspond to a zone. Within each zone the trading ratio would be 1 to 1 and between zones it would be accorded to the fixed trading ratio which would be based on the average impact of the emissions of each zone on pollution concentration.

Implementation of policies

A lively debate on the advantages and disadvantages of the different policy instruments was found in the economic literature. Adding institutional and administrative aspects or transaction costs to the consideration of efficiency suggests that there is no single instrument that is superior in all dimensions. However, there is an overall claim for a wider use of flexible incentive-based policies (Cropper and Oates, 1992; Goulder and Parry, 2008).

EU member states rely on standard and effluent fees. The basis of the taxation varies greatly between the different member states. In the Netherlands one pollution equivalent is defined by the amount of effluent that an individual produces, in France by the amount of effluent that reaches a water body on a normal day of a month at the maximum, or in Germany by the toxicity or damage equivalent that effluents cause. In Denmark for instance effluent fees are based on biological oxygen demand (1.48€/kg), nitrogen (N) (2.69€/kg) and phosphorus (P) (14.78€/kg), while in Germany effluent fees are based on one unit of noxiousness (35.79€) (Mattheiss et al., 2010). This unit can consist for example of 25kg of N, or 3kg of P or 50kg of chemical oxygen demand. The different bases of a pollution equivalent makes it extremely complicated to compare effluent fees between member states and presents a significant obstacle for an integrated policy.

The taxes collected in most member states were well below the marginal environmental damages and did not provide incentives to decrease water pollution beyond the standard. In fact the tax revenue essentially often served to raise revenues but was not sufficient to recover costs or to provide incentives to reduce emissions (Glachant, 2002; OECD, 2005). With the translation of the WFD into action programs, in particular the realization of river basin management plans from 2009-2015, member states are obliged to recover the full cost of water services including environmental and resource scarcity costs, thus one would expect taxes to rise substantially in a number of countries.

In contrast to the EU, US policy relies on standards that must be met. They are enforced by taking legal action to bring polluters into compliance with the law. Different authorities within the US also implemented emission trading programs. One of them is the Long Island Sound Nitrogen Credit Exchange Program (DEEP, 2013) effective from January 2011 to December 2015. It regulates the discharge of total nitrogen from 79 "publicly owned treatment works" in Connecticut. In the program definition, the geographic location is considered by an equivalency factor – similar to the definition of a vulnerability index used in this chapter. It means a ratio of the unit response of

Table 9.3 National summary of probable sources of impairment in assessed rivers and streams, reporting year 2010

Probable source group	Miles threatened or impaired
Agriculture	114,714
Unknown	94,795
Atmospheric deposition	91,774
Hydromodification	53,568
Urban-related runoff/storm water	51,573
Municipal discharges/sewage	51,347
Natural/wildlife	49,659
Unspecified nonpoint source	42,779
Habitat alterations (not directly related to hydromodification)	31,090
Resource extraction	22,611

Source: http://www.epa.gov/waters/ir/index.html.

Table 9.4 National summary of probable sources of impairment in assessed lakes, reservoirs, and ponds, reporting year 2010

Probable source group	Acres threatened or impaired
Atmospheric deposition	4,139,795
Unknown	3,125,304
Agriculture	1,317,892
Municipal discharges/sewage	743,914
Natural/wildlife	730,263
Urban-related runoff/storm water	691,395
Hydromodification	588,411
Other	547,900
Legacy/historical pollutants	529,686
Resource extraction	380,524

Source: http://www.epa.gov/waters/ir/index.html.

dissolved oxygen to nitrogen based on the location of the discharge point divided by the unit response of the location with the highest impact (DEEP, 2010). Another trading program, implemented outside the US, is the Australian Hunter River Salinity trading scheme which has been operating since 2002 (DEC, 2006). The monitoring of the salinity occurs in real time and trading of the overall limited credits (cap) takes place online. The exchange of credits depends on the location of the emitter and the river flow conditions.

Nonpoint source pollution

As mentioned above, emissions where neither the origin nor the quantity can be observed with sufficient accuracy, are defined as nonpoint source emissions or diffuse pollution. The underlying reason may be that it is technically not possible to meter the emissions, or the associated costs are prohibitive. Nonpoint source emissions usually result from precipitation, atmospheric deposition, runoff from urban or agricultural land, infiltration, drainage, or seepage from hydrologic modifications. Tables 9.3 and 9.4 show the ten most important sources of water impairment for US rivers and streams, and lakes, reservoirs and ponds respectively. In line with the data published by the EEA (2012), agriculture, urban runoff, storm water and atmospheric deposition are the principal causes for nonpoint source pollution in Europe.

Since the emissions cannot be metered the regulator cannot apply standard policy instruments that relate to the emissions. Instead, the regulator needs to design alternative instruments where the compliance base can be observed.

The economic literature covering the management of nonpoint source pollution has identified three strategies. The earliest literature suggested using observation of inputs or technology choice since they are highly correlated with the emissions. These instruments can be designed as standards or based on economic incentives. The second approach is based on the water quality of the water body itself. Often it is referred to as an ambient scheme. The third approach, the most recent one, is voluntary and information-based. It is based on voluntary compliance either by following the guidelines of eco-labels or by disclosing private information on emissions in exchange for a reduction in tax payment. Likewise, it may provide incentives to change the modus operandi of the firm leading to lower nonpoint source emissions.

Inputs or technology-based instruments

These instruments aim to target choices that affect nonpoint source pollution. In the case of runoffs from agricultural land, they could target the reduction of mineral and organic fertilizers, animal waste, pesticides, or the adoption of minimum tillage practices, strip cropping or the creation of vegetative buffer strips. In the case of urban runoff, policy measures could focus on the reduction of impervious surfaces or the construction of percolation trenches, infiltration basins, green roofs or porous pavements. With respect to atmospheric deposition, basically PCB, PAB and toxic metals, the possible policy instruments are very specific for each pollutant and for every site and for the sake of brevity they are not discussed in this review. For more details consult the EPA-905-R-08-001 report published with the IADN (2008).

Input-based instruments can be designed in the form of standards, taxes, subsidies or permit trading. In the case of an input-based tax, the term $D(s)$ of the decision problem (9.5) needs to be replaced by $\tau_i x_i$ which leads to the private first-order condition $\dfrac{p\partial y}{\partial x_i} - c_i - \tau_i = 0$. The later one coincides with the socially optimal first-order condition (9.6) if the tax is set equal to

$$\tau_i(t,v) = \lambda(t)\frac{\partial e(\overline{x}^{\star}(t,v),v)}{\partial x_i}, \tag{9.12}$$

that is, the tax per unit of input is a function of time and the vulnerability of the water body with respect to emissions from site v. In the case where x_i is employed for pollution control $\tau_i(t, v)$ is negative and presents a subsidy.

So far we assumed implicitly that all production units at location v are utilized. Yet, it may be optimal to leave some production units idle. Let the share of production units at location v dedicated to the production of the good at time *t* be presented by $n(t,v)$. Hence, the decision problem (9.5) has to be modified by taking account of the new decision variable, $n(t,v)$., and the additional constraint $1 - n(t,v) \geq 0$. The new decision variable leads to an additional first-order condition given by

$$\left(py - \sum_i^I c_i x_i - \lambda e \right) h + \mu_1 - \mu_2 = 0, \tag{9.13}$$

where μ_i, $i = 1, 2$ are the Lagrange multipliers related to the lower and upper limit of the variable $n(t,v)$.

The tax $\tau_i(t,v)$, defined in equation (9.12), would establish the socially optimal use of the input related to the intensity of the production but not related to the extension of the production, that is, the number of production units dedicated to the production of the good. In other words τ_i does correct the private first-order condition with respect to x_i but not with respect to n. For this purpose another tax $\hat{\tau}$ has to be established. This production unit-use tax is given by

$$\hat{\tau}(t,v) = -\tau_i(t,v) - \lambda e \left(\overline{x}^*(t,v), v \right). \tag{9.14}$$

The introduction of a tax at the intensive margin causes a distortion at the extensive margin. For productive inputs, Goetz et al. (2006) and Xabadia et al. (2006) have shown that $\hat{\tau}$ is positive (proper tax) if the function $e(\cdot)$ is concave in x_i, and $\hat{\tau}$ is negative (subsidy) if $e(\cdot)$ is convex. If $e(\cdot)$ is linear in x_i, $\hat{\tau}$ is equal to zero. A tax on the proportional increment in production units is positive if the value of the marginal damages of pollution resulting from a more extensive production outweighs the marginal damages from an intensification of the production. In other words, if the emission function is concave the taxes favor an increase in production by a more intensive use of the production units and less by the utilization of more production units. Similar arguments allow the sign of $\hat{\tau}$ to be determined if *e* is convex or linear.

The input tax τ_i and the production unit-use tax $\hat{\tau}$ are specific for the location and vary over time. Alternatively, one can propose subsidies in order to reduce emissions. Subsidies are often based on the "provider gets principle" which forms part of agricultural policy programs. The "provider gets principle" can be applied if firms undertake activities which avoid environmental damage or provide environmental amenities and go beyond required good practices.

Yet, in practice it is difficult to draw the exact line between the application of the "provider gets principle" and the offense of the "polluter pays principle" advocated by the EU in the Single European Act 1987 or by the OECD Council Recommendation in 1974 (Rosso Grossman, 2007). In any case, the application of subsidies to reduce emissions may provide incentives for existing firms to expand production or for new firms to enter into production. Hence, as in the case of a Pigouvian tax, the design of the subsidies should be such that it does not provoke distorting effects at the extensive margin.

Implementation of policies

The implementation of taxes and subsidies differentiated over location and time is extremely complex. On one hand it requires information about input and technology for every firm and every location. Moreover, even if this information were fully available, the implementation of differentiated taxes requires the regulator to be able to separate markets completely. Otherwise firms with a higher tax rate would buy the input from a firm with a lower tax rate. Only the complete separation of the markets would impede the development of a black market which would fully undermine any differentiation of taxes. Moreover, input taxes/subsidies on polluting inputs are likely to lead to substitution processes between all inputs. Consequently only the far-fetched case where all inputs are either taxed or subsidized would allow the replication of the first-best solution.

For these reasons second-best approaches are proposed which do not consider all inputs, technologies and production units. Regulation focuses on a subset of easily observable choices that are highly correlated with nonpoint source emissions (Brady, 2003; Iho, 2010).

Nevertheless, even if the selected inputs are highly correlated with emissions there is no guarantee that a significant reduction in emissions will be achieved. Taxes on an input may lead to a reduction in output which in turn may drive up output prices. Rising prices may encourage increased production at the intensive and extensive margins, offsetting, in part or completely, the reduction of the emissions caused by the incentive effect of the tax. Besides the market effects on the firms, the reaction of the indirectly affected firms needs to be accounted for. For instance Austria introduced a tax on the N content of fertilizer in 1986 which was abolished in 1994 when it joined the EU. Yet, despite a rising tax rate the price of nitrogen fertilizers actually went down since the tax burden was absorbed by the fertilizer manufacturers and the distributors of mineral fertilizer (ECOTEC Research and Consulting, 2001; Bel et al., 2002). The latter group not only feared a decrease in the sale of nitrogen fertilizers but also in complementary products like pesticides, herbicides or growth regulators. Since distributors often calculate the profit margin for each product as a function of their sales mix, regulators cannot easily predict the effect of a tax on a particular product. Thus, intermediate dealers most likely found it profitable to renounce the profits from fertilizer sale and compensate the reduced profits with a higher profit margin on other products.

Despite the possible adversities, some of the OECD states levy taxes and charges on mineral fertilizers (Sweden, Italy and some states of the US) or on pesticides (Denmark, France, Belgium, Italy and Norway) (Vojtech, 2010). Other policies aim at inputs or technologies where market effects are likely to be moderated given that they affect some farmers but not the whole agricultural sector. This could be the case for payments on farm fixed assets like waste management facilities covered by the Environmental Quality Incentives Program in the US or by the EU Rural Development Regulation 1698/2005.[8] Special care needs to be taken that the supported installations replace old ones and do not favor the construction of additional ones.

Instead of choosing particular inputs it is also possible to choose a constructed compliance base that takes account of the aggregate choice of inputs and technologies. In this way the compliance base is akin to an environmental performance index. Although it is frequently used for the regulation of point sources, it is more difficult to implement in the context of nonpoint source pollution (OECD, 2010b). A scheme which comes close to the idea of performance indices was the Dutch scheme MINAS which combined nutrient accounting with a tax on nutrient surplus. However, it was replaced by the manure application standard in 2006 to comply with the EU Nitrogen Directive (Wright and Mallia, 2008).

Another policy option to reduce emissions from nonpoint source pollution is related to the regulation of the authorization of potential pollutants in the production and the subsequent use of the produced goods. Examples are the international Stockholm Convention on Persistent Organic Pollutants in 2001 and its subsequent amendments. Since these pollutants accumulate in living organisms or different spheres of the environment, where water is only one of them, the reduction or elimination of the risks associated with these pollutants is promoted more generally outside the policy area of water management. A similar trend can be observed for the regulation of bioaccumulable heavy metals, for instance mercury, lead, tin or cadmium. A tighter standard for the admissible presence of these metals in food will also reduce the presence of these metals in water bodies (EU Regulation 1881/2006).

Over recent years several new market-based approaches in form of point–nonpoint trading programs have been developed, basically within the US. One of them is the Chesapeake Bay Nutrient Trading Program. The complexity of the tradable commodity and the singularity of each case suggests that interchange might be complicated. For this reason careful attention has to be paid to the program design. Specific criteria for successful pollution permit trading schemes are discussed by Fisher-Vanden and Olmstead (2013).

Ambient scheme

The input tax and technology-based approach is based on the emissions from the different sources. In contrast, the ambient scheme focuses on the water quality of the receptor, i.e., the receiving water body.

Based on Holmstrom's work (Holmstrom, 1982) on moral hazards in teams, Segerson (1988) proposed a scheme where the compliance base is not given by individual emissions but by collective emissions, i.e., the contamination of the receiving water body. The tax is imposed if the measured contamination exceeds a previously defined acceptable level $\tilde{s}$. In the case that the measured contamination is below the cutoff level $\tilde{s}$ the tax turns into a subsidy. Formally, the ambient tax scheme τ_j for firm j or source of pollution j, $j = 1,\ldots,J$ is given by (Horan et al., 1998)

$$
\left.
\begin{aligned}
\tau_j(s) &= \tau(s-\tilde{s}) + k_j, s > \tilde{s}, \text{ with } \tau = D'(s^*), \forall j \\
\tau_j(s) &= \tau(s-\tilde{s}), \ s \le \tilde{s}
\end{aligned}
\right\},
\tag{9.15}
$$

where k is specific for each firm. The value of k is chosen such that it corrects the distortionary effect of the tax on entry and exit decisions.

Equation (9.15) presents a linear ambient tax as proposed by Segerson (1988). It achieves a first-best outcome if polluters are risk-neutral, if regulators and firms have identical expectations about random events, if any firm can predict the water quality actions of all other firms and their impact on the water quality, and if all firms take the actions of the other firms as given. Horan et al. (1998) and Hansen (1998, 2002) extended the approach by considering a non-linear ambient tax as in equation (9.16).

$$
\tau_j(s) = D(s) - k_j \ .
\tag{9.16}
$$

Other extensions of this approach consider situations where firms are not risk-neutral (Horan et al., 2002), where firms and regulator have heterogeneous expectations about random events (Cabe and Herriges, 1992), and where polluters do not act independently and may form coalitions (Hansen, 2002). Xepapadeas (1992) also analysed the case where the pollutant accumulates in the receiving water body so that the optimal ambient tax is defined for every moment of time.

What both the linear and non-linear versions of the tax have in common is that the approach is not budget balancing. Tax τ_j is equal to the marginal damage of the contamination s, i.e., the pollution of all potential polluters and not just of the individual contribution. Since all firms pay τ_j, the collected taxes exceed the social damages.

Corrective measures to balance the budget have been proposed by Xepapadeas (1991) and Jones and Corona (2008) in the form of fines/payments from a randomly selected small subset of the potential polluters. Although these proposals achieve budget balancing they still do not overcome a principal problem of ambient schemes. There is no clear link between individual behavior and tax payments. For this reason its acceptability might be low among the affected firms and among politicians who have to defend it in public.

The consideration of time underlines an implementation problem for the ambient tax scheme. In the case where the pollutant does not accumulate in the water body and the regulator does not permanently monitor water quality, high emissions and contamination peaks may go unnoticed, i.e., without any costs for the polluters. Another implementation problem for ambient tax schemes is related to the definition of the cutoff level $\tilde{s}$ for rivers. Should it be defined uniformly over the course of the river or specific for different sections of the course? In the former case, firms have incentives to locate at sections where the probability that emissions exceed the cutoff level is low. The problems related with the implementation of an ambient tax scheme and its most likely low political acceptability might explain why this approach has hardly ever been used for actual policy design to date.

Implementation of policies

Although water quality has not been used as a compliance base, the idea of triggering off policy measures as a function of water quality has been applied. The Clean Water Act introduced the regulatory term Total Maximum Daily Load. Its value includes a margin of safety to take account of the stochastic nature of emissions and determines the maximum acceptable pollution limits, including point and nonpoint sources. In the case of a body water being classified as impaired or threatened, the load allocation for all sources is reassigned in order to meet water quality criteria in the future. While point sources are regulated on permit bases, nonpoint sources are generally regulated on a voluntary compliance base. In the European Union, for instance, the Nitrate Directive 91/676/EEC established that member states can designate vulnerable zones, where agricultural activities are limited, if the nitrate concentration of the water is above 50mg/l or if the water is classified as eutrophic.

Voluntary and information-based approaches

Over recent decades an extensive number of voluntary and information-based approaches have emerged and often complement standard instruments of water management. These approaches can be classified as a) product or site related b) emission related and c) technology related.

PRODUCT OR SITE RELATED

The first strategy often bundles environmental characteristics (public goods or bads) of the produced goods with the characteristics of the goods that are private (Kotchen, 2013). Examples are eco-labelling programs like Salmon Safe[9] or trademarks like Processed Chlorine Free[10] which are related to a product. An extensive list of eco-labels can be found at www.ecolabelindex.com.

EMISSION RELATED (MONITORING)

The second category is of great interest for the management of water quality as it allows regulators to obtain information about individual emissions in the context of nonpoint source pollution. Within the development and application of new technologies like remote sensing, georeferenced data processing, and geographic information system, the border line between point and nonpoint sources becomes blurred if the cost–benefit analysis of investing in the monitoring technology is positive for the decision maker.

Initially the economic literature looked at the regulator's decision to invest in monitoring equipment. Dinar and Xepapadeas (1998, 2002) developed a model that allows the determination of the optimal input tax on water use and the optimal investment in monitoring equipment which increases the informational base with respect to individual water use and emissions. Yet, the decision about investment in the monitoring equipment is taken by the regulator and not by the individual firm. A similar approach was presented by Farzin and Kaplan (2004) for the problem of sedimentation as it increases the risk of flooding and reduces salmon spawning sites in the Redwood National Park. The regulator and the polluter have complete information about the cost of monitoring but the polluter has more information about the future benefits of the acquired information than the regulator. As a result of this asymmetric distribution of information, efficiency can be increased if the choice of investing in monitoring equipment is endogenous for the firm and not exogenously taken by the regulator.

Millock et al. (2002) considered this possibility by allowing individual polluters to install monitoring equipment in order to reduce payment for emissions. Since the regulator cannot distinguish between the different sources, payments for emissions are uniform. Yet, some firms are more efficient and less polluting than others and thus the investment in monitoring may reduce their payment since it is based on individual metering and not on average data. Millock et al. (2012) extended this approach by explicitly considering the aspect of time, i.e., the fact that pollutants often accumulate at the receptor. In a way this approach is similar to self-reporting where polluters voluntarily disclose information and do not rely on monitoring equipment (Innes, 1999). However, self-reporting schemes require polluters to have full information about pollution processes: a prerequisite which might be difficult to meet in the context of pollution processes subject to stochastic variations. Moreover, the regulator needs to design a compensation scheme that leads firms to report the true value of their emissions and not report false values for strategic considerations. Hansen and Romstad (2007) drafted a self-reporting scheme that is robust to strategic cooperation among the polluters in the context of ambient taxes.

The third approach provides incentives to alter the modus operandi of the firm. In agriculture, the amount of inputs that are related to emissions or the technology in itself cannot often be classified as clean or dirty but rather the way in which the inputs are applied or the technology used leads to higher or lower emissions. In other words, the modus operandi is not embodied in the input or the technology but in management practices. Unfortunately, management practices can hardly be observed by the regulator and it would be difficult to define a compliance base in terms of the application of best management practices.

To overcome this problem Goetz and Martínez (2013) proposed to make use of an accredited firm that either validates that the input or technology has been used in accordance with the best management practices or provides this service to the firm. In this way the regulator creates a market for best management practices and the resulting market transaction can be used to form part of policy instrumentation. Similar to the work by Fullerton and Wolverton (2000) and in accordance with the work by Segerson and Wu (2006) and Millock et al. (2002, 2012) this instrument would consist of a combination of voluntary and mandatory policies. Firms would be obliged to pay a tax on the inputs or estimated emissions that correspond to the situation where best management practices are not followed. Obviously firms are only willing to contract an accredited firm if they receive something in exchange. By following best management practices firms reduce their emissions. The monetary equivalent of the reduced marginal damage can be reimbursed to the firm. The firms thus have incentives to voluntarily contract an accredited firm if the costs of the contract are less than the benefits (cost savings for the work now carried out by the accredited/external firm and reduction in payment for emissions). The regulator may implement this policy if the transaction costs plus the costs of the accredited firm are less than the social net benefit of this policy. If the farm produces different outputs and uses a variety of inputs this approach can be implemented in form of a two-part instrument. Goetz and Martínez (2013) present a variant of the so-called deposit-refund approach.

Another voluntary scheme related to farming practices, but not specifically to nonpoint source pollution, is payments to farmers in exchange for environmentally friendly farming practices like integrated pest management, extensive crop production, green set-aside, cover crop in winter, etc. (US: Environmental Quality Incentives Program 1996 and Food, Conservation and Energy Act, 2008; EU: Rural Development Regulation 1257/1999, 1698/2005). Since 2005 all farmers in the EU receiving direct payments are subject to compulsory compliance, regulation 1783/2003, with different legislative acts that refer to the environment, animal and plant health and animal welfare. In the US, farmer support programs are widely linked to the compliance with the environmental protection of environmentally sensitive land (OECD, 2010a). Albeit that most of these direct payments are not targeted to improve water quality, they may contribute to it at certain locations.

Conclusions

The review of the literature shows that the quality of water bodies in the developed world has been improved over recent decades for some chemical compounds while for others little improvement can be perceived. The most significant advancement has been observed in connection with the control of point source emissions while relatively little progress can be observed for the control of nonpoint sources. For example, nitrate emissions from agricultural production are still a prominent problem and if the current trend of nitrate emissions is not substantially changed, the presence of nitrate in water bodies will continue to pose a problem in future decades. For this reason policy makers are likely to be called upon to implement further instruments to control emissions. This chapter has shown the difficulties in designing and implementing well targeted polices, leading to a gap between optimal and feasible policies. Instruments based on economic incentives and tailored to site-specific conditions are not able to establish the first-best optimum but they may provide additional headway and either substitute or complement many of the instruments currently in place. Likewise, water management policies need to be coordinated with urban waste water and regional development policies in order to design an efficient and coherent set of policy instruments.

Besides nitrogen, persistent organic pollutants and heavy metal compounds which accumulate in living organisms present an important threat for plants, animals and humans. These substances form part of the nutrient chain as they are present in the water, air and soil. Thus, the emission of these substances does not correspond to the traditional concept of nonpoint source pollution since its origin cannot even be traced back to a particular region or particular production process. In this respect the control of these pollutants calls for coordinated action far beyond water management. Therefore water-related policy instruments should be coordinated with product or process-related instruments as well as with air and soil management instruments. A first attempt in this direction is the REACH initiative by the European Commission (Registration, Evaluation and Restriction of Chemical Substances, phased in from 2007–2018). Another important step is the continuous update of the list substances to be monitored (Priority Substance Directive 2008/105/EC). From 2014 onward it will most likely include pharmaceutical substances for the first time.

In addition to the chemical status of the water bodies, water quality is related to ecological and quantitative characteristics. Economists have considered these three aspects for a long time as separate aspects. So far the simultaneous consideration of all three aspects, including the cost–benefit consideration, presents a challenge for economic analysis that should be accepted in the near future by economists.

Notes

1 The production of PCB was banned in the US in 1979 and Europe in 1985 (Directive 85/467/EEC), but the use of equipment which contains PCB was allowed until 2010 (phase out period). The persistence of this organic pollutant is the reason why it still presents a serious threat to water quality.
2 For surface waters, for instance, priority substances should be monitored every month, salinity every three months, phytoplankton every six months and fish every three years.
3 For an explicit modeling of the fate and transport processes or the hydrological processes see Dosi and Tomasi (1994), Aftab et al. (2007), Aftab et al. (2010) and Lankoski et al. (2010).
4 We assume that the utility function of the consumers is quasilinear with respect to the traded good and the externality. Given that prices are influenced neither by regional production decisions nor by the regional production of the externality, the optimal level of the externality is independent of consumer expenditures, and it is possible to derive a utility function which

depends only on the externality *s* (Mas Collell et al., 1995). To discuss the results of our model in a practical setting, we propose that the derived utility function is represented by the damage function $D(s)$. The assumption of quasilinearity of the utility function helps to keep the model simple and allows us to concentrate our analysis on policy issues.

5 Corner solutions could emerge if the marginal damage tends to infinity after a critical level of emissions has been exceeded. For certain pollutants the critical level is considered to be zero as the current legislation in the US and Europe completely prohibits their emission. Unless the marginal damages do not tend to infinity with relatively low emissions, corner solutions are not optimal from an economic point of view. Alternatively, the application of the precautionary principle may lead to corner solutions as well if damages are unknown or monitoring costs are very high. Yet, from an economic point of view the trade-off between marginal benefits and marginal costs for the vast majority of the pollutants is more interesting. For this reason, this chapter concentrates on the case of interior solutions.

6 Shortle and Horan (2013) respond to the question for the case where emissions are stochastic. In this case the answer to the question is still yes, however, only if the monetary loss function $D(s)$ is linear in *s*. This would guarantee that a variation of *s* does not affect the expected monetary loss, i.e., $E(D(s)) = D(E(s))$.

7 The superindex ★ indicates that the variable is evaluated along its optimal trajectory.

8 Another example would be payments based on land retirement programs implemented in many OECD countries.

9 Salmon Safe is an independent nonprofit organization which has certified more than 400 US West Coast farms, vineyards, dairies and other sites based on riparian habitat protection.

10 Certification is undertaken in accordance with environmental management audits (ISO 17011, 17021, 14011 or Environmental Audit and Management System III EC 121/2009).

References

Aftab, A., Hanley, N. and Baiocchi, G. (2010) "Integrated regulation of nonpoint pollution: Combining managerial controls and economic instruments under multiple environmental targets", *Ecological Economics*, vol 70, no 1, pp. 24–33

Aftab, A., Hanley, N. and Kampas, A. (2007) "Co-ordinated environmental regulation: Controlling non-point nitrate pollution while maintaining river flow", *Environmental and Resource Economics*, vol 38, no 4, pp. 573–593

Bel, F., d'Aubigny, G., Lacroix, A. and Mollard, A. (2002) "Fertilization Taxation and Regulation of Nonpoint Water Pollution: A Critical Analysis after European Experiences", Laboratoire d'economie et de sociologie rurales, no 2002-01, Grenoble

Brady, M. (2003) "The relative cost-efficiency of arable nitrogen management in Sweden", *Ecological Economics*, vol 47, no 1, pp. 53–70

Cabe, R. and Herriges, J. A. (1992) "The regulation of non-point-source pollution under imperfect and asymmetric information", *Journal of Environmental Economics and Management*, vol 22, no 2, pp. 134–146

Cropper, M. and Oates, W. (1992) "Environmental economics: A survey", *Journal of Economic Literature*, vol 30, no 2, pp. 675–740

DEC (2006) "Hunter River Salinity Trading Scheme: Working together to protect river quality and sustain economic development", http://www.environment.nsw.gov.au/resources/licensing/hrsts/hrsts.pdf, accessed 24 March 2014

DEEP (2010) "General Permit for Nitrogen Discharges", http://www.ct.gov/deep/lib/deep/water/municipal_wastewater/2011_2015_nitrogen_gp.pdf, accessed 24 March 2014

DEEP (2013) "Nitrogen Control Program for Long Island Sound", http://www.ct.gov/deep/cwp/view.asp?a=2719&q=325572&deepNav_GID=1654, accessed 24 March 2014

Dinar, A. and Xepapadeas, A. (1998) "Regulating water quantity and quality in irrigated agriculture", *Journal of Environmental Management*, vol 54, no 4, pp. 273–289

Dinar, A. and Xepapadeas, A. (2002) "Regulating water quantity and quality in irrigated agriculture: Learning by investing under asymmetric information", *Environmental Modeling and Assessment*, vol 7, no 1, pp. 17–27

Dosi, C. and Tomasi, T. (1994) *Nonpoint Source Pollution Regulation: Issues and Analysis*, Kluwer, Boston, MA

Duriancik, L. F., Bucks, D., Dobrowolski, J. P., Drewes, T., Eckles, S. D., Jolley, L., Kellogg, R. L., Lund, D., Makuch, J. R., O'Neill, M. P., Rewa, C. A., Walbridge, M. R., Parry, R. and Weltz, M. A. (2008) "The first five years of the Conservation Effects Assessment Project", *Journal of Soil and Water Conservation*, vol 63, no 6, pp. 185A–197A

ECOTEC Research and Consulting (2001) "Study on the Economic and Environmental Implications of the Use of Environmental Taxes and Charges in the European Union and its Member States", no C1653/PtB/DH/MM, Brussels, Birmingham

EEA (2012) "European Waters — Current status and future challenges", European Environment Agency, no 9/2012, Copenhagen

European Commission (2000) "Water Framework Directive", 2000/60/EC, European Parliament and Council, http://eur-lex.europa.eu/legal-content/EN/ALL/?uri=CELEX:32000L0060, accessed 24 March 2014

European Commission (2012) "Implementation of the Water Framework Directive (2000/60/EC)", Report from the Commission to the European Parliament and the Council, Vol 1/2, Brussels

Farzin, Y. H. and Kaplan, J. (2004) "Nonpoint source pollution control under incomplete and costly information", *Environmental and Resource Economics*, vol 28, no 4, pp. 489–506

Fisher-Vanden, K. and Olmstead, S. (2013) "Moving pollution trading from air to water: Potential, problems, and prognosis", *Journal of Economic Perspectives*, vol 27, no 1, pp. 147–172

Fullerton, D. and Wolverton, A. (2000) "Two generalizations of a deposit-refund system", *American Economic Review*, vol 90, no 2, pp. 238–242

Glachant, M. (2002) "The political economy of water effluent charges in France: Why are rates kept low?", *European Journal of Law and Economics*, vol 14, no 1, pp. 27–43

Goetz, R. U. and Martínez, Y. (2013) "Nonpoint source pollution and two-part instruments", *Environmental Economics and Policy Studies*, vol 15, no 3, pp. 237–258

Goetz, R. U., Schmid, H. and Lehmann, B. (2006) "Determining the economic gains from regulation at the extensive and intensive margins", *European Review of Agricultural Economics*, vol 33, no 1, pp. 1–30

Goetz, R. U. and Zilberman, D. (2007) "The economics of land-use regulation in the presence of an externality: A dynamic approach", *Optimal Control Applications and Methods*, vol 28, no 1, pp. 21–43

Goulder, L. H. and Parry, I. W. H. (2008) "Instrument choice in environmental policy", *Review of Environmental Economics and Policy*, vol 2, no 2, pp. 152–174

Hansen, L. (1998) "A damage based tax mechanism for regulation of non-point emissions", *Environmental and Resource Economics*, vol 12, no 1, pp. 99–112

Hansen, L. (2002) "Regulation of non-point emissions – A variance based mechanism", *Environmental and Resource Economics*, vol 21, no 4, pp. 303–316

Hansen, L. G. and Romstad, E. (2007) "Non-point source regulation — A self-reporting mechanism", *Ecological Economics*, vol 62, no 3–4, pp. 529–537

Hoag, D. L. and Hughes-Popp, J. S. (1997) "Theory and practice of pollution credit trading in water quality management", *Review of Agricultural Economics*, vol 19, no 2, pp. 252–262

Holmstrom, B. (1982) "Moral hazard in teams", *Bell Journal of Economics*, vol 13, no 2, pp. 324–340

Horan, R. D., Shortle, J. S. and Abler, D. G. (1998) "Ambient taxes when polluters have multiple choices", *Journal of Environmental Economics and Management*, vol 36, no 2, pp. 186–199

Horan, R. D., Shortle, J. S. and Abler, D. G. (2002) "Ambient taxes under m-dimensional choice sets, heterogeneous expectations, and risk-aversion", *Environmental and Resource Economics*, vol 21, no 2, pp. 189–202

IADN (The Integrated Atmospheric Deposition Network) (2008) "Atmospheric Deposition of Toxic Substances to the Great Lakes: IADN Results through 2005", Environment Canada and the United States Environmental Protection Agency, no EPA-905-R-08-001

Iho, A. (2010) "Spatially optimal steady-state phosphorus policies in crop production", *European Review of Agricultural Economics*, vol 37, no 2, pp. 187–208

Innes, R. (1999) "Remediation and self-reporting in optimal law enforcement", *Journal of Public Economics*, vol 72, no 3, pp. 379–393

Jones, K. R. and Corona, J. P. (2008) "An ambient tax approach to invasive species", *Ecological Economics*, vol 64, no 3, pp. 534–541

Kneese, A. and Bower, B. (1968) *Managing Water Quality: Economics, Technology, Institutions, Resources for the Future*, Johns Hopkins University Press, Washington, DC

Kotchen, M. J. (2013) "Voluntary- and information-based approaches to environmental management: A public economics perspective", *Review of Environmental Economics and Policy*, vol 7, no 2, pp. 276–295

Lankoski, J., Lichtenberg, E. and Ollikainen, M. (2010) "Agri-environmental program compliance in a heterogeneous landscape", *Environmental and Resource Economics*, vol 47, no 1, pp. 1–22

Mas Collell, A., Whinston, M. D. and Green, J. R. (1995) *Microeconomic Theory*, Oxford University Press, New York

Mattheiss, V., Goral, F., Volz, P. and Strosser, P. (2010) "Economic instruments for mobilising financial resources for supporting Integrating Water Resource Management", Report to the OECD, ACTeon, Colmar

Millock, K., Sunding, D. and Zilberman, D. (2002) "Regulating pollution with endogenous monitoring", *Journal of Environmental Economics and Management*, vol 44, no 2, pp. 221–241

Millock, K., Xabadia, A. and Zilberman, D. (2012) "Policy for the adoption of new environmental monitoring technologies to manage stock externalities", *Journal of Environmental Economics and Management*, vol 64, no 1, pp. 102–116

OECD (2005) "Reducing Water Pollution and Improving Natural Resource Management", Sustainable Development in OECD Countries: Getting the Policies Right, Organisation of Economic Co-operation and Development, Paris

OECD (2008) "Environmental Outlook to 2030", Organisation of Economic Co-operation and Development, Paris

OECD (2010a) "Environmental Cross Compliance in Agriculture", http://www.oecd.org/tad/sustainable-agriculture/44737935.pdf, accessed 24 March 2014

OECD (2010b) "Guidelines for Cost-effective Agri-environmental Policy Measures", Organisation of Economic Co-operation and Development, Paris

OECD (2012) "Water Quality and Agriculture: Meeting the policy challenge", OECD Studies on Water, Organisation of Economic Co-operation and Development, Paris

Richardson, C. W., Bucks, D. A. and Sadler, E. J. (2008) "The conservation effects assessment project benchmark watersheds: Synthesis of preliminary findings", *Journal of Soil and Water Conservation*, vol 63, no 6, pp. 590–604

Rosso Grossman, M. (2007) "Agriculture and the Polluter Pays Principle", *Electronic Journal of Comparative Law*, vol 11 no. 3, pp. 1–66. http://www.ejcl.org/113/article113-15.pdf, accessed 19 August 2014

Segerson, K. (1988) "Uncertainty and incentives for nonpoint pollution control", *Journal of Environmental Economics and Management*, vol 15, no 1, pp. 87–98

Segerson, K. and Wu, J. (2006) "Nonpoint pollution control: Inducing first-best outcomes through the use of threats", *Journal of Environmental Economics and Management*, vol 51, no 2, pp. 165–184

Sharpley, A. N., Weld, J. L., Beegle, D. B., Kleinman, P. J. A., Gburek, W. J., Moore, P. A. and Mullins, G. (2003) "Development of phosphorus indices for nutrient management planning strategies in the United States", *Journal of Soil and Water Conservation*, vol 58, no 3, pp. 137–152

Shortle, J. and Horan, R. D. (2013) "Policy instruments for water quality protection", *Annual Review of Resource Economics*, vol 5, no, pp. 111–138

Spulber, D. (1985) "Effluent regulation and long-run optimality", *Journal of Environmental Economics and Management*, vol 12, no 2, pp. 103–116

Tietenberg, T. (1995) "Tradable permits for pollution control when emission location matters: What have we learned?", *Environmental and Resource Economics*, vol 5, no 2, pp. 95–113

US EPA (2013) "Watershed Assessment, Tracking & Environmental Results", http://www.epa.gov/waters/ir/index.html, accessed 24 March 2014

US Government (1972) "Clean Water Act", http://www2.epa.gov/laws-regulations/summary-clean-water-act, accessed 19 August 2014 .

Vojtech, V. (2010) "Policy measures addressing agrienvironmental issues", Food, Agriculture and Fisheries Papers. OECD, Paris.

Wright, S. and Mallia, C. (2008) "The Dutch approach to the implementation of the Nitrate Directive: Explaining the inevitability of its failure", *Journal of Transdisciplinary Environmental Studies*, vol 7, no 2, pp. 1–16. http://www.journal-tes.dk/vol_7_no_2/No_5_Stuart_Wright.pdf, accessed 19 August 2014.

Xabadia, A., Goetz, R. U. and Zilberman, D. (2006) "Control of accumulating stock pollution by heterogeneous producers", *Journal of Economic Dynamics and Control*, vol 30, no 7, pp. 1105–1130

Xabadia, A., Goetz, R. U. and Zilberman, D. (2008) "The gains from differentiated policies to control stock pollution when producers are heterogeneous", *American Journal of Agricultural Economics*, vol 90, no 4, pp. 1059–1073

Xepapadeas, A. P. (1991) "Environmental policy under imperfect information: Incentives and moral hazard", *Journal of Environmental Economics and Management*, vol 23, no 2, pp. 113–126

Xepapadeas, A. P. (1992) "Environmental policy, adjustment costs, and behavior of the firm", *Journal of Environmental Economics and Management*, vol 23, no 3, pp. 258–275

10
GROUNDWATER USE AND IRRIGATED AGRICULTURE IN CALIFORNIA

Dynamics, uncertainty, and conjunctive use

Keith Knapp and Kurt Schwabe

Introduction

Future global food demand will likely require an increase in irrigated agricultural production worldwide given that irrigated land is three times more productive than non-irrigated land (Anderson et al., 1997; Postel, 2000; Rosegrant et al., 2001). More irrigated production will require additional water, possibly 30 percent more than current use to meet global food production by 2025 (Shiklomanov, 1999). While both surface water and groundwater supplies contribute to meeting irrigated agricultural demands, each confronts challenges in terms of meeting future water demand.

The future sustainability of numerous groundwater systems worldwide as a viable irrigation source is threatened due to groundwater overdraft and degradation (Tanji and Kielen, 2002; Thayalakumaran et al., 2007). Many of these groundwater systems are characterized with shallow saline water tables due to a lack of adequate drainage disposal; consequently, substantial tracts of agricultural land are threatened. Surface water supplies, alternatively, are subject to additional scarcity as demand for environmental and urban uses increasingly competes with agricultural uses, while freshwater supplies become less certain due to climate change (Doll, 2002; Milly et al., 2005). Results from Shindell et al. (2006) suggest less precipitation and further reductions in freshwater supplies in many of the already water-stressed semi-arid and arid regions worldwide that are reliant on irrigation, including parts of the Middle East, North Africa, Australia, and the U.S.

The challenge, then, seems to be the following: how will various agricultural regions respond to increases in global food demand at a time when the availability of two major inputs—surface water and groundwater—seem to be contracting? The objective of this research is to evaluate how one particular region—the Kern County Basin in California's San Joaquin Valley (SJV)—might respond to such challenges with a focus on groundwater and how it is affected under different biophysical and management scenarios. With concern over falling water tables largely attributed to the unregulated nature of these common property resources, we evaluate how agricultural rents and the groundwater system respond to efficient management, under both deterministic and stochastic surface water supplies. A second issue we address is how potential changes in water supply characteristics that might be expected under climate change influence irrigated agricultural productivity and groundwater sustainability and use. In particular, we evaluate how rents and the groundwater system respond to decreases in water supply availability and reliability. Finally, we analyze an issue of interest in much of today's research—how does strategic behavior among groundwater users impact groundwater resources? We compare the solutions from two commonly-used game-theoretic approaches to the solutions under a common property regime and present-value optimality to illustrate how groundwater systems might evolve across these different analytical regimes.

Conceptual Framework

Conceptually, the problem can be illustrated as follows (Figure 10.1). Consider an irrigated agricultural region with two sources of water—a high quality surface water source and a (potentially) lower quality (more saline) groundwater source. A portion of applied irrigation water goes to crop evapotranspiration (ET); however, irrigation system non-uniformities and salt leaching requirements imply deep percolation flows from agricultural production to the aquifer. Decision variables in the model are crop area, irrigation technology, applied water depth, and aggregate ground and surface water

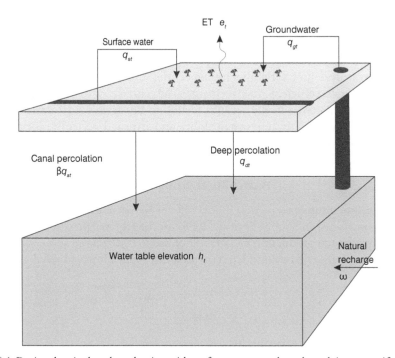

Figure 10.1 Regional agricultural production with surface water supply and overlying an aquifer system. Symbols are defined in the text

quantities. Throughout the chapter we will generally consider both common property (CP) usage, as well as efficient usage defined by present value (PV) optimization.

Our empirical focus is Kern County, California, with nearly a million acres of irrigable farmland. The Kern aquifer is essentially an unregulated common property resource,[1] and surface water allocations to irrigation districts within the region have been reduced substantially in recent years due to both environmental concerns and drought. Indeed, over the past two years surface water allocations from the main water supplier in the area—the federal government via the Central Valley Project (CVP)—have been only 40 percent and 50 percent of their full allocations. The challenges confronting Kern County, then, are not uncommon elsewhere as most irrigated agricultural regions worldwide are threatened by lower and less reliable surface water supplies, and overdraft of their aquifers. Many of these regions, furthermore, consist of a common property aquifer whose unregulated use can lead to potentially costly inefficiencies. As such, the lessons learned from this chapter may be useful in understanding issues surrounding groundwater use and irrigated agriculture elsewhere.

The next section sets up the general irrigated agricultural production model and highlights some of the advantages of our particular modeling approach in which the marginal value of water and deep percolation flows are nonlinear functions of applied water. The fourth section develops a dynamic hydro-economic model of irrigated agricultural production to estimate the impacts on groundwater use, water table height, and annual net benefits from managing the aquifer as an unregulated common property resource relative to the efficient solution developed using a dynamic programming algorithm. Analyzing groundwater management using a game-theoretic framework is the focus of the fifth section, with particular attention to what can be learned from a policy perspective about groundwater use using game theory relative to what can be gleaned from a more traditional, and simple, common property framework. The sixth section adds a stochastic surface water supply along with artificial recharge to the general framework to analyze how variability influences the solutions under common property management versus efficient management. The surface water supply probability distribution is then altered in the seventh section to illustrate the potential impacts of climate change on groundwater sustainability and irrigated agricultural productivity. The final section concludes with a summary of the qualitative and quantitative insights.

Agricultural production

The regional agricultural production model follows Kan et al. (2002) for the crop-water production functions, and Schwabe et al. (2006) for the regional programming model. A base-level agricultural production model is described in this section; some modifications and extensions of this will be considered in the actual applications (coupled models) that follow.

Model

Annual net benefits from crop production in year *t* are defined as:

$$\pi_t = \sum_j \sum_k \pi_{jkt} x_{jkt} - p_{wt} q_{wt} \tag{10.1}$$

where indice j denotes crop type, k denotes irrigation system type, and per-acre net returns are $\pi_{jkt} = (p_{cj} y_{jkt} - \gamma_{jk} - \gamma_{wjk} w_{jkt})$. Variables are y_{jkt} = crop yield, w_{jkt} = applied water depth (feet), x_{jkt} = cropped area (acres), and q_{wt} = total water supply purchased in the region. Parameters are p_{cj} = crop price (US\$/ton), γ_{jk} = nonwater production cost (US\$/acre), γ_{wjk} = pressurization cost (US\$/acre-foot), and p_{wt} = water price (US\$/acre-ft).

Crop-water production functions give crop yield and deep percolation flows as a function of applied water depth and salt concentration. The functions used here follow the development in Kan et al. (2002), and are specified as:

$$y_{jkt} = \sum_{i=1}^{2} \phi_{yjk}^i [e_{jkt}(w_{jkt}, c_t) - \underline{e}_{jk}]^i \qquad (10.2)$$

$$e_{jkt} = \frac{\bar{e}_j}{1 + \phi_{ejk}^0 (c_t + \phi_{ejk}^1 w_{jkt}^{\phi_{ejk}^2})^{\phi_{ejk}^3}} \qquad (10.3)$$

where e_{jkt} is evapotranspiration (with lower and upper bounds given by the parameters $\underline{e}_{jk}$ and $\bar{e}_j$, respectively), irrigation water salt concentration is represented by c_t (deciSiemens per meter (dS/m)), and ϕ_{yjk} and ϕ_{ejk}^i for $i \in \{0,1,2,3\}$ are estimated parameters.

As the S-shaped functions described in (10.3) are influenced by the spatial distribution of applied water, the parameters ϕ_{yjk} and ϕ_{ejk}^i are estimated for each specific crop-irrigation systems. Surface and groundwater salt concentration is a constant over time (i.e., $c_t = \bar{c}$). Finally, the mass-balance conditions, $d_{jkt} \equiv w_{jkt} - e_{jkt}$ and $c_{jkt}^d \equiv c_t w_{jkt} / d_{jkt}$, give deep percolation flows and salt concentrations for each land use under rootzone steady-state conditions.

Land constraints are defined so the sum of crop-irrigation system areas in any period t,

$$\sum_j \sum_k x_{jkt} \leq \bar{x} \qquad (10.4)$$

cannot exceed total land available for irrigated production $\bar{x}$ (million acres). Rotational constraints are also imposed on individual crops,

$$\underline{x}_j \leq \sum_k x_{jkt} \leq \bar{x}_j \qquad (10.5)$$

where $\underline{x}$ and $\bar{x}$ are lower and upper bounds, respectively. These values are specified within historical ranges observed over the past 20 years, which includes a period in the 1990s in which the region confronted drought conditions somewhat similar to those currently faced.

The regional water constraint in this model for immediate purposes is

$$\sum_j \sum_k w_{jkt} x_{jkt} \leq q_{wt} \qquad (10.6)$$

This implies that water use for irrigation is less than the total supply available. In some instances, there will be constraints on some sources of water (e.g., surface water). This constraint will be modified as needed in the subsequent coupled aquifer models.

Data

A production function for cotton under a furrow 1/2 mile irrigation system is illustrated in Figure 10.2. The yield-water relation is convex-concave, implying that a certain amount of water is necessary for any yield to occur at all. Conversely, yield is bounded

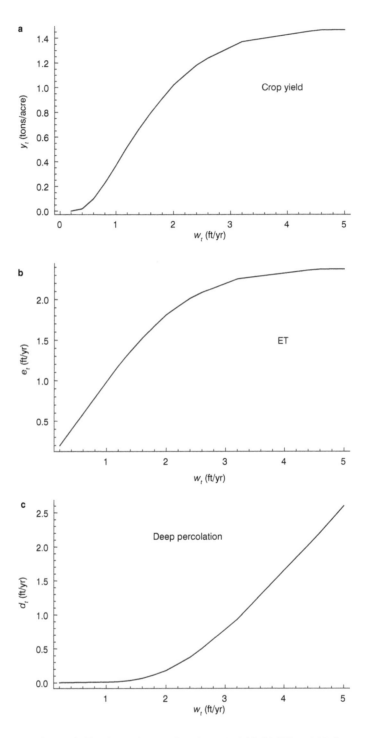

Figure 10.2 Cotton-furrow half-mile production function. (a) yield, (b) ET, and (c) deep percolation as a function of applied water depth (salt concentration of irrigation water = 0.7 dS/m)

above by maximum possible yield regardless of the amount of water applied. This simply reflects the fact that at some point, water is no longer the limiting factor in crop production. The deep percolation-water relation is convex which has implications for modeling as described below.

The six crops considered are cotton, tomatoes, wheat, lettuce, alfalfa, and Bermuda grass. Cost, price, and production data come from a variety of sources, including Kan et al. (2002), Schwabe et al. (2006), and Knapp and Baerenklau (2006). Market prices for each cropping system are derived from county agricultural commissioner crop reports, while surface water costs are a weighted average of water prices in the region. Nonwater production costs account for planting, land preparation, weed cultivation, fertilizer, and tile and tile and drainage systems. Harvest costs include both a yield-related variable component and a fixed per acre component. Irrigation system data are generally from Posnikoff and Knapp (1996), with adjustments for inflation. The six irrigation systems considered include furrow with 1/2 mile runs, furrow with 1/4 mile runs, low-energy precision application (LEPA), linear move, sprinkler, and drip.

Water demand and deep percolation

This basic setup can be used to derive regional water demand and deep percolation flows which will be helpful in understanding the dynamics of the system. For a given water price p_w, the model is solved by maximizing annual net benefits subject to the constraints and definitions in (10.1)–(10.6) and various bounds on the variables. Water demand and deep percolation flows are then generated by solving the model for various water prices. The annual net benefits, water demand, and deep percolation functions are illustrated in Figure 10.3. As can be seen, water demand is very elastic at low water prices, implying that substantial water savings are possible even with reasonably small increments in water price. As water price increases, there comes a point when demand becomes relatively inelastic implying an effective minimum water supply. Finally, at sufficiently high water prices, irrigated agriculture is no longer profitable and water use falls to zero.[2]

Regional deep percolation flows as a function of regional water use are illustrated in Figure 10.3c. These are a convex function of water applications. At low water applications, a high fraction of applied water goes to meeting crop ET and the fraction percolating below the rootzone is relatively small. After some point, however, the slope becomes constant implying that a constant fraction of incremental water applications goes to deep percolation. Note also that this is an economic relation and not a physical relation. For instance, a change in crop prices implies a change in crop mix, which in turn implies a different level of deep percolation flows for a given total regional water use.

These relations have implications for groundwater economics modeling. Linear water demands have been frequently used in the groundwater economics literature; it seems clear from Figure 10.3, however, that water demand is highly nonlinear. This could have impacts on several issues in the literature, including management benefits from aquifer management (e.g., the Gisser-Sanchez effect) as described in the succeeding sections.

In the literature it is also standard to model deep percolation flows as a constant fraction of applied water (typical values for the percolation coefficient are around 20 percent). The analysis here demonstrates that this assumption may be limiting and

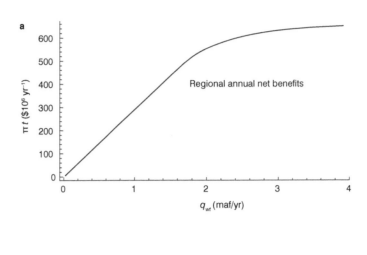

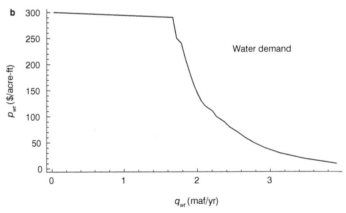

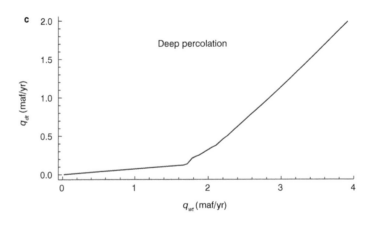

Figure 10.3 Regional agricultural programming model for Kern County, California. (a) Annual net benefit function, (b) water demand, and (c) deep percolation emission function. Variables are $\pi_t =$ annual net benefits, $p_{wt} =$ water price, $q_{dt} =$ deep percolation, and $q_{wt} =$ applied water

that deep percolation flows are actually a nonlinear function of applied water, and may vary with the economic parameters of the system as well as hydrologic parameters. The relationship between applied water rates and deep percolation flows can be important in situations where return flows, high water tables, and integrated drainwater management are concerns. These considerations are an advantage of including an explicit agricultural production model in the coupled aquifer model; something that would be more difficult to achieve with the use of aggregate functions (e.g., derived water demand or constant percolation rates) as is generally done.

Groundwater usage and management

This section begins with the standard lumped-parameter groundwater model as initially developed in the classic studies of Burt (1964), Brown and Deacon (1972), and Gisser and Sanchez (1980), and subsequently pursued in an extensive literature of theoretical and applied studies. The model considers only the quantity dimensions of the aquifer and not quality. This provides a baseline of theoretical, empirical, and policy implications. Subsequent sections extend this model to include game-theoretic equilibria with a finite number of users, conjunctive use with stochastic surface supplies, and a variety of policy issues. Interestingly enough, for reasons to be explicated as we proceed, it turns out that the standard model is still quite informative despite its many years of existence and a myriad of extensions in the literature.

Biophysical model

Three general adjustments need to be made to the model developed in the previous section. First, equation (10.1) is modified to include the costs associated with pumping groundwater as follows:

$$\pi_t = \sum_j \sum_k \pi_{jkt} x_{jkt} - \gamma_e (\bar{h} - h_t) q_{gt} - \gamma_{sw}(q_{st}). \qquad (10.7)$$

The second expression in (10.7) is groundwater pumping costs where γ_e is a pumping cost parameter, $\bar{h}$ is height of the land surface, h_t is water table elevation relative to sea-level, and q_{gt} is the quantity of groundwater extractions from the aquifer. The third expression in (10.7), replaces the simple linear surface water costs (constant price) with a convex function of total surface water imports to the region, q_{st}, as in Knapp and Baerenklau (2006). The convex function reflects the fact that growers and districts often rely upon a combination of different surface water sources, each with a different pricing structure.

Second, the regional water constraint in this model is amended to:

$$\sum_{j=1}^{nc} \sum_{k=1}^{K} w_{jkt} x_{jkt} \leq q_{st} + q_{gt} \qquad (10.8)$$

implying that water use for irrigation is less than or equal to total supply, where total supply is combined surface and ground water. Surface water deliveries are subject to the constraint that $q_{st} \leq \bar{q}_{st}$ where $\bar{q}_{st}$ is the maximum amount of surface water available to the region.

Finally, the equation of motion describing the water table elevation, h_t, response to extractions and deep percolation flows is defined as:

$$h_{t+1} = h_t + \frac{1}{As^y} \left(\omega + \beta_s q_{st} + \sum_j \sum_k d_{jkt} x_{jkt} - q_{gt} \right) \qquad (10.9)$$

where A is the regional aquifer area, s^y is the aquifer specific yield, ω is natural recharge, and β_s is the surface water infiltration coefficient. Here the water table rises with surface water imports (canal losses) and deep percolation, and falls with extractions for irrigation. The water table elevation is constrained by $\underline{h} \le h_t \le \overline{h}$, where $\underline{h}$ is determined by the lower confining layer and $\overline{h}$ is determined by the rootzone depth. The lower bound limits ground water extractions to the available supply and the upper bound limits net deep percolation flows to the maximum storage capacity that is feasible for crop production.

Aquifer characteristics for the Kern County empirical application come from Knapp and Olson (1995). The aquifer specific yield (s^y) is 0.13, with aquifer lower and upper bounds at -233 feet below MSL (mean sea level) and 375 feet above MSL, respectively. The surface water infiltration coefficient is 0.3, while natural recharge in the region is 0.052 ac-ft per year. Surface water salinity concentrations are 0.7 dS/m, while groundwater salinity concentration is also assumed to be 0.7 dS/m. Finally, aquifer area extends beyond the regional irrigated area at 1.29 million acres.

Common property usage

When there are many small users in an infinite-transmissivity basin, it is reasonable to suppose that individual users perceive that their water use decisions have a negligible impact on their own future welfare. In Kern County, as in many other if not most regions with aquifers, there are a very large number of irrigators—over 2,000. Coupled with the fact that most groundwater pumping is not metered, we assume that under CP usage, irrigators select their decision variables to maximize annual net benefits as defined in (10.7) for a unit area of land given our estimated production functions and the water and land constraints (10.4), (10.5), and (10.8). While a rigorous approach can be given utilizing game theory, and the infinite-transmissivity assumption is not very realistic, as we will show later these considerations will not be limiting for our purposes here.

Figure 10.4 illustrates annual net benefits and groundwater table heights over a period of 500 years under CP usage assuming an aquifer that is initially full. As shown, the aquifer declines over time until it reaches a steady-state level. Two possibly useful observations are (a) this process potentially takes a long time, and (b) the aquifer is not physically exhausted even under CP (unregulated) usage. The latter result is due to increasing marginal extraction costs as the water table falls. In policy discussions it is common to describe groundwater usage (and other CP resources for that matter) as non-sustainable, presumably due to the falling water table. However, a falling watertable as evidence of unsustainable usage is incorrect, in part because it can be—and typically is—efficient to draw down the water table from an initially high elevation. The inefficiency is the *rate* of drawdown as demonstrated next.

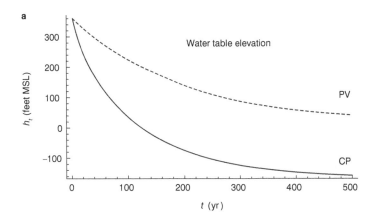

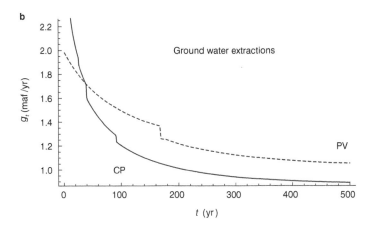

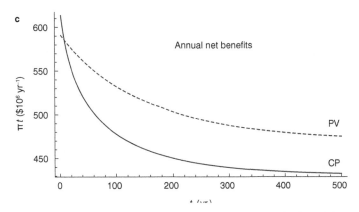

Figure 10.4 Common property (CP) and efficiency (PV) for the standard model. (a) Water table elevation, (b) regional groundwater extractions, and (c) regional annual net benefits

Present value optimality

Efficient basin management is achieved by selecting the decision variables to maximize the PV of annual net benefits

$$\sum_{t=1}^{\infty} \alpha^t \pi_t \tag{10.10}$$

where $\alpha = \dfrac{1}{1+r}$ is the discount factor and r is the interest rate, which is assumed to be 4 percent. This problem is solved utilizing a dynamic programming (DP) algorithm. The optimal value function satisfies Bellman's equation

$$v(h) = Max \ \pi(h,z) + \alpha v[g(h,z)] \tag{10.11}$$

where $z = \{w_{jk}, x_{jk}, q_g, q_s\}$ is the vector of decision variables, annual net benefits $\pi(h,z)$ are defined by (10.7) and the production functions, and $g(h,z)$ is the equation of motion for water table height (10.9). The maximization problem is over the decision variables z and subject to the land and water constraints ((10.4), (10.5) and (10.8)) and non-negativity conditions.

The value function (not illustrated) is convex; this is both intuitive and matters for theoretical analysis as discussed in Knapp and Olson (1995). The optimal decision rule for groundwater extractions (also not illustrated) is increasing in the water table height as would be expected. (A later section illustrates the optimal decision rule in a more general setting of conjunctive use.) As a methodological point, this analysis demonstrates that DP problems can be solved with a large number of controls (there are 74 in this problem).

The time-series for efficient water table levels, groundwater extractions, and annual net benefits are illustrated in Figure 10.4 along with the corresponding CP values. As can be seen, efficient water table levels decline over time, eventually reaching a steady-state, although this process, again, takes a considerable amount of time. In comparison to CP, efficient management results in a slower decline of the water table and a considerably higher steady-state level. Groundwater management as an investment is illustrated in the middle panel of Figure 10.4. Extractions are reduced in the early periods under efficient management, and annual net benefits are correspondingly less than CP. Eventually, however, the higher water tables lead to significantly lower pumping costs, and annual net benefits are significantly greater under efficiency.

Management benefits and sustainability

Benefits from groundwater management are the difference in present values between the CP and PV-optimality solutions. Beginning from an initially high water table, annualized net benefits per-acre of farmland are US$650 versus US$673 under CP and PV solutions, respectively, resulting in per acre annualized management benefits of US$23, or 3.5 percent. While somewhat small, the actual advantage of efficient management may be even smaller given that we have only accounted for the benefits of management, and not the management or information, contracting, monitoring (ICM) costs associated with setting up and maintaining an appropriate institutional framework. Also to be noted is that the analysis here assumes perfect knowledge of the underlying parameters, functions, and equations of motion. Since this would not be possible in an operational setting, the low management benefits associated with efforts to efficiently manage the aquifer could evaporate quickly, and even turn negative.

However, the percentage calculation is subject to the criticism that it depends on the production costs including *all* costs; this may include various types of fixed costs that can be harder to obtain. An alternate approach is to measure income improvement. The gain of US$23/acre-yr implies an income increase of US$14,467 annually for a 640-acre farm, which may seem more substantial. From this perspective, groundwater management might be judged as beneficial.

The finding that groundwater management benefits may not necessarily be large is perhaps the currently biggest dilemma in groundwater quantity economics. It was originally pointed out by Gisser and Sanchez (1980) and is sometimes termed the *Gisser-Sanchez effect* (GSE). It has also been the subject of a large literature, with a particularly informative review by Koundouri (2004). Without larger management benefits, there would be little reason to manage groundwater basins. Interestingly, groundwater basins in California largely have been unmetered and unmanaged except in rare adjudicated basins, although recent legislation (SBx7-6) now requires the California Department of Water Resources to monitor groundwater levels in all of its groundwater basins.

This analysis also suggests some insights into sustainability. As mentioned before, sustainability appears to be equated to declining water tables in policy arenas (explicit definitions are generally not given, therefore this is a reverse-engineered statement). However, as noted above, declining water tables themselves are not the problem. As the efficiency analysis demonstrates, the problem is that the *rate* of decline can be (and typically is) excessive under CP. In addition, sustainability is defined over household consumption, whereas annual net benefits in the groundwater economics literature (as here) are actually an income stream. Household consumption can then differ from this depending on investments in other forms of capital and the degree to which these substitute for income streams generated by the aquifer.

A full analysis of sustainability thus requires an expanded framework from the models in this chapter as well as the extant literature. This involves coupling the agricultural production/aquifer model to other forms of capital (human, physical, biological). Since this involves some issues of a more macro-nature and not currently part of the groundwater economics literature, we do not pursue this further here. The main conclusion here is that one cannot assess sustainability by simply observing the physical system alone, which is what seems to be widely done: one needs to assess a wider system that includes both human-generated and natural capital stocks.

Game-theoretic analysis

Here we consider a game-theoretic model closely related to the standard model just explored. Early dynamic game models for natural resources include Levhari and Mirman (1980) and Eswaran and Lewis (1984); Long (2010) provides a useful review. Negri (1989) was the first paper to apply dynamic games to groundwater, while other game-theoretic treatments for groundwater include Provencher and Burt (1993) and Rubio and Casino (2001). Of particular interest here is Dixon (1988) who demonstrated that even with a small number of users (as small as ten), payoffs in the game-theoretic equilibrium differ from payoffs in the standard (Gisser-Sanchez) CP (myopic) model by just 2 percent. Since there are typically hundreds if not thousands of users in groundwater basins, this raises questions about the relevance of game theory in these situations. We will explore the same issue as Dixon, with qualitatively similar findings.

This analysis extends Negri (1989) to include surface and deep percolation flows consistent with the standard model set out in the previous section, and differs from Dixon (1988) in that we consider an infinite horizon. The analysis is continuous time. We suppose that there are a units of land available for production, n identical users in the basin, and we consider a linear (derived) demand curve for water instead of the explicit agricultural production model in the previous model.

For an individual user, annual net benefits per-unit area are

$$\pi[h(t), w_i(t)] = \int_0^{w_i(t)} b'[w_s + w] \mathrm{dw} - \gamma_e[\bar{h} - h(t)] w_i(t) \tag{10.12}$$

where $b'(w) = b_1 + b_2 w$ is a linear water demand curve, $w_s = (1 - \beta_s) q_s / a$ is applied surface flows per-unit area (exogenous in this model), and $w_i(t)$ is applied groundwater depth. The aquifer equation of motion is defined as

$$\dot{h}(t) = \frac{1}{As^\gamma} [(\beta_s + \beta_d(1 - \beta_s)) q_s - (1 - \beta_d) \frac{a}{n} \sum_{j=1}^n w_j(t) + \omega] \tag{10.13}$$

where we are again maintaining the assumption of infinite transmissivity in the aquifer. Similar to equation (10.9), β_s and β_d are infiltration coefficients for surface water and groundwater, respectively, ω is natural recharge, and the other factors are defined previously.

With this setup, there are four regimes: pure common property as before (CP), open and closed-loop game-theoretic equilibria, and social optimum. Under CP, users just maximize annual net benefits, so optimal withdrawals satisfy $b'[w_s + w_i(t)] = \gamma_e[\bar{h} - h(t)]$. This defines a function $w_i(t) = \phi^{cp}[h(t)]$ giving CP withdrawals as a function of water table height. Substituting into (10.13) gives the equation of motion for the CP aquifer.

The open-loop Nash equilibrium (OLNE) is a Nash equilibrium in decision sequences; that is, equilibrium occurs where the time-series of each user's decisions are optimal given the decision sequences of all other users. Mathematically, individual user i maximizes the PV of annual net benefits $\int_0^\infty e^{-rt} \frac{a}{n} \pi[h(t), w_i(t)] dt$ subject to the equation of motion (10.13), where the decision sequences of other users are taken as given.

Formulating the Hamiltonian with co-state $\lambda_i(t)$ and differentiating yields a function $w_i(t) = \phi^{ol}[h(t), \lambda_i(t)]$. This gives OLNE withdrawals for an individual user as a function of water table height and the co-state. The co-state equation of motion is

$$\dot{\lambda}_i(t) = r\lambda_i(t) - \frac{\gamma_e a w_i(t)}{nAs^\gamma} \tag{10.14}$$

where r is the interest rate. Substituting the decision rule $w_i(t) = \phi^{ol}[h(t), \lambda_i(t)]$ into both equations of motion leaves a differential equation system in $h(t)$ and $\lambda_i(t)$. Note that the pure CP problem above can be obtained as the limit of the OLNE as $n \to \infty$.

The third regime is a closed-loop Nash equilibrium (CLNE). Here equilibrium is defined as a set of user decision rules $\{\phi_1^d[h(t)], \dots \phi_n^d[h(t)]\}$ giving water withdrawals as a function of the water table, and such that each is optimal given the decision rules for all the users. Following Negri (1989), this can be solved by formulating the same individual optimal control problem as OLNE except that the decisions of all other users in the aquifer equation of motion (10.13) are replaced by $\varphi_j, j \neq i$. The first-order condition defines the same withdrawal function – $w_i(t) = \phi_i^d[h(t), \lambda_i(t)]$ – as in the OLNE. However, the co-state equation is now

$$\dot{\lambda}_i(t) = r\lambda_i(t) - \frac{\gamma_e aw_i(t)}{nAs^y} + \lambda_i(t)[\frac{a(n-1)}{n}\frac{\partial\varphi^d(h)}{\partial h}] \tag{10.15}$$

noting the common decision rule φ^d since users are identical. The co-state equation (10.15) is what distinguishes the CLNE from the OLNE. In particular, the co-state equation of motion now includes a third expression on the right-hand side representing strategic action among aquifer users. Social optimum, finally, is defined as the optimized basinwide present value of net benefits. It can be obtained from either the OLNE or CLNE by letting $n = 1$.

We now compare steady states under the four regimes. We first note that the water table equation of motion pins down an explicit value for steady-state withdrawals w_{ss} which is determined solely by the physical parameters of the system and is invariant across the regimes. The steady-state co-state values are determined from the respective co-state equations of motion. These steady-state co-state values are

$\{0, \frac{\gamma aw_{ss}}{nr}, \frac{\gamma aw_{ss}/n}{r+\theta\gamma}, \frac{\gamma aw_{ss}}{r}\}$ for CP, OLNE, CLNE, and efficiency, respectively, and

where $\theta = (-1 + n(1-\beta_d) + \beta_d)/(nb_2As_y)$ enters the denominator for CLNE. Given steady-state withdrawals w_{ss}, the associated steady-state water table heights can be readily computed from the respective functions $w(t) = \phi[h(t), \lambda(t)]$.

Some theoretical analysis is possible utilizing the above steady-state co-state values (Negri 1989). Under the reasonable assumption that $\partial\varphi(h)/\partial h > 0$ (Negri 1989), then the above values are increasing from CP to efficiency. As the first-order conditions are the same across regimes, this implies that the equilibrium steady-state water table levels are lowest under CP, lower under CLNE than OLNE, and highest under the efficient social optima solution. Thus CLNE results in a higher level of inefficiency in the aquifer than OLNE. It can also be seen from the λ_{ss} values that OLNE and CLNE equilibria will approach CP as $n \to \infty$.

Table 10.1 illustrates how the steady-state water table levels vary with the number of users under the four behavioral regimes. As can be seen, while CLNE has theoretical justification over OLNE, it doesn't necessarily imply a quantitatively significant difference from OLNE. More importantly, with even as few as 40 users, both OLNE and CLNE are virtually indistinguishable from CP. This result is analogous to Dixon (1988). There are a large number of irrigators—well over 2,000 (Knapp and Baerenklau 2006)—pumping water from the Kern County aquifer, a characteristic that likely represents most basins of public policy interest. Given this characteristic, a question naturally arises as to the relevance of applying (classical) game theory to groundwater use and management, sustainability, and policy.

Conjunctive use and artificial recharge

This section extends the standard model from the fourth section to stochastic surface water flows and artificial recharge. Burt (1964) and Provencher and Burt (1993) consider

Table 10.1 Steady-state water table heights (ft MSL) as dependent on number of users and behavioral regime

n	CP	CLNE	OLNE	PV-opt
1	−222.7	−116.8	−116.8	−116.8
10	−222.7	−215.1	−212.1	−116.8
20	−222.7	−219.0	−217.4	−116.8
40	−222.7	−220.8	−220.8	−116.8

stochastic recharge to the aquifer but without explicit recharge. Knapp and Olson (1995, 1996) include stochastic surface flows, artificial recharge, and risk aversion with recursive utility. Theoretical results utilizing lattice programming (non-concave value function) in the latter two papers demonstrate monotonic convergence to a limiting distribution. Empirical results, alternatively, do not find much of a role for artificial recharge, an outcome that Knapp and Olson (1995, 1996) suggest could differ in a shallow aquifer. While incorporating risk aversion does increase management benefits, the result is still small and thus is not analyzed further below.

Model

The model here extends Knapp and Olson (1995) to include the explicit agricultural production economics model as set out above. Annual net benefits are as in (10.7) with the associated definition of π_t and the production functions ((10.2)–(10.3)). The water constraint remains as (10.8) except that we replace q_{st} by q_{1t} where q_{1t} is the portion of surface water devoted to irrigated agricultural production. The land and rotation constraints ((10.4)–(10.5)) remain as before.

Surface water allocation is now

$$q_{1t} + q_{2t} = q_{st} \tag{10.16}$$

where q_{2t} is artificial recharge to the aquifer. In this problem, maximum surface water availability $\bar{q}_{st}$ to the region is a state variable and stochastic. As before, $q_{st} \le \bar{q}_{st}$; however, $\bar{q}_{st}$ now evolves according to

$$\bar{q}_{s,t+1} = \rho \bar{q}_{st} + \varepsilon_t \tag{10.17}$$

where ε_t is a random variable and ρ is a correlation coefficient (Knapp and Olson 1995). The water table equation of motion is now modified to

$$h_{t+1} = h_t + \frac{1}{As^y}\left(\omega + \beta_s q_{st} + q_{2t} + \sum_j \sum_k d_{jkt} x_{jkt} - q_{gt}\right) \tag{10.18}$$

to account for artificial recharge.[3]

Common property

CP decisions follow the standard model in which users maximize annual net benefits after observing the water table height and surface flow upper bound for that year. However, the stochastic flows imply that the time-series are now random sequences. A Monte-Carlo simulation (500 iterations) was used to estimate the probability distributions for the model variables over a 300-year period.

Figure 10.5a reports the mean and standard deviation over time for the water table height. As can be seen, the mean water table is apparently converging to a value reasonably consistent with that of the (deterministic) standard model (Figure 10.4). Inspection of the standard deviation (not reported) suggests that it has still not converged even over this relatively long time span.

Tsur (1990) theoretically investigates aquifer development as a means of stabilizing water availability to regions when surface flows exhibit variability. The model here can be used to quantitatively investigate this. Figure 10.5b reports the histogram for applied irrigation water (surface plus ground water) in year 300, which can be contrasted with the surface water histogram (Figure 10.5c) which is constant over time (no serial correlation). As can be seen, the aquifer has a very strong buffering effect: the aquifer acts to stabilize water flows to agricultural production, and this stabilization is quite large.

Efficiency

Efficient basin management is achieved by selecting a decision rule to maximize the expected PV of annual net benefits

$$E[\sum_{t=1}^{\infty} \alpha^t \pi_t] \qquad (10.19)$$

where the expectation is over the random variable ε_t as identified in (10.17). This problem is solved utilizing a dynamic programming algorithm. The optimal value function satisfies Bellman's equation

$$v(h, \overline{q}_s) = Max \; \pi(h, z) + \alpha \, E[v[g(h, \overline{q}_s, z, \varepsilon)]] \qquad (10.20)$$

where $z = \{w_{jk}, x_{jk}, q_1, q_2, q_s, q_g\}$ is the vector of decision variables, annual net benefits $\pi(h, z)$ are defined by (10.7) and the production functions, and $g(h, \overline{q}_s, z, \varepsilon)$ is now a vector equation of motion for water table height (10.9) and surface flows (10.17).

The maximization problem is over the decision variables z and subject to the land and water constraints and non-negativity conditions. Bellman's equation and the associated optimal decision rule are solved using a value function iteration algorithm. Once the value function and associated optimal decision rule have been found, then time-series for the probability distributions of the states, controls, and annual net benefits are estimated utilizing a Monte-Carlo simulation.

The optimal decision rule for groundwater extractions is displayed in Figure 10.6a. This shows that groundwater extractions are generally increasing in hydraulic head, and decreasing in surface water availability. Also, the decision rule is not linear—consistent with the fact that the water demand function is not linear, as discussed in the third section.

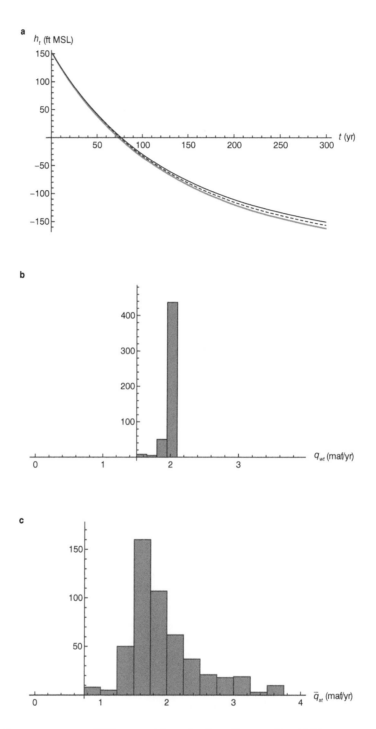

Figure 10.5 Common property (CP) usage with stochastic surface supplies. (a) Mean water table height plus/minus 1 standard deviation. (b) Histogram for applied water in year 300. (c) Histogram for surface water supplies

Optimal decision rules are compared in Figure 10.6b for the deterministic model in the fourth section and the stochastic model of this section. It can be seen that for surface water availability equal to 1.9 million acre-feet per year (maf/yr) as in the deterministic model, the decision rules are essentially identical. Formally, certainty equivalence (CE) only holds strictly for linear-quadratic problems. The formal conditions for CE are not met here as the annual net benefit function is not quadratic, the equation of motion for the aquifer is not linear in the controls, and there are various bounds and constraints not found in a formal linear-quadratic model. Nevertheless, the results suggest that the stochastic groundwater problem might be approximated reasonably well with an appropriately formulated deterministic model even though the theoretical conditions for CE are not met and the stochastic variables have a large coefficient of variation. Note,

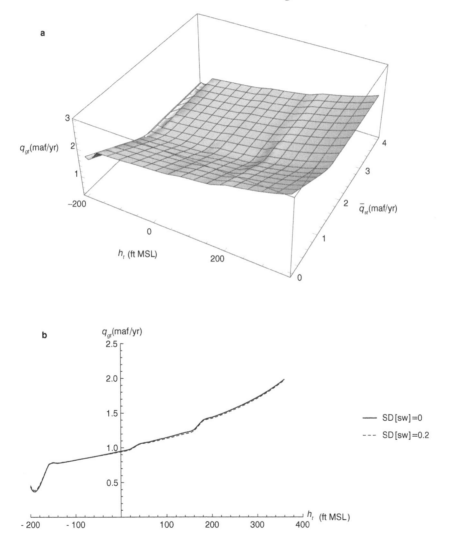

Figure 10.6 Optimal decision rules for the conjunctive use model. (a) Groundwater extractions as a function of water table elevation and surface water availability. (b) Comparison of optimal groundwater extractions as a function of water table height with $\bar{q}_{st}$ =1.9 maf for the deterministic and stochastic models

however, that the decision rule is not linear as in a pure linear-quadratic problem, and that the deterministic analysis needs to be modified to account for the actual surface water availability prevailing in a given year.

Artificial recharge is widely discussed in policy forums. Inspection of the results here does not indicate that artificial recharge is an efficient policy, even for surface flows 0.4 maf/yr greater than the mean surface flow value. This is consistent with Knapp and Olson (1995), but very surprising in view of substantial policy interest. This is a topic deserving of considerable further attention.

Management benefits

Expected value of annualized net benefits are US$650 per acre-year under CP, and US$672 per acre-year under efficiency, for annualized management gains of US$22 per acre-year when h_0 =360 feet MSL. This is not so different from the deterministic analysis in the fourth section as would be expected from the near CE of the model discussed above. Thus introducing uncertainty and artificial recharge does not significantly increase the gains from management, at least in the lumped-parameter/ infinite transmissivity model of this chapter. Note also that Knapp and Olson (1995, 1996) reached a similar conclusion for both risk neutrality as here, and an extension to recursive utility with intertemporal elasticity of substitution (IES) and risk parameters consistent with macro-data. As with artificial recharge, this is overall a surprising conclusion in view of substantial surface water availability, and therefore also deserving of additional analysis.

Climate change

This section addresses how climate change—via its potential impact on the mean and reliability of surface water supply deliveries—might affect agricultural sustainability. As such, this section is intended to provide a better understanding of groundwater resources under alternative climate-related impacts and the associated economic consequences. Such analyses as this and those mentioned below are necessary to develop informed policies that can properly weigh the benefits and costs of efforts intended to address or remediate the possible impacts of climate change on agriculture not only in California, but worldwide as well.

Literature review

A number of studies have investigated the impacts of climate change on agriculture using mathematical programming models, including Adams et al. (1995; 1999), Chen et al. (2001), Mejias et al. (2004), Connor et al. (2009; 2012), Rowan et al. (2011).[4] A common result from these studies is that there exists low-cost adaptation opportunities, especially for low to moderate climate change scenarios, but that the impacts can be significant for severe scenarios. Furthermore, as emphasized in a number of studies (e.g., Quiggin et al., 2010; Rowan et al., 2011; Connor et al., 2012), models and analyses that overlook the climate-related changes on water supply reliability and quality can significantly underestimate the overall impacts from climate change.

Few studies, though, have combined a hydro-economic model that accounts for two sources of water—a surface water supply and a groundwater supply—to investigate the role groundwater might play under a future with less reliable and lower surface water supplies as might be expected under climate change. Wang et al. (2011) develop a hydro-economic model of the potential impacts of climate change on irrigated agriculture above the Ogallala aquifer within a dynamic framework. In their analysis, groundwater is the single source of irrigation water with a recharge term unrelated to climate. Through incorporation of different climate change model outputs into an Environmental Policy Integrated Climate (EPIC) model, they illustrate how climate impacts irrigated crops via temperature effects and dryland crops via precipitation effects. Alternatively, Heidecke and Heckelei (2010) develop a sophisticated static hydro-economic model of irrigated agricultural production in the Draa River, Mococco, for analyzing climate change impacts with variable surface water supplies and an increasingly saline groundwater supply. They find that while groundwater will be increasingly relied upon in the future to deal with climate change, the rise in groundwater salinity will result in the aquifer developing into more of a complementary relationship with surface water supplies rather than to serve as a substitute.[5]

Model

This section utilizes the model developed in the previous section which includes a stochastic surface water supply. Agriculture in the Kern region in California is highly reliant on both groundwater and highly variable surface water supplies for irrigation (Knapp and Olson, 1995). As such, including a stochastic surface water supply is a reasonable and potentially more accurate representation of the resource problem confronting growers. Determining an accurate distribution of surface water supplies to agriculture under climate change is difficult given that such a distribution depends on future trajectories and interactions among biophysical, socioeconomic, political, and demographic factors. Accordingly, it seems reasonable to assume decreased mean surface water deliveries to agriculture, and/or increased variability of those supplies.

With this in mind, we choose a simple representation of the impacts of climate change on irrigated agriculture by taking the parameter values characterizing surface water deliveries in Knapp and Olson (1995) as the baseline values under no climate change, and evaluate the impacts of lower and less reliable supplies relative to this baseline. In particular, we evaluate the impacts on groundwater levels and annual net benefits over a 100-year time horizon from a 15 percent, 30 percent, and 50 percent change in the mean and variance of the baseline distribution—e.g., a 15 percent decrease in mean and a 15 percent increase in variance of water supplies relative to the baseline levels.[6] Besides accounting for variability (something that is often overlooked), this can also be used to assess the potential benefits of more secure water supplies, an issue of prominence in many agricultural and urban settings.[7]

Here, because of the short-run nature of the problem, the model also assumes a constant surface water price as in Knapp and Olson (1995). Finally, given that most of California's aquifers, and many worldwide, are not regulated and not metered, the analysis assumes the aquifer is an unregulated common property resource. In particular, the impacts of changes in the mean and variability of surface water supplies on groundwater depth, surface water application rates, and annual net benefits is analyzed in a period-by-period framework representative of an unregulated common property resource.

Common property usage under alternative water supply scenarios

Figure 10.7 illustrates how changes in the mean and variance of surface water supplies impact groundwater table elevation, applied groundwater use, and annualized net benefits over a 100-year time horizon. Included in each figure are the results from a deterministic surface water supply, represented by the label, *Certainty*; a baseline value with a mean and variance from Olson and Knapp (1995) which is used in the fifth section (represented by *Baseline*); followed by results from changes in water supply characteristics of 15 percent, 30 percent, and 50 percent in the mean (a decrease) and variance (an increase) of the distribution relative to the baseline. The graphs show the time profile of the mean levels and the whiskers above and below each line represents one standard deviation from those means.

As shown in Figure 10.7a, over the 100-year time horizon, water tables increase under the Certainty and Baseline scenarios, and decline under each of the climate change scenarios as surface water supply and reliability decrease. The baseline scenario closely mimics the Certainty scenario, both of which reach a steady-state well below the crop rootzone. As shown in Figure 10.7b, one of the main responses to the lower and less reliable surface water supplies is, initially, an increase in groundwater use. Another response (not shown) is a movement towards more efficient but costly irrigation technology using the lower surface water supplies, which together with the reduction in surface water allocations results in lower deep percolation flows contributing to the groundwater table. The lower deep percolation flows and greater groundwater use combine to lower groundwater tables which, subsequently, increase pumping costs. These increased costs lead to an increase in irrigation efficiency and deficit irrigation on those lands using groundwater; consequently, groundwater use declines over time under the various climate change scenarios. For the Certainty and Baseline scenarios, groundwater use increases over time in response to the lower pumping costs that arise from a higher watertable. Also shown in Figure 10.6b are standard deviations above and below the mean levels for a subset of periods for each scenario which do not change significantly over the 100-year horizon.[8]

The last graph in Figure 10.7 shows how annual net benefits change over time under the various scenarios. Consistent with the above discussion surrounding losses in surface water and changes in pumping costs from differences in water table levels, under the Certainty and Baseline scenarios annual net benefits stay relatively constant over time, whereas under the climate change scenarios annual net benefits decline. While the divergence between the annual net benefits under the Certainty outcome relative to the climate change outcomes diverge over time, after accounting for discounting the differences are minimal. For instance, the present value net benefits (PVNB), assuming a 5 percent discount rate under the Baseline solution for the 100-year simulation are 3 percent less than under the Certainty case (US$11,800 versus US$12,160). The PVNB under the 15 percent, 30 percent, and 50 percent solutions are 5 percent, 7 percent, and 11 percent lower than the Certainty case. Adjustments to changes in the mean and

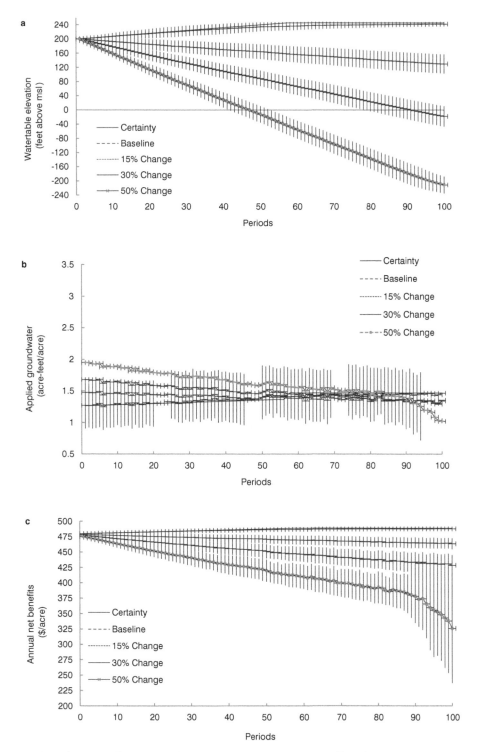

Figure 10.7 Alternative climate-related water supply impacts on time profiles of (a) water table elevation, (b) applied groundwater use, and (c) annual net benefits

reliabilities of the surface water supplies—which include mostly deficit irrigation, but some crop switching, and irrigation efficiency improvements—result in very minor profit losses in the initial periods. As those losses increase with time, they occur further out and thus the discounting leads to the consequent minimal impact, even though the biophysical differences would seem to justify something greater.

Management benefits and sustainability

The results suggest that as water supply characteristics worsen due to climate change (as measured by lower mean and higher variability), growers will engage in more groundwater usage as a buffer against these changes; consequently, water tables decline. As water table levels decline, though, the associated increase in pumping costs lead to less groundwater usage through both increased deficit irrigation and a switch to more efficient irrigation systems. Similar to much of the literature evaluating water supply reductions, if irrigated agriculture can engage in numerous mitigative procedures to reduce the negative impacts from changes in water supplies, the overall impacts on profitability can be minimal. Of course, if there are limited opportunities to respond due to demand hardening over time or a more capital intensive production regime, such responses will be more limited and the impacts will be greater. This was illustrated in Connor et al. (2012) in the case of a sub-region in the Lower Murray Basin with somewhat salt sensitive perennial crop production.

Conclusions

California's agricultural production, similar to many irrigated agricultural regions worldwide, is highly reliant on groundwater. Within its most productive area—the Central Valley, in which the Kern County aquifer is located—irrigated agriculture relies on groundwater for nearly 30 percent of the region's applied water on average, a number that escalates to 60 percent during drought (Dale et al. 2013). Concerns exist, though, regarding current extraction rates and management regimes, as well as the future sustainability of the Kern County aquifer given that climate change models suggest arid and semi-arid regions are likely to experience greater aridity and more frequent drought (Seagar et al. 2007). Indeed, the aquifers in the regions in and around Kern County have been identified recently as one of the hotspots for groundwater depletion. As discussed in Scanlon et al. (2012), groundwater depletion in the Central Valley from 1900 to 2010 has accounted for 15 percent of the total groundwater storage loss in the entire U.S.

This chapter evaluates policy-relevant issues regarding groundwater use in California, issues that are certainly germane to other regions with irrigated agriculture and groundwater, and along the way illustrates some methodological issues of importance. The model identifies salient characteristics of water demand and deep percolation flows that are important to the responsiveness of irrigated agriculture to changes in water supply or water supply prices, and also demonstrates a nonlinear link between the aquifer and applied water rates. The latter condition has implications for issues surrounding returns flows, water savings from more efficient irrigation management schemes, and water quality.

The inefficiency associated with use of an unregulated common property aquifer is relatively minor, a result consistent with the Gisser-Sanchez effect and which has been found in a number of other studies. An outcome of this analysis is that declining water tables levels do not necessarily mean unsustainable use as the efficient solution also exhibits declining water tables from an initially full aquifer. Common property and efficient usage both converge to a steady-state due to increased pumping costs. On the issue of groundwater sustainability, it is likely more useful to discuss rates of groundwater use rather than simply focus on whether water table levels have declined.

The fifth section investigates game-theoretic models of groundwater use. As many have noted, use likely involves strategic interactions among users of any particular aquifer. Within the empirical application, which is likely similar to many basins of policy interest worldwide, there are many users—over 2,000 irrigators in Kern County. As the number of users increase, the solutions under OLNE and CLNE theoretically approach the common property solution (Negri, 1989). In the empirical application here, the solutions are nearly identical when the number of users is as few as 40. As such, a natural question arises as to the relevance of applying more complex game theoretic models to understanding the problem for policy purposes. That is, while recent studies of game theory are intellectually interesting, the results here and elsewhere (Dixon, 1988) suggest that the policy relevance is not yet established, and that the lump-parameter model still serves as a useful reference. Of course, more attention is warranted on this issue as the results are limited to a computational analysis of steady-state conditions in a classical (global) game setting where all players play against one another. Evolutionary and localized-interaction games might lead to different results and conclusions, transition time-paths have not been investigated, and the quantitative results to date are limited to simplified functional forms.

The sixth section finds a similar result as that from Tsur and Graham-Tomasi (1991)—the aquifer has a very strong buffering effect in that the aquifer acts to stabilize water flows to agricultural production, and this stabilization is quite large. We also compare the solution under a stochastic surface water supply relative to a deterministic one and find that the stochastic groundwater problem might be approximated reasonably well with an appropriately formulated deterministic model even though the theoretical conditions for certainty equivalence are not met and the stochastic variables have a large coefficient of variation. Of course caution is suggested again due to the fact that the decision rule is not linear as in a pure linear-quadratic problem, and that the deterministic analysis needs to be modified to account for the actual surface water availability prevailing in a given year. Somewhat surprising in view of the large surface water variability is that there is a minimal role for artificial recharge, and that aquifer management benefits don't increase. These findings merit additional investigation in view of substantial policy interest in these topics.

Finally, the seventh section evaluates the impacts of decreases in water supply availability with and without increases in its variability. The analysis provides insight into the dynamics of how irrigated agricultural production in regions with substantial groundwater resources might respond to the water-related effects of future climate change. What we find is that for any particular scenario, simply focusing on a reduction in mean supplies and overlooking the likely increases in variability will result in an underestimate of the groundwater withdrawals and an overestimate of the aquifer volume and annualized net benefits. Also, and as expected, as climate change effects become more severe, groundwater systems are increasingly adversely affected through larger extraction rates. Yet, at least in the application here, the multitude of mitigative strategies available to irrigated agriculture, including deficit irrigation, crop switching, and improvements in irrigation efficiency, all serve to minimize the negative impacts from changes in water supply availability and reliability. Of course, the magnitude of the result will also depend on numerous economic, biophysical, and institutional factors, including the volume of the aquifer, the degree of demand hardening that has already taken place, and whether water banks and/or water trading is available and functioning, to name a few.

Notes

1 While there is progress in California legislation to begin monitoring groundwater use, for most aquifers in the basin neither management nor monitoring exists.
2 A related outcome of this is that mild to moderate reductions in water supply can be met at minimal impact irrigated agriculture, a finding consistent with numerous previous studies, including Adams et al. (1995; 1999), Chen et al. (2001), Mejías et al. (2004), Schwabe et al. (2006), and Connor et al. (2009).
3 A statistical analysis of Kern county surface water supplies found zero serial correlation (Knapp and Olson (1995)). The analysis here accordingly assumes no serial correlation. This is certainly worthy of further investigation utilizing new data since the time of that study.
4 Alternative research uses hedonic methods to assess the impacts of climate change on irrigated agriculture (e.g., Mendelsohn et al. 1994; Mendelsohn and Dinar 2003; Deschenes and Greenstone 2011; Schlenker et al. 2007; Mukherjee and Schwabe 2013). While the majority of these studies find that changes in mean water supply levels can negatively impact farmland values, Mukherjee and Schwabe (2013) find that water supply reliability also can play a significant role.
5 A similar concern and discussion surrounding salinization of irrigation sources appears in other studies, including Schwabe et al. (2006), Knapp and Baerenklau (2008), and Connor et al. (2012).
6 Admittedly, our representation of the climate-related impacts on water supply deliveries is simplistic, although qualitatively likely in the correct direction. Analyses by Heidecke and Heckelei (2010), Quiggin et al. (2010), Wang et al. (2011), Rowan et al. (2011), and Majone et al. (2012), for instance, incorporate results from global climate change models to guide how such impacts are represented.
7 Significant attention is devoted to increasing water supply reliability independent of mean levels in California and elsewhere. For instance, one of the two main objectives of the proposed US$24.5 billion California Bay-Delta Conservation Plan is to improve water supply reliability, not overall water supplies (the other primary objective is to preserve ecosystem habitat). Alternatively, numerous water agencies in Southern California have implemented or are implementing waste water recycling programs for indirect potable reuse more so as a measure to increase water supply reliability than increase overall water supplies.
8 The variance associated with each of these scenarios does not change appreciably over time in Figure 10.6b; a subset of periods was chosen for ease of exposition.

References

Adams, R.M., Fleming, R.A., Chang, C.C., McCarl, B.A., and Rosenzweig, C. (1995) "A reassessment of the economic effects of global climate change on US agriculture", *Climatic Change,* vol 30, pp. 147–167.

Adams, R.M., McCarl, B.A., Segerson, K., Rosenzweig, C., Bryant, K.J., Dixon, B.L., and Conner, R. (1999) "The economic effects of global climate change on US agriculture", in Mendelsohn, R., and Neumann, J. (eds.) *The Impacts of Climate Change on the United States Economy*, Cambridge University Press, Cambridge.

Anderson, K., Dimaranan, B., Hertel, T., and Martin, W. (1997) "Asia-Pacific food markets and trade in 2005: a global, economy-wide perspective", *Australian Journal of Agricultural and Resource Economics,* vol 41, no 1, pp. 19–44.

Brown, G., and Deacon, R. (1972) "Economic optimization of a single-cell aquifer." *Water Resources Research*, vol 8, no 3, pp. 557–564.

Burt, O.R. (1964) "The economics of conjunctive use of ground and surface water", *Hilgardia*, vol 36, pp. 31–111.

Chen, C., Gillig, D., and McCarl, B.A. (2001) "Effects of climatic change on a water dependent economy: A study of the Edwards Aquifer", *Climatic Change*, vol 49, no 4, pp. 397–409.

Connor, J., Schwabe, K., Kirby, M., Kaczan, D., and King, D. (2009) "Impacts of climate change on Lower Murray irrigation", *Australian Journal of Agricultural and Resource Economics*, vol 53, no 3, pp. 437–456.

Connor, J.D., Schwabe, K., King, D., and Knapp, K. (2012) "Irrigated agriculture and climate change: the influence of water supply variability and salinity on adaption", *Ecological Economics*, vol 77, pp. 149–157.

Dale, L., Dogrul, E., Brush, C., Kadir, T., Chung, F., Miller, N., and Vicuna, S. (2013) "Simulating the impact of drought on California's Central Valley hydrology, groundwater, and cropping". *British Journal of Environment and Climate Change* vol 3, pp. 271–291.

Deschenes, O., and Greenstone, M. (2004) "The economic impacts of climate change: evidence from agricultural profits and random fluctuations in weather", *American Economic Review,* vol 97, no 1, pp. 354–385.

Dixon, LS. (1988) *Models of Groundwater Extraction with an Examination of Agricultural Water Use in Kern County, California*. PhD Dissertation, UC Berkeley.

Doll, P. (2002) "Impact of climate change and variability on irrigation requirements: a global perspective", *Climate Change*, vol 54, pp. 269–293.

Eswaran, M. and Lewis, T. (1984) "Appropriability and the extraction of a common property resource", *Economica*, vol 51, no 204, pp. 393–400

Gisser, M., and Sanchez, D. (1980) "Competition versus optimal control in groundwater pumping", *Water Resources Research*, vol 16, pp. 638–642.

Heidecke, C., and Heckelei, T. (2010) "Impacts of changing water inflow distributions on irrigation and farm income along the Draa River in Morocco", *Agricultural Economics*, vol 41, pp. 135–149.

Kan, I., Schwabe, K.A., and Knapp, K.C. (2002) "Microeconomics of irrigation with saline water", *Journal of Agricultural and Resource Economics*, vol 27, pp. 16–39.

Knapp, K.C., and Baerenklau, K.A. (2006) "Groundwater quantity and quality management: agricultural production and aquifer salinization over long time scales", *Journal of Agricultural and Resource Economics*, vol 31, no 3, pp. 616–641.

Knapp, K.C., and Olson, L. (1995) "The economics of conjunctive groundwater management with stochastic surface supplies", *Journal of Environmental Economics and Management*, vol 28, pp. 340–356.

Knapp, K.C., and Olson, L. (1996) "Dynamic resource management: intertemporal substitution and risk aversion", *American Journal of Agricultural Economics*, vol 78, pp. 1004–1014.

Koundouri, P. (2004) "Potential for groundwater management: Gisser-Sanchez effect reconsidered", *Water Resources Research*, vol 40, no 6, pp. 1–13.

Levhari, D., and Mirman, L.J. (1980) "The great fish war: an example using a dynamic Cournot-Nash solution", *Bell Journal of Economics*, vol 11, pp. 322–334.

Long, N.V. (2010) *A Survey of Dynamic Games in Economics*. Singapore: World Scientific Publishing Co.

Majone, B., Bovolo, C., Bellin, A., Blenkinsop, S., and Fowler, H. (2012) "Modeling the impacts of future climate change on water resources for the Gallego river basin (Spain)", *Water Resources Research*, vol 48, pp. 1–18.

Mejías, P., Ortega, C.V., and Flinchman, G. (2004) "Integrating agricultural policies and water policies under water supply and climate uncertainty", *Water Resources Research*, vol. 40, W07S03, DOI:10.1029/2003WR002877.

Mendelsohn, R., and Dinar, A. (2003) "Climate, water, and agriculture", *Land Economics*, vol 79, pp. 328–341.

Mendelsohn, R., Nordhaus, W.D., and Shaw, D. (1994) "The impact of global warming on agriculture: a Ricardian analysis", *American Economic Review*, vol 84, pp. 753–771.

Milly, P., Dunne, K., and Vecchia, A. (2005) "Global pattern of trends in streamflow and water availability in a changing climate", *Nature*, vol 438, pp. 347–350.

Mukherjee, M., and Schwabe, K. (2013) "Valuing a Water Portfolio in Irrigated Agriculture", Working Paper Series 01-2012. Riverside, CA: Water Science and Policy Center, UC, Riverside.

Negri, D. (1989) "The common property aquifer as a differential game", *Water Resources Research*, vol 29, pp. 9–15.

Posnikoff, J., and Knapp, K.C. (1996) "Regional drainwater management: source control, agroforestry, and evaporation ponds", *Journal of Agricultural and Resource Economics*, vol 21, no 2, pp. 277–293.

Postel, S.L. (2000) "Entering an era of water scarcity: the challenges ahead", *Ecological Applications*, vol 10, no 4, pp. 941–948.

Provencher, B., and Burt, O. (1993) "The externalities associated with the common property exploitation of groundwater", *Journal of Environmental Economics and Management*, vol 24, pp. 139–158.

Quiggin, J., Adamson, D., Chambers, S., and Schrobback, P. (2010) "Climate change, uncertainty, and adaptation: the case of irrigated agriculture in the Murray-Darling Basin in Australia", *Canadian Journal of Agricultural Economics*, vol 58, pp. 531–554.

Rosegrant, M., Paisner, M., Meijer, S., and Witcover, J. (2001) "2020 Global Food Outlook: Trends, Alternatives, Choices", International Food and Policy Research Institute, Washington, DC.

Rowan, T., Maier, H., Connor, J., and Dandy, G. (2011) "An integrated dynamic framework for investigating the impact of climate change and variability on irrigated agriculture", *Water Resources Research*, vol 47, no 7, pp. 1–13.

Rubio, S.J., and Casino, B. (2001) "Competitive versus efficient extraction of a common property resource: the groundwater case", *Journal of Economic Dynamics and Control*, vol 25, pp. 1117–1137.

Scanlon, B. R., Faunt, C. C., Longuevergne, L., Reedy, R., Alley, W., McGuire, V., and McMahon, P. (2012) "Groundwater depletion and sustainability of irrigation in the U.S. High Plains and Central Valley", *Proceedings of the National Academy of Sciences of the United States of America*, vol 109, no. 24, pp. 9320–9325.

Schlenker, W., Hanemann, W.M., and Fisher, A.C. (2007) "Water availability, degree days, and the potential impact of climate change on irrigated agriculture in California", *Climate Change*, vol 81, pp. 19–38.

Schwabe, K., Kan, I., and Knapp, K.C. (2006) "Integrated drainwater management in irrigated agriculture", *American Journal of Agricultural Economics*, vol 88, no 1, pp. 133–149.

Seagar, R., Ting, M., Held, I., Kushnir, Y., Lu, J., Vecchi, G., Huang, H., Harnik, N., Leetmaa, A., Lau, N., Li, C., Velez, J. and Naik, N. (2007) "Model projections of an imminent transition to a more arid climate in southwestern North America". *Science*, vol 316, pp. 1181–1184.

Shiklomanov, I. (1999) World Water Resources and Their Use. International Hydrological Programme, UNESCO's Intergovernmental Scientific Program in Water Resources, available at: http://webworld.unesco.org/water/ihp/db/shiklomanov/. Accessed 2/28/2009.

Shindell, D., Faluvegi, G., Lacis, A., Hansen, J., Ruedy, R., and Aguilar, E. (2006) "Role of tropospheric ozone increases in 20th century climate change", *Journal of Geophysical Research*, vol 111. D08302, doi:10.1029/2005JD006348

Tanji, K., and Kielen, N. (2002) *Agricultural Drainage Water Management in Arid and Semi-Arid Areas.* Food and Agricultural Organization. Rome, Italy.

Thayalakumaran, T., Bethune, M.G., and McMahan, T. (2007) "Achieving salt balance—should it be a management objective?", *Agricultural Water Management*, vol 92, pp. 1–12.

Tsur, Y. (1990) "The stabilization value of groundwater when surface water supplies are uncertain: the implications for groundwater development", *Water Resources Research*, vol 26, pp. 811–818.

Tsur, Y. and Graham-Tomasi, T. (1991) "The buffer value of groundwater with stochastic surface water supplies", *Journal of Environmental Economics and Management,* vol 21, no 3, pp. 201-224.

Wang, W., Park, S., McCarl, B., and Amosson, S. (2011) "Economic and Groundwater Use Implications of Climate Change and Bioenergy Feedstock Production in the Ogallala Aquifer Region", Agricultural and Applied Economics Association Selected paper. AAEA Annual Meetings, Pittsburgh, Pennsylvania.

PART III

Institutions and information

11

INSTITUTIONS FOR MANAGING GROUND AND SURFACE WATER AND THE THEORY OF THE SECOND-BEST

Karl Jandoc, Richard Howitt, James A. Roumasset, and Christopher A. Wada

This chapter derives principles of water pricing using a simple model of conjunctive use. We begin with a review of first-best principles for the joint management of ground and surface water, i.e. in abstraction from transaction costs. Inasmuch as information and enforcement costs are pervasive in water management, however, how can we extend the conventional water management theory to account for transaction costs?

In what follows, we recognize two types of second-best analysis. The first involves adding a constraint to an otherwise first-best problem. For example, optimal taxation is concerned with the optimal markup of the tax-inclusive price above marginal cost in the presence of a minimum constraint on tax revenue and, because of a missing market, the inability to tax labor. In the conjunctive-use context, this corresponds to the problem of how to price irrigation water when groundwater use is unregulated.

The second type of second-best analysis involves the comparative analysis of alternative institutions regarding their ability to simultaneously enhance the benefits of economic cooperation and economize on transaction costs (Roumasset 1978; Williamson 1985). We consider the case of optimal governance of groundwater.

Second-best pricing of surface water with unregulated groundwater

There is an abundant literature on the importance of water pricing to implement demand-side conservation. This has led to the spread of markets for irrigation water in several parts of the U.S., Australia, Israel, and several developing countries (Tsur et al. 2004). While most of the literature on the economics of surface and groundwater management is at the first-best level of analysis, groundwater is often left unpriced due to the administrative costs of regulatory arrangements.

Pricing surface water at its full opportunity cost in such situations may induce some farmers to switch to groundwater thereby exacerbating the open access problem. There remains an "unpaid factor" externality because farmers are not paying for the marginal user cost of groundwater. Once this externality is taken into account, how is the optimal surface-water price affected, given that the costs of directly pricing or regulating are excessive? Just as taxing complements of untaxed leisure higher than other goods is second-best (Corlett and Hague 1953), one suspects that pricing water lower than its first-best level will partially ameliorate that externality. It is even conceivable that creating a market for surface water, without groundwater regulation, could decrease welfare relative to leaving surface water unpriced. This section extends the Chakravorty–Umetsu (2003) model of the first-best boundary between groundwater and surface water to the case of second-best.

Model setup and optimal conjunctive use in the absence of transaction costs

We begin with a simplified version of the model in Chakravorty and Umetsu (2003). Farms are located along a continuum with the headworks located upstream. These farms differ only by their distance from the source, which we denote by x, and are similar in all other respects. We assume that the command area is a fixed distance $\bar{x}$. Water sent from the headworks to a farm located at distance x is denoted by $Q(x)$. Some of the water that flows through the canal is lost in conveyance because of seepage and percolation. Therefore the amount of *received water* at distance x is given by $q(x) = Q(x)h(x)$ where $h(x)$ is the conveyance efficiency $(0 < h(x) < 1)$ which is a decreasing function of distance, that is, $h'(x) < 0$. The constant marginal cost of *sent* water is denoted by c_s, and hence the marginal cost of received water at distance x is given by $c_s / h(x)$. Apart from surface water, farms also have access to groundwater. The use of groundwater at distance x is given by $g(x)$. The marginal extraction cost of groundwater, c_g, as well as the marginal user cost of groundwater, λ, are assumed to be constant.[1] The benefit of water use is given by the function, $B(\omega(x)) = \int_0^{\omega(x)} p(\gamma) d\gamma$, where $\omega(x) = q(x) + g(x)$ and $p(\omega)$ is the inverse demand function $\left(\dfrac{dp}{d\omega} < 0\right)$. The benefit function is assumed to be concave.

The water authority wishes to maximize the net benefit of water use by choosing the boundary of surface water use. This is given by the equation:

$$\max_{q(x),g(x)} V = \int_0^{\bar{x}} \left[B(\omega(x)) - \frac{q(x)c_s}{h(x)} - g(x)(c_g + \lambda) \right] dx \tag{11.1}$$

As in Chakravorty and Umetsu (2003) and Chapter 5 of this Handbook, farmers closer to the surface water headworks have a comparative advantage in using surface water. That is, there is a unique boundary x^* at which farmers located before x^* exclusively use surface water and those farther away exclusively use groundwater. Hence, we can write (11.1) by decomposing it into farms using surface water exclusively and those who use groundwater exclusively:

$$\max_{x^*} V = \int_0^{x^*} \left(B(\omega(\theta)) - c_s \frac{q(\theta)}{h(\theta)} \right) d\theta + \int_{x^*}^{\bar{x}} \left(B(\omega(\varphi)) - g(\varphi)(c_g + \lambda) \right) d\varphi \tag{11.2}$$

Note that $\omega(x) = q(x)$ for $x \leq x^*$ and $\omega(x) = g(x)$ for $x > x^*$

As in Coasean economics (e.g. Coase 1960), it is instructive to abstract from transaction costs as a point of departure. In this case, the marginal benefit (inverse demand function) of all those who exclusively use surface water equals the full marginal cost of surface water at distance x, that is, $B'(q(x)) = \dfrac{c_s}{h(x)}$. Also, the marginal benefit of those who exclusively use groundwater equals the marginal extraction cost plus the marginal user cost, that is, $B'(g) = c_g + \lambda$. The unique boundary is found by the equality of marginal benefits at that point. This means that at the boundary:

$$\frac{c_s}{h(x^*)} = c_g + \lambda \tag{11.3}$$

The first best solution might be implemented, for example, by setting the wholesale price of water released from the headworks at c_s and pricing groundwater at $c_g + \lambda$. Alternatively, the same solution could be implemented by quantity controls.

Surface water markets with open access in groundwater

As motivation for the case of optimal surface water pricing in the face of unregulated groundwater, consider the equilibrium allocation under a market for surface water but unregulated groundwater. In this case, the first-best solution cannot be implemented. Farmers will not internalize the full marginal cost of groundwater, but, rather, will use groundwater up to the point where its marginal benefit equals the (constant) marginal extraction cost, that is, groundwater is extracted up to $\tilde{g}$ where $B'(\tilde{g}) = c_g$. The boundary $\tilde{x}$ in this case is determined by the equality of marginal benefits of *received* water and groundwater at that point. Since the marginal benefit of surface water at $\tilde{x}$ is given by the equality $B'(q(\tilde{x})) = \dfrac{c_s}{h(\tilde{x})}$, it follows that the boundary $\tilde{x}$ is determined by the equation

$$c_g = \frac{c_s}{h(\tilde{x})}. \tag{11.4}$$

This equation says that at $\tilde{x}$, the marginal extraction cost of groundwater and the marginal cost of received water at that distance are equal. Clearly, since in open access farmers draw more groundwater than in the first-best solution (at open access marginal user cost is ignored), the boundary $\tilde{x}$ should be closer to the surface water source than the first best boundary, x^*. But since the marginal social cost of groundwater, $c_g + \lambda$, is greater than marginal benefits under open access, groundwater is overused.

Second-best pricing of surface water in the absence of groundwater regulation

The problem is now to choose the price of surface water in order to maximize net benefits, subject to the constraint that groundwater is unregulated. The water authority in this situation can induce people to economize on groundwater by subsidizing the price of surface water.[2] In this second-best scenario, the water authority maximizes (11.1) subject to the constraint that at the boundary the marginal benefit of surface water use equals the marginal extraction cost. With the subsidy, the marginal benefit of surface water use is equal to the marginal cost of surface water minus the subsidy, that is:

$$B'(q(x)) = \frac{c_s - s}{h(x)}. \tag{11.5}$$

where s is the per-unit subsidy of surface water. Therefore, at the second-best boundary x^{**}, we have that

$$\frac{c_s - s}{h\left(x^{**}\right)} = c_g \tag{11.6}$$

Thus, the water authority wishes to find the boundary x^{**} at which net benefits are maximized subject to the constraint that at the boundary the marginal benefit of surface water use equals the marginal extraction cost of groundwater, that is, the problem is:

$$\max_{x^{**}} V = \int_0^{x^{**}} \left(B(q(\theta)) - c_s \frac{q(\theta)}{h(\theta)} \right) d\theta + \int_{x^{**}}^{\bar{x}} \left(B(g(\phi)) - g(\phi)(c_g + \lambda) \right) d\phi \tag{11.7}$$

subject to equation (11.6).

However, raising the revenue required for the subsidy to surface water users creates "tax friction," due to the distorted incentives imposed by the taxes needed to finance the subsidy (Ballard, Shoven and Whalley, 1985). Hence, we modify equation (11.7) to include the excess burden of tax friction as follows:

$$\max_{x^{**}} V = \int_0^{x^{**}} \left(B(q(\theta)) - c_s \frac{q(\theta)}{h(\theta)} - \alpha s \frac{q(\theta)}{h(\theta)} \right) d\theta + \int_{x^{**}}^{\bar{x}} \left(B(g(\phi)) - g(\phi)(c_g + \lambda) \right) d\phi \tag{11.8}$$

subject to equation (11.6). Here α is the estimated tax friction and $S = \int_0^{x^{**}} s \frac{q(\theta)}{h(\theta)} d\theta$ is the total subsidy given to surface water users.

Numerical example

We illustrate the results with a numerical example. Suppose that $c_s = 1$, $c_g = 2$ and $\lambda = 4$. Assume that the benefit function for water use is $B(\omega(x)) = \omega(x) \left(10 - \frac{\omega(x)}{2} \right)$. Let the conveyance efficiency function be $h(x) = e^{-.02x}$ for each x km. Let the command area be $\bar{x} = 100$ km. In the first-best scenario, the boundary $\hat{x}$ is determined as the solution to equation (11.3)

$$\frac{c_s}{h(\hat{x})} = c_g + \lambda \Rightarrow h(\hat{x}) = \frac{1}{6} \Rightarrow \hat{x} = \frac{\ln(6)}{.02} \Rightarrow \hat{x} \approx 89.6 \ km$$

In the case of open access, the boundary $\tilde{x}$ is likewise computed as the solution to equation (11.4)

$$\frac{c_s}{h(\tilde{x})} = c_g \Rightarrow h(\tilde{x}) = \frac{1}{2} \Rightarrow \tilde{x} = \frac{\ln(2)}{.02} \Rightarrow \tilde{x} \approx 34.7 \ km$$

We see in this example that the open access boundary is closer to the source than the first best boundary, which implies that more farmers use the groundwater resource under open access. Intuitively, since farmers do not have to pay the full opportunity cost of groundwater under open access, more farmers use groundwater than the fully priced surface water.

In computing the second-best boundary x^{**} we do the following steps: First, we guess a value of the second-best boundary, say 50 km, and we compute the subsidy s using equation (11.6). Second, we use this subsidy and find an expression for $q(x)$ using the condition that the marginal benefit of surface water should be equal to the marginal extraction cost (minus the subsidy). Third, we use the expression of $q(x)$ and plug this into the net benefit function in equation (11.7) or equation (11.8), along with the corresponding amounts of groundwater obtained by the solution to $B'(\tilde{g}) = c_g$, to compute the corresponding net benefit. We do this step for the case where we do not assume zero tax friction and for the case where tax friction is estimated to be 30 percent of the total subsidy. Finally, we iterate this process by choosing another boundary and compute the corresponding net benefit. The boundary x^{**} is at the level where net benefits are maximized.

Case 1 in Table 11.1 illustrates the boundaries in this system, the optimal quantities of sent water and groundwater extraction as well as the total subsidy and subsidy per unit of sent water in the scenario with no tax friction. Case 2 illustrates the case where tax friction is 30 percent of the total subsidy.

For the tax friction case, the subsidy of surface water (US$0.54 per unit) is slightly more than half of the wholesale price and cuts the welfare loss of open access by slightly more than half as well (from US$2500 million–US$1270 million to US$2500 million–US$1921 million). The relatively large subsidy of surface water moves the second-best boundary (74 km) most of the way from the open access boundary of 34.7 km towards the first-best boundary of 89.6 km. In this particular example, each unregulated groundwater farmer gains rents that are exactly offset by the excess burden associated with ignoring the marginal user cost. By subsidizing surface water, more farmers end up in the surface water category and garner rents in excess of even the full cost.

Now suppose that groundwater is depleted over time, prior to reaching a steady state. In this case, both the marginal extraction cost and the marginal user cost of groundwater extraction increase. Suppose the new marginal extraction costs and marginal user costs increase to $c_g = 2.25$ and $\lambda = 4.5$ (from 2 and 4 respectively). As seen in Case 3 in Table 11.2, this increase in the scarcity value of groundwater moves the boundaries for the open access, first-best and second-best cases to 40.5, 95.5 and 79 km, respectively. The subsidy of surface water slightly decreases to US$0.53 per unit. There is a significant

Table 11.1 Open access, first best and second best with and without tax friction

	Case 1			Case 2		
	Open access	First best	Second best	Open access	First best	Second best
Total quantity of:						
Sent water, Q	425	1,625	2,022	425	1,625	1,488
Groundwater extraction, g	523	42	112	523	42	208
Net welfare (in 10^6US$) from:						
Surface water	1,270.37	2,416.90	2,229.61	1,270.37	2,416.90	1,921.18
Groundwater	0	83.29	0	0	83.29	0
Total	1,270.37	2,500.19	2,229.61	1,270.37	2,500.19	1,921.18
Boundary (km)	34.7	89.6	86	34.7	89.6	74
Total subsidy (US$)			1,298			811
Subsidy per unit Q (US$)			0.641			0.544

Note: In Case 1, we disregard tax friction. In Case 2, tax friction is equal to 30 percent of total subsidy.

Table 11.2 Open access, first best and second best when groundwater is scarce

	Case 3		
	Open access	First best	Second best
Total quantity of:			
Sent water, Q	523	1,761	1,666
Groundwater extraction, g	461	15	163
Net welfare (in 10⁶US$) from:			
Surface water	1,453.11	2,455.89	1,955.26
Groundwater	−287.97	23.88	−101.72
Total	1,165.14	2,479.77	1,853.54
Boundary (km)	40.5	95.5	79
Total subsidy (US$)			894
Subsidy per unit Q (US$)			0.536

Note: In Case 3, we include the same value of tax friction (30 percent) but let marginal extraction cost and marginal user cost of groundwater rise to 2.25 and 4.5, respectively.

loss in welfare from groundwater extraction for all the cases, but it is ameliorated by the welfare gain of farmers switching to surface water.

The evolution of optimal groundwater governance

In this section, we portray a different level of second-best analysis by explicit treatment of transaction costs. For this purpose, we focus only on optimal groundwater extraction.

In many regions around the world, groundwater is pumped from private wells for irrigation of crops. When a common pool resource such as groundwater faces overuse by multiple users with unrestricted extraction rights, additional governance may be warranted if the gains from governance exceed the costs (Demsetz 1967; Ostrom 1990). As discussed in the previous section, the optimal first-best solution may be unattainable if one of the resources such as groundwater cannot be directly managed. However, even if no such restriction is in place, the first-best solution may remain out of reach when enforcement and information costs are considered. Which of several institutions (e.g. privatization, centralized ownership, user associations) maximizes the net present value of the groundwater resource depends on the benefits generated from each option net of the governance costs involved in establishing and maintaining the candidate institution. For example, if the demand for water is small relative to the size of the aquifer, the gains from management are likely to be small, and open access might be preferred initially. As demand grows over time and water becomes scarcer, however, a common property arrangement (e.g. a user association) or a water market may become efficient.

Augmenting the standard groundwater optimization problem to allow for governance costs is fairly straightforward. Let the head level (h_t), which is the vertical distance between mean sea level and the top of the groundwater aquifer, be an index for the volume of stored freshwater. The groundwater stock increases with exogenous natural recharge (R) and decreases when water discharges from the aquifer to adjacent water bodies (d) or is pumped for irrigation (q_t), as described by the following equation of motion:

$$\dot{h}_t = R - d(h_t) - q_t \tag{11.9}$$

Pumped groundwater, which is an input to the production of crops, generates marginal benefit equal to P. The marginal pumping cost (c) is a function of the head level because more energy is required to lift groundwater a longer distance when the head is lower. Given q_t, the net benefit of resource use for society at time t is given by consumer surplus net of extraction costs:

$$NB_t = \int_0^{q_t} P(x)dx - c(h_t)q_t \tag{11.10}$$

Without governance, harvest will continue to the point where rent diminishes to zero. The associated open access harvest q_{oa} satisfies $P(q_{oa}(h)) = c(h)$. Note that the open access harvest at time t depends on the stock level in that period. Governance, which limits harvest to a level below q_{oa}, is costly. Governance costs may be incurred once or may be recurrent, but we begin by focusing on variable (recurring) costs of governance. Once the institution is established, the variable cost at each time t depends on the difference between open-access harvest and actual harvest:

$$G(q_t; q_{oa}) = g(q_{oa}(h_t) - q_t) \tag{11.11}$$

The planner's problem is to choose extraction to maximize the present value of net benefits from resource use (equation 11.10) less governance costs (equation 11.11) subject to the aquifer's equation of motion (equation 11.9), i.e.

$$\max_{q_t} \int_0^{\infty} e^{-rt} [\int_0^{q_t} P(x)dx - c(h_t)q_t - g(q_{oa}(h_t) - q_t)]dt \tag{11.12}$$

subject to $\dot{h}_t = R - d(h_t) - q_t$ and $0 \leq q_t \leq q_{oa}(h_t)$ for all t.

To simplify the analysis, we further assume that the price (marginal benefit) of the resource is constant and the initial resource stock is above its steady state level. One can show that if the optimal solution involves positive governance (i) the groundwater stock decreases monotonically to its steady state level, (ii) the shadow price of groundwater increases monotonically to its steady state level, and (iii) the steady state head level is lower than the first best level that would prevail if $g = 0$ (Roumasset and Tarui 2010). Figure 11.1 illustrates hypothetical dynamic paths of the full marginal net benefit (FMB) and shadow price (λ) of a groundwater aquifer in transition to governance.

The FMB of extraction is defined by $P - c(h_t) + g$, and the shadow price (λ) is the co-state variable associated with the resource stock under the optimal solution. The non-instantaneous convergence of the FMB and λ appears to contradict the tendency to associate scarcity with net price or marginal user cost, the justification being that they are all equal along the optimal path in the standard management problem (without governance). However, this puzzle is resolved by examining one of the necessary conditions for the maximization problem (equation 11.12):

$$P - c(h_t) + g = \lambda_t + \theta_t \tag{11.13}$$

where θ is the Lagrangian multiplier associated with the governance constraint. When g and θ are equal to zero in every period, the condition can be described as net price =

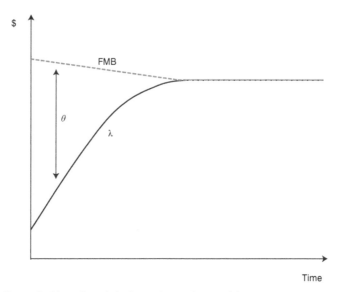

Figure 11.1 Full marginal benefit and shadow price under transition to governance

MUC. When transition to governance is optimal, however, the FMB may be declining over time while the shadow price (the true scarcity value) is rising, and the difference between the two values is the shadow value of the binding governance constraint. Once the constraint is no longer binding, i.e. the optimal extraction quantity is less than open access, FMB and λ converge.

The speed of transition depends on the scarcity of the resource, the net price, and the cost of governance. As governance costs decrease due to advances in technology, e.g. improved satellite monitoring capabilities, we are able to move closer to the solution where the full marginal benefit is equal to the full marginal cost of groundwater.

From open access to common property to markets

In an open access regime where groundwater extraction is unregulated, farmers pump water until their rent from doing so is exhausted. In many countries there have been several forms of groundwater management that are designed to curb over-extraction of the resource. Examples of common property institutions are community-managed systems such as water users associations. A case study from Minquin County in China (Aarnoudse et al. 2012) revealed that water users associations have been instrumental in reducing excessive groundwater extraction. Such common property institutions rely on specific systems of reward and punishment, often including community monitoring. Market institutions, on the other hand, rely on price or quantity mechanisms to allocate water and utilize public enforcement mechanisms. In Hawai'i, for example, the Board of Water Supply (BWS) chooses the price per unit of groundwater extracted, consumers determine use, and the BWS extracts accordingly. Similarly in West Bengal, volumetric pricing is implemented by metering tubewell extraction (Mukherji 2013). One could imagine cases where the management authority sets quantities extracted and allows trading such that the market determines price. Due to conveyance costs,

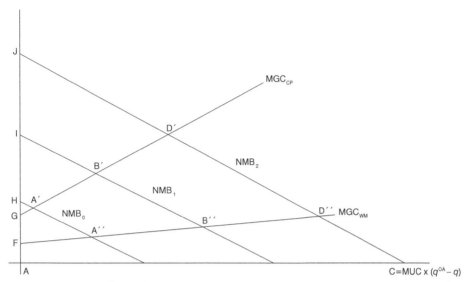

Figure 11.2 Evolution of groundwater governance (open access to common propery to water markets)

however, distribution charges would have to be added to the market-established wholesale price.

Figure 11.2 presents a heuristic illustration of how institutions of groundwater management could efficiently evolve from open access to common property to water markets. The downward sloping lines are the net marginal benefits of conservation, i.e. the marginal extraction cost plus the marginal user cost saved from not using the last unit minus the forgone marginal benefit of that unit, for three periods (0, 1, and 2). The upward sloping lines are the marginal costs of governance, i.e. the monitoring and enforcement costs as a function of conservation value, defined as the amount actual extraction is reduced from its open access level times the marginal user cost. That is, the greater is the value of the resource (in terms of its shadow price), and the more pumping is restricted away from its open access level, the higher are requisite governance costs. We also assume that the marginal governance cost schedule for water markets, G_{WM}, has a lower intercept and slope than that of common property, but that these advantages come only at the higher fixed cost of setting up markets as opposed to that water user associations, G_{CP}.

Beginning at time zero, note that water markets have the highest net gains before consideration of fixed costs. But assuming that these net gains are less than the requisite fixed costs, the optimal governance is at zero, i.e. open access is second-best optimal (point A). At time 1, the optimal institution is at point B′ (common property) if and only if the net gains from common property (triangle IB′G minus G_{CP}) is positive and larger than the net gains from water markets (triangle IB″F minus G_{WM}). At time 2, the optimal institution is at point D″ (regulation) if and only if the net gains from regulation (triangle JD″F minus G_{WM}) is positive and is larger than the net gains from common property (triangle JD′G minus G_{CP}). Intuitively, when the resource is sufficiently abundant, open access may be optimal. When the resource is sufficiently scarce in the sense of high marginal opportunity cost of extraction, water markets may be warranted.

In the intermediate range, common property institutions (e.g. water users associations or informal customs limiting extraction) may (or may not be) efficient. This exercise in comparative institutional analysis illustrates that one cannot rank the efficiency of institutions apart from their circumstances. In particular, where groundwater is still plentiful, it may be prudent to delay restrictions, regulations, and market institutions. Moreover, planners should not assume that one institution will always be appropriate.

Satellite imaging to reduce measurement costs: the barbed wire of irrigation

The fixed costs of water market institutions

In the previous section, the optimal timing of the establishment of water management institutions was based on the evolution of benefits and costs of those institutions. In the analysis, the variable costs were defined as being continuous and well-behaved. The optimal timing of the switch from one institution to another is said to maximize net benefits, inclusive of institutional governance costs as depicted by equation (11.12). However, this equation abstracts from the fixed costs of starting, stopping, or switching institutions.

But as anticipated by the discussion of Figure 11.2, the question remains as to how do we determine when it is worth committing the fixed costs of establishing water market institutions? This section will address this question which is fraught with difficulties of non-convexity of establishment costs, uncertainties of the net benefits of the institution and its cost of operation, and the irreversible nature that seems to dog most institutional shifts. This section has two veins: first, we argue that the question of discrete changes in property rights should be analyzed using capital asset pricing theory pioneered by Dixit and Pindyck (1994), which leads to the conclusion that there is a fixed total cost for the establishment of new institutions which needs to be compensated by additional benefits from its establishment over and above the expected returns. The second part of this section concentrates on some operational concepts of institutions with low transaction costs that may be able to compensate for the costs expected in institutional establishment.

Institutional innovation as a discrete capital investment

Shifts in the nature of property rights for natural resources invariably involve a discrete switch point as illustrated by Figure 11.2, and a significant investment in the lump sum of establishing the institutional monitoring system and enforcement. Clearly this fixed cost of institutional establishment will impede optimal solutions that consider only marginal costs.

In the classic paper on brand registration in the establishment of cattle property rights in the American West between 1850 and 1900, Anderson and Hill (1975) show that individuals and institutions increased time and resources devoted to the definition and enforcement of rights as the benefits increased and the costs (e.g. fencing with barbed wire) decreased. For example, cattle brands fluctuated with the prosperity of the industry which was significantly changed by the blizzards of the 1870s. The registration of brands in the local courthouses may have been quite comprehensive, but the enforcement of these brands by the regional Sheriff was very variable as demonstrated by many Western movies.

When applied to the question of when comprehensive institutions can be justified for water markets and groundwater management, the standard continuous marginal analysis breaks down under the pressure of the discrete non-convexity of establishment costs, uncertainties about the net benefits to overlying groundwater pumpers, and the costs of committing to a fixed institutional structure that has high costs of reversibility. The situation in several states in the Western US is the surface water rights are allocated based on prior appropriation which was formally recognized in 1855 in California in the case of *Irving* v. *Phillips*. The concept of reasonable use was first clarified in terms of groundwater and surface rights in 1903 in the case of *Katz* v. *Wilkinshaw*. Reasonable use was finally required for all water uses in California by a modification of the Constitution in 1928. A further modification occurred in the interpretation of reasonable use by a challenge in 1967 in the case of *Joslin* v. *Marin Municipal Water District* (Gray 1993). In this case the property rights had to define between types of reasonable use and not merely rely on the concept of "first in time first in right" which had been the guiding principle up until that point. In the Joslin case the court had to distinguish between different types of values of use, essentially choosing between municipal uses of water from a diversion dam, versus the value of water washing gravel onto adjoining property from whence it was sold.

While these developments in the establishment of surface water rights and reasonable use were evolving, the situation with California and groundwater in other Western states such as Texas remained with the laissez-faire concept of correlative property rights in which the overlying pumper has no restrictions on pumping unless injury can be proved by an adjoining groundwater pumper. In several Western states, for example California and Texas, there is a natural inclination by groundwater users against the establishment of fixed property rights. Users there strongly resist efforts to measure and monitor groundwater use, which is a prerequisite for operational groundwater management. In California, a recent multibillion-dollar water bill was held up in the state legislature until the provision for mandatory measurement of groundwater pumping measurement was removed and, as of this writing, California is the only major Western irrigating state to still have no statewide groundwater management legislation enacted.

These ill-defined property rights for groundwater in the Western U.S. is hardly surprising given the uncertainty about how groundwater moved through the aquifers, its source of recharge, and the ability to accurately measure pumping extraction. An additional factor that makes the prospect of spending financial and political capital on establishing groundwater management less attractive is the relatively low cost of extracting groundwater in most Western states. Until recently, the relative abundance of groundwater in most areas makes the default position of remaining with undefined correlative rights one that was probably socially optimal.

The problem of establishing groundwater rights that enable more precise management can be characterized as a stochastic dynamic problem with discrete lump sum costs of institutional establishment in both fiscal and political terms. The economic benefits of managed groundwater can be summarized as resulting from excessive overdrafting of the aquifer and resulting increases in pumping cost to the overlying users. A second and possibly more valuable source of benefit has been characterized by Tsur and

Graham-Tomasi (1991) as the buffer stock value. This buffer stock value can be thought of as resulting from a difference in groundwater extraction capacity among different pumpers. Usually this is due to differences in well size and capacity, and in particular, the depth at which the screens on the wells are set to draw from different strata in the aquifer. Analysis of well log data shows that the depth of the top of well screens in California's central valley is distributed normally with a significant set of outlying wells having screens at higher levels. These small shallow wells are usually associated with small farms but these often grow very valuable crops. It follows that a given reduction in groundwater depth during drought years will have a differential impact on different groundwater pumpers. However, a small farmer is unlikely to have a portfolio of wells of different depths, and thus the social impact of groundwater is closer to a switching function rather than a continuous function in terms of the damage for any one business, or group of businesses by size. In this case the benefit function from stabilization of groundwater depth during dry years is a stochastic event with a discontinuous cost function.

During the 1991 California drought the buffer value of groundwater in California, in particular in Butte County, was brought into sharp contrast by a lawsuit brought by a group of small farmers in the area termed the Cherokee Strip. The Cherokee Strip region is situated on a bench area on the edge of the Sacramento Valley which was significantly higher than the farms with more alluvial soil on the valley floor. As the drought progressed, an increasing number of small farm wells went dry in the Cherokee area and the Cherokee farmers petitioned to prevent farmers further down the hydraulic gradient on the valley floor from pumping their wells to the detriment of the Cherokee Strip groundwater depth. While the evidence for the hydraulic linkages between the groundwater in the two areas was disputed by the California Department of Water Resources, and was not found to be conclusive in the court hearings, the case had a significant effect on water markets and exports, and it spawned many locally enacted county ordinances that prevented the export of water from many counties in the Sacramento Valley of California (Hanak 2003).

A formal quantitative specification of the decision to adopt water market institutions is shown in Howitt (1995). Unlike the spatially-based analysis earlier in this chapter where surface water cost is a function of distance from the source, the approach taken by Howitt defines the optimal timing of institutions over a uniform set of water market opportunities. The value of a water market is defined as depending on the state variable which characterizes the gains from trade in water from the difference between value marginal products (excess demand) within the potential market. This state variable is defined as a stochastic continuous time Ito process which when combined with a Weiner process, leads to more manageable differential equations describing the change in social value of the resource. The optimal switch time for establishment of the market institution is characterized by two sets of dynamic first order conditions, the value matching and smooth pasting conditions. The value matching position states that at the optimal switch point, the expected present value of the flow of costs of unmet excess demands for water trades, minus the value of the option to switch in the future, must be equal to the lump sum institutional cost. The value of the option to switch in the future, which is given up by taking action in the establishment of an institution, is generally termed the hurdle rate for any discrete capital investment. This condition is in contrast to the standard marginal trade conditions, which do not take into account the cost of losing the option to switch in the future once the fixed financial and political costs of institutional establishment have been spent. The cost of the lost option to switch in the future will drive a further wedge between the optimal benefit and cost conditions derived in equation (11.12). Howitt uses some typical parameters from the California water market to calculate the average delay in the optimal establishment of a market due to the hurdle rate cost of market establishment. The results show that the average delay varies from 4 to 16 years depending on the stochastic properties of the excess demand driven by fluctuations in water supply and the discount rate of the alternative risk-free asset. These latter parameters affect the value of the option to switch in the future that is forgone by the decision to introduce an institution at a given time. This increase in the cost of market establishment due to uncertainty, fixed costs, and irreversibility, further emphasizes the point made earlier, that as the transaction costs of a water market institution are reduced, the probability that the market institution will be adopted increases.

Institutions and methods to reduce water market transaction costs

Several economic historians have pointed out that land allocation by the Homestead Act was initially ineffective on the treeless western prairies and ranges due to the impossibility of building fences out of local timber. While settlers had clear and legal title to their land, it was ineffective as an incentive to develop crop farming, due to the excessive costs of effective fencing in many treeless parts of the high plains. It was not until the development and perfection of barbed wire in 1874 by Joseph Glidden that agriculture became a practical alternative to ranching in the fertile plains of the Midwest (Hornbeck 2010). So rapid was the adoption of barbed wire that Glidden was not prepared for the response. Hornbeck recounts that when Glidden received an order for a hundred tons of barbed wire he was dumbfounded and telegraphed the purchaser asking if his order should not read 100 pounds. The order was in tons not pounds. In addition, the price of barbed wire dropped substantially from US$20 per hundred pounds in 1874 to US$1.80 per hundred pounds in 1897 further stimulating its widespread adoption. The enclosure of farmland on the High Plains was met with opposition from ranchers which initially resulted in destruction of some fences; however, by 1880 this new way of enforcing property rights was reluctantly accepted by the common property interest groups. This example illustrates the role that transaction costs of defining, monitoring, and enforcing may play in controlling groundwater extraction. Without accepted and enforced measurement there can be no property rights or trade, so economizing on the implementation cost of these institutions is the key to their more widespread adoption.

Many economic analyses of water markets are predicated on either fully defining property rights for groundwater in quantitative terms, or a conceptually optimal set of Pigouvian taxes to internalize the user cost of groundwater extraction. The first of these institutions requires the adjudication and allocation of pumping capacity of a given basin. If the basin is to be used correctly as a capital asset whose scarcity value fluctuates with the climate, groundwater pumpers must be able to optimally draw down on this asset in times of drought without incurring excessive pumping costs or, more importantly, buffer stock costs of lost wells. This optimal drawdown function of groundwater is essential in most Mediterranean climates but requires flexible institutions to allow optimal asset use without imposing externalities on other users. Many advocates of groundwater management presume that without full adjudication of the basin and allocation of full property rights in quantitative terms such management cannot take place. This desire for full adjudication of the basins before management is a serious impediment to its implementation given the costs and incredible time delays of basin adjudication. In California, for example, the adjudication of the waters of the Klamath River basin was finally settled in 2013 after 38 years in which it was under legal dispute. The cost was extraordinary, but the delay of 38 years in resolving the case was even more remarkable.

There is, however, a very effective alternative institution for groundwater management and marketing that does not require any regulation of pumping or legal adjudication of the existing groundwater. It is essentially a "cap and trade" method for allocating groundwater, and has been successfully implemented in Orange County, California, for the past 40 years and probably in many other parts of the West. The essential idea is that all that is required is the measurement of the quantity of water pumped and a determination of the average hydrologic safe yield to the basin. Given these two pieces of information, groundwater users are allowed to pump whatever quantity they wish, subject to the condition that they pay the replacement costs of the groundwater if they exceed their share of the average safe yield of the basin. The average safe yield quantity is prorated across overlying land owners in accordance with the correlative groundwater rights to the aquifer. This is essentially a "cap and trade" solution to groundwater institutions given that there are no Pigouvian costs, fines, or charges levied.

In the case of Orange County there have, for most years, been adequate supplies of expensive but available recharge water. The cost of recharge water effectively sets the price on replacing any pumping above a landowner's share of the average safe yield of the basin. However, in many basins where outside sources of recharge are either unavailable, or it is too costly to inject into the groundwater aquifer, users who pump more than that prorated share will have to purchase additional shares of water from their neighbors at the opportunity cost. This institution is so simple and clear that the equity of the system is apparent to all users, and the transaction costs of its implementation are those of measuring groundwater extraction and enforcing the recharge cost. An advantage of the institution is that the cost of replacing excessive pumping is internalized in the system. Orange County has successfully managed three major droughts since the implementation of this groundwater institution, each time drawing down the aquifer to substitute for shortage of surface water supplies and increased evapotranspiration demand. In each instance the water management agency has successfully recharged the aquifer and restored groundwater levels after the drought was over. In addition, Orange County has a potential problem of seawater intrusion into the aquifer if a preventive mound of water is not maintained along the coastal strip. The maintenance of this mound is achieved by the differential pricing of groundwater and surface water. The cost difference encourages those pumpers along the coastal strip to substitute cheaper surface water for groundwater in this special area, and thus maintains the integrity of the aquifer over a long period of time.

Just as barbed wire revolutionized the transaction costs of enforcing property rights in the High Plains, cheap and consistent measurement of total water use, and by inference, groundwater use by remote sensing sources from satellites offers an alternative to the costly and politically difficult process of on the ground monitoring and measurement of wells, capacities, and water use.

Over the past 15 years a line of research has concentrated on using the energy information from the Landsat satellite to estimate net water consumption. The researchers have been able to use six bands of thermal energy readings at a 40 m² pixel basis to calculate the net evapotranspiration on a pixel-by-pixel basis for each two weekly pass of the satellite. This research has, in the U.S., advanced most rapidly in the state of Idaho, driven by researchers at the University of Idaho (www.idwr.idaho. gov/GeographicInfo/METRIC/et.htm). Such is the acceptance of the system in the State of Idaho that several cases of water curtailments in 2009 and disputes over water rights in 2006 have been settled on the basis of remotely-sensed data measurements. These decisions by the director of the Idaho Department of Water Resources have been adjudicated and verified up to the High Court of the state. In addition to the greater accuracy and frequency of remotely-sensed evapotranspiration data, the costs of measurement by remote sensing in Idaho have been shown to be one third of the cost of those achieved by on-site visits. The system which is called Metric is now being applied in several Western states, and promises over time to significantly reduce the cost of monitoring water use in agriculture by increasing its precision.

A test of a similar system called Surface Energy Balance Algorithm for Land (SEBAL) was performed for the California Delta by Medellín-Azuara and Howitt (2013). They explored the potential of remote sensing technology using the SEBAL method (Bastiaanssen et al. 2012) to provide an accurate estimate of consumptive use of water in crop production on five islands in the Sacramento San Joaquin Delta. The SEBAL-based data set on energy based measures of evapotranspiration for year 2007 were compared to evapotranspiration (ET) estimates using methods from the Irrigation Training and Research Center (ITRC) of California Polytechnic State University-San Luis Obispo. The ITRC measurements were based on a spreadsheet approach that extrapolated the results of field trials. A second method used to compare the SEBAL measurements was developed by the California Department of Water Resources; this approach is called the Simulation of Evapotranspiration of Applied Water (SIMETAW) model. SIMETAW uses historical climate data to determine a daily soil water balance for individual cropped fields within a watershed region having one set of reference evapotranspiration (ETo) estimates. The alternative approach uses historical climate data and batch files of soil and climate data to perform daily soil water balance for individual cropped fields, for 20 crops, and four land-use categories over the period of record by combinations of detailed analysis.

The results showed that the SEBAL-based remotely-sensed estimates of evapotranspiration were more precise and consistent than those based on field measurements and spreadsheet extrapolations. However, the main advantage of remote sensing is that it removes the need for costly and intrusive ground measurement, and has a consistent and legally justifiable basis for the measurement of total net water use and thus net groundwater use. If adopted on a statewide basis, similar remotely-sensed water measurement methods should significantly reduce the transaction costs of managing water and water markets and thus possibly play the role that barbed wire played in implementing landowning institutions on the High Plains.

Notes

1 In a dynamic setting, the marginal user cost would change over time in response to changes in various factors that affect water scarcity such as depth to groundwater and current precipitation.
2 Later in the chapter, we discuss the case of Orange County which uses this subsidization mechanism to maintain their coastal ground water mound (stock).

References

Aarnoudse, E., B. Bluemling, P. Wester and W. Qu (2012) "How WUAs Facilitate Direct Groundwater Regulation: A Case Study of Minqin County, China." IWMI Water Policy Research Highlight 23. Coloombo, Sri Lanka: International Water Management Institute

Anderson, T.L. and P.J. Hill (1975) "The Evolution of Property Rights: A Study of the American West." *Journal of Law & Economics*, 18: 163–179.

Ballard, C.L., J.B. Shoven and J. Whalley (1985) "General Equilibrium Computations of the Marginal Welfare Costs of Taxes in the United States", *The American Economic Review*, 75(1): 128–138.

Bastiaanssen, W.G.M., M.J.M. Cheema, W.W. Immerzeel, I.J. Miltenburg and H. Pelgrum (2012). "Surface Energy Balance and Actual Evapotranspiration of the Transboundary Indus Basin Estimated from Satellite Measurements and the ETLook Model". *Water Resources Research* 48: doi: 10.1029/2011WR010482.

Chakravorty, U. and C. Umetsu (2003) "Basinwide Water Management: A Spatial Model", *Journal of Environmental Economics and Management*, 45(1): 1–23.

Coase, R.H. (1960) "The Problem of Social Cost", *Journal of Law and Economics,* vol 3(1) 1–44.

Corlett, W. and D. Hague (1953). "Complementarity and the Excess Burden of Taxation." *Review of Economic Studies*, 21(1): 21–30.

Demsetz, H. (1967) "Toward a Theory of Property Rights." *American Economic Review*, 57(2): 347–359.

Dixit, A.K. and R.S. Pindyck (1994) *Investment under Uncertainty*. Princeton, NJ: Princeton University Press.

Gray, B.E. (1993) "The Modern Era in California Water Law." Working Paper, Hastings College of Law, San Francisco, CA: University of California.

Hanak, E. (2003) "Who Should Be Allowed to Sell Water in California? Third-Party Issues and the Water Market." Public Policy Institute of California: San Francisco, CA.

Hornbeck, R. (2010) "Barbed Wire: Property Rights and Agricultural Development..*The Quarterly Journal of Economics*, 125(2): 767–810.

Howitt. R.E. (1995) "Malleable Property Rights and Smooth-Pasting Conditions." *American Journal of Agricultural Economics,* 77(5): 1192–1198.

Medellín-Azuara, J. and R. Howitt (2013) "Comparing Consumptive Agricultural Water Use in the Sacramento-San Joaquin Delta: A Proof of Concept Using Remote Sensing." http://www.delta. ca.gov/res/docs/landscapes/UCD_ET_Report_9-2013.pdf

Metric Landsat www.idwr.idaho.gov/GeographicInfo/METRIC/et.htm

Mukherji, A. (2013) "Strategies for Managing India's Groundwater." http://wle.cgiar.org/ blogs/2013/02/12/strategies-for-managing-indias-groundwater/

Ostrom, E. (1990) *Governing the Commons: The Evolution of Institutions for Collective Action*. Cambridge: Cambridge University Press.

Roumasset, J. (1978) "The New Institutional Economics and Agricultural Organization." *Philippine Economic Journal*, 17(3): 331–348.

Roumasset, J. and N. Tarui (2010). "Governing the Resource: Scarcity-Induced Institutional Change." University of Hawaii, Department of Economics Working Paper No. 10-15.

Tsur, Y., T. Roe, R. Doukkali and A. Dinar (2004) *Pricing Irrigation Water: Principles and Cases from Developing Countries*. Washington, DC: Resources for the Future.

Tsur, Y. and T. Graham-Tomasi (1991) "The Buffer Value of Groundwater with Stochastic Surface Water Supplies". *Journal of Environmental Economics and Management*, 21(3): 201–224.

Williamson, O. (1985) *The Economic Institutions of Capitalism: Firms, Markets, Relational Contracting*. New York: The Free Press.

12

TOWARDS AN ECONOMICS OF IRRIGATION NETWORKS

Karl Jandoc, Ruben Juarez, and James A. Roumasset

Introduction

According to the 2012 UN World Water Development Report (WWDR), there will be a sharp growth in demand for water resources in the next two decades: food demand is predicted to increase by 50 percent and demand for energy, which includes hydropower, is predicted to increase by 60 percent (WWAP [2012]). Inasmuch as water is fungible, pressure from these various sectors is said to require an estimated additional 6 percent increase in water withdrawals (FAO [2012]). Faced with the prospect of an impending global water crisis, governments, water authorities, and academics are trying to find innovative ways of managing water resources.

The challenge, therefore, is to efficiently allocate water across time and space. Modeling water distribution along these dimensions is complicated by considerations such as conveyance losses, water recharge to the aquifer, and the like. Several studies that explored efficiency in a dynamic setting include Roumasset and Wada (2010) where they examine the optimal path of groundwater extraction in the face of extraction costs and groundwater recharge. If extraction costs were sufficiently convex, then the extraction path of groundwater will approach a steady state above the maximum sustainable yield level of groundwater stock. This line of research in efficiently allocating water across time has provided insights on the optimal ordering of extracting water resources such as groundwater, recycled water, and desalination (see Roumasset and Wada [2011]).

On the other hand, several papers dealt with water allocation across space. For instance, Chakravorty and Roumasset (1991) develop a spatial model of irrigation in the presence of conveyance losses. In their linear network model, they find that efficient water distribution implies that farms farther away from the source receive less water and that farmers at the tail end of the system may be required to pay a larger total charge for receiving less water.[1] In another paper, Chakravorty, Hochman and Zilberman (1995) model the choice of reducing these conveyance losses through investment, for instance, in improved canal lining or other conservation technology.

In this context, network economics may provide a useful approach along several dimensions. First, since most of the recent studies that deal with water allocation issues assume that the water network is already known (that is, the connection of sources to farms is given), network economics may extend conventional principles of irrigation design. For instance, the previously mentioned studies on spatial efficiency only deal with a single water source with farms arrayed in a linear network. In a model with several alternative water sources (e.g. multiple dams, lakes, and/or availability of deep and shallow tubewells), efficiency should require some knowledge of which of these sources should be developed and how they should or should not be linked together. Certain principles from network economics may provide a guide to achieve desirable characteristics of water distribution, such as efficiency, over a general network setting. Second, since engineering design principles of water networks do not fully take into account optimal allocation, allocation principles obtained from network economics may help to increase the net present value of these irrigation projects.

There is some literature on water networks, specifically on cost sharing. Aadland and Kolpin (1998) discussed the appeal of several cost sharing variants of serial and average cost sharing in 25 irrigation ditches located in Montana. They provided an axiomatic characterization of these mechanisms that underpinned the attractiveness of these cost-sharing mechanisms. Márkus, Pintér and Radványi (2011) discussed a solution concept in cost sharing called the Shapley value in a class of "irrigation games". There is also a strand in this literature that discusses ways to divide a cost of maintaining a waterway (say, a river) with several agents benefitting from that waterway (see Ambec and Sprumont [2002]; Ni and Wang [2007]; Ambec and Ehlers [2008] for the context of a linear network and Dong, Ni and Wang [2012] for the context of a general network). There is also a related problem in graph theory where we want to connect a set of agents to a single source at the minimal cost. Several algorithms have been proposed to find the minimal cost spanning tree of a connected network. The minimal cost spanning tree is the network that connects all the nodes together at the least cost (see Claus and Kleitman [1973]; Bird [1976]; Bergantiños and Vidal-Puga [2007]). A related problem in this regard is to efficiently select a location for a water source. There are also many rules or algorithms proposed in the literature that cover this class of problems (see, for instance, Vygen [2005]).

In this paper we aim to provide principles for efficient allocation of water over a network and to illustrate some algorithms to operationalize those principles. It takes into account transportation or conveyance losses in bringing water from the source to the farm. In several cases developed throughout the paper, we examine how marginal costs of holding water at the source[2] and conveyance efficiency affect the efficient distribution of water in the network.

We develop the spatial model of a water network in the next section. The third section examines the case of a linear network with one source. The fourth section extends to the case of a general network with identical costs for the different sources. The fifth section looks at the case when the sources have different cost functions. The sixth section looks at the related problem in locating a source among several potential locations that will achieve the highest social net benefit. The seventh section reconciles equity and efficiency objectives in water allocation. The final section concludes the paper and presents some open questions for further research.

The model

A water network consists of sources and users with links or connections between them.[3] Examples of sources of water are dams, deep tubewells, and other such structures. Users can be farms or households. Each of the k sources holds a stock of water, Z_k, that can be released to the different farms that are connected to them. There is some water lost in conveyance from source to the farm due to seepage, evaporation and percolation.[4] That is, when source k releases Q_i^k units of water, farm i only receives. $q_i^k = Q_i^k h(d_{ik})$ The function $h(d_{ik})$ denotes the percent of water from source k designated for farm i which actually reaches that farm. This function is decreasing in distance d_{ik} (here distance is interpreted to be the *distance of the shortest path* from source i to farm k), that is, more distant farms require greater amounts of water to be sent in order to receive the same amount of water as nearer farms.

The farms use the water to produce output, which is given by a production function $f(q_i)$, common to all farms (where $q_i = \sum_k q_i^k$), The price of output is given at P per unit. For every extra unit of water the farm receives, we can define the value of how much output is increased. This is given by the *value of marginal product for received water* for farm i from source k:

$$VMPS_i^k = Pf'(q_i)h(d_{ik}) \tag{12.1}$$

It is easy to show that the value of equation (12.1) decreases with distance d_{ik} from the source. The reason is that farms farther away from the source require larger release of water from the source to receive a unit of water. This makes the *VMPS* smaller for farms that are farther away.

The water authority wishes to maximize net social benefits, that is,

$$\max_{Q,Z} \left[\sum_{k=1}^{K} \sum_{i=1}^{n} \int_0^{Q_i^k} VMPS_i^k(\theta)d\theta \right] - \sum_k C_k(Z_k) \tag{12.2}$$

where $Z_k = \sum_i Q_i^k$ is the total amount of water released from source k and C_k is the cost function of source k.

In the fourth and fifth sections (exogenous and endogenous matching of sources and demands, respectively) deal with cases where the cost functions are different between sources. In the section with exogenous matching of sources and demands, we examine the case where marginal cost is constant, that is, the cost of storing an extra unit of water is not increasing. In the section with endogenous matching, we examine the case where marginal cost of storing an extra unit of water is increasing.

In what follows, we provide several interesting network structures to characterize water distribution efficiency.

A linear network with one source

This section extends Chakravorty and Roumasset (1991) who consider a single source called headworks (e.g. a dam) and farms that are linked along a line with fixed distances between each farm. Consider the network illustrated as in Figure 12.1 and assume at the outset that the demand at each node is known. In this case, the maximization problem in equation (12.2) simplifies to:

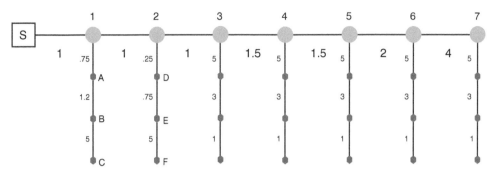

Figure 12.1 Linear network with one source and seven farms

$$\max_{Q,Z} \sum_{i=1}^{N} \int_{0}^{Q_i^1} VMPS_i^1(\theta)d\theta - C_1(Z) \tag{12.3}$$

subject to $Z = \sum_i Q_i^1$ where the i's denote nodes instead of farms.

Solving this problem requires constructing the Lagrangian and obtaining the corresponding first order conditions with respect to Q_i^1 and Z. There are two important results that will characterize efficient water distribution in this network. First, the first order conditions imply that

$$VMPS_i^1(q_i^*) = VMPS_j^1(q_j^*) \tag{12.4}$$

for all farms i,j that receive positive amounts of water.

This equation tells us that for the allocation to be optimal, the value of marginal product of source water should be equal across nodes. If this were not the case, then it would be possible to increase aggregate payoff by reallocating source water to nodes with higher marginal productivity. Note from equation (12.1) that the only way in which *VMPS* can be equalized for farms in the network is for the value $Pf'(q_i)$ to increase for nodes that are farther away from the source since the function h is decreasing with distance. This implies that nodes that are farther away should receive less water.

Second, efficient allocation of water should equate the value of marginal product of source water to the marginal cost of producing the water. That is,

$$VMPS_i^1(q_i^*) = C_1'(Z) \tag{12.5}$$

Farms receiving zero water must satisfy $VMPS_i^1(q_i^*) \leq C_1'(Z)$. That is, the benefit of sending marginal unit of water to farm i is less than the cost of producing the same unit. The first order conditions also imply that $VMPS_i^1(q_i^*) > VMPS_j^1(q_j^*)$ if farm i receives a positive amount of water and farm j does not.

If there are $n \leq N$ farms receiving positive amounts of water, this gives us n equations and n unknowns. This result says that the water authority should equate the extra benefit of releasing a unit of water against the cost of "producing" it. If the former were higher than the latter, then it would make sense for the water authority to increase the amount of water to be stored and released, since the cost of doing so is less than the benefits.

Equations (12.4) and (12.5) characterize the *equimarginal principle* of efficiency. Thus, for this case an allocation that solves the water authority's problem in (12.3) is efficient if and only if it satisfies equations (12.4) and (12.5), where the *VMPS* are equal for farms being served and the *VMPS* should be equal to the marginal cost.

Table 12.1 Summary of results from the linear network in Figure 12.1

Farm	h	Optimal q	Optimal Q	VMPS	Benefit	Consumer surplus
1	0.98	0.82	0.83	9.00	7.85	0.33
2	0.96	0.63	0.66	9.00	6.13	0.20
3	0.94	0.44	0.47	9.00	4.34	0.10
4	0.91	0.15	0.17	9.00	1.51	0.01
5	0.89	0.00	0.00	8.87	0.00	–
6	0.85	0.00	0.00	8.52	0.00	–
7	0.79	0.00	0.00	7.87	0.00	–
Total		2.04	2.13		19.83	0.64

Consider a numerical illustration where marginal cost is constant and the distances between nodes/farms are given in Figure 12.1.

Following Chakravorty and Roumasset (1991), suppose that $h(d_{i1}) = e^{-.02d_{i1}}$ where $e \approx 2.718$ and d_{i1} is the distance of the farm from the source. The numerical value of h is given in the second column in Table 12.1. The VMPS function is given by $VMPS_i^1 = (10 - q_i)h(d_{i1})$ and the marginal cost function is given by $C_1'(Z) = 9$. Table 12.1 gives the values of the functions for each farm and summarizes the results.

In this example, only farms 1 to 4 receive an allocation of water from the source while farms 5, 6 and 7 receive no water. Here the farms that received water have their VMPS equal to each other (which is also equal to the constant marginal cost of 9) and the VMPS of the farms that receive zero water is less than the VMPS of the farms that receive a positive amount of water. Moreover, farms that are farther away from the source receive less water.

An important thing to note is that if we assume demands to be nonlinear and cost functions to have some flat parts in relevant ranges, then it may be possible for the sent water (Q) to be increasing with distance but the received water (q) will remain to be decreasing with farms farther from the source (Chakravorty and Roumasset [1991]).

Now suppose that MC is non-constant, say increasing in the relevant range. In this case, one cannot simply equate the VMPS function for each farm with the constant MC and solve independently for each Q_i. Rather we have n equations and n unknowns that must be solved simultaneously. One solution procedure is to equate the aggregate demand for water as a function of price with the increasing MC function. This yields the optimal MC and aggregate quantity. Then setting each VMPS to the optimal MC yields the optimal allocation of sent water among farms.

So far we have assumed that we know the demand curves at each node. In order to determine those demands for the case where there are farmers along the secondary canals as illustrated in Figure 12.1, we know to simply apply the procedure for determining the aggregate demand at the source, as just described, except this time we treat each node as a source for the farms along the secondary canals.

As an illustration, suppose that the conveyance efficiency along the main canal is the same as above and the conveyance efficiency for the secondary canals is given by the formula $h(d_{i0}) = e^{-.02d_{j0}} e^{-.043d_{ij}}$.[5] Thus, the percentage of water that reaches farm i from the source is just the water that remains after it has reached node j (when travelling through the main canal) net of water lost as it travels from node j to farm i in the secondary canal. Thus for node 1 this is equal to h_1 =−[0.95, 0.90. 0.73] for farms A, B and C, respectively, and h_2 = [0.95. 0.92. 0.74] for farms D, E and F (ignore nodes 3,4,5,6,7 since in this example farms in these nodes will not receive water). Assume that the *VMPS* is the same as above and the marginal cost function is $C'(Z) = 9$. Applying the same procedure as described above, the efficient allocation implies that the first two farms from node 1 receive positive amount of water (q_A = 0.52 and q_B = 0.02, respectively) and also the first two farms in node 2 (q_D = 0.3 and q_E = 0.03). Notice that even though farm B is closer to source 1 in terms of distance than farm E, the severity of losses in the secondary canals implies that more water is allocated to farm E since the distance that the water travels through the secondary canals is shorter.

Multiple-source networks: exogenous matching of sources and demands

In this section, we extend the case of the linear network into more general networks. Suppose that there is more than one source and farms are connected to different sources via a water network. Figure 12.2 gives a simple example of a network with two sources and three farms. This network gives all the potential paths for which water can flow from the sources to the farms. It is useful then to distinguish between this "potential water network" and the "economic water network" which contains only the optimal links where water flows. An economic water network may be the whole potential network or a distinct subnetwork. As discussed in the section with endogenous matching of sources and demands, all the sources in an integrated economic network have the same optimal marginal costs and all the nodes have the same *VMPS*.

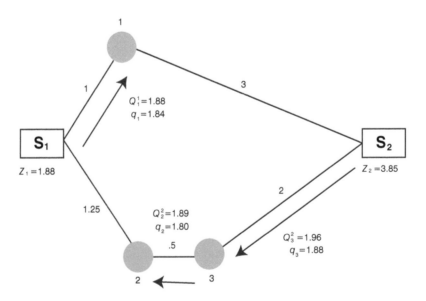

Figure 12.2 A network with two sources and three farms and two economically integrated subnetworks

If an economic network contains one or more subnetworks, then the marginal costs within each subnetwork equal the *VMPS* but the marginal costs across subnetworks are unequal. As an illustration, the arrows in Figure 12.2 represent optimal water flows. We will show that source 1 serving node 1 and source 2 serving nodes 3 and 2 are distinct economic subnetworks.

Consider the case where water sources have different cost functions. For instance, farms may have a choice between sourcing water from a dam or from deep tubewells located in different parts of the aquifer and those sources will have different cost functions.[6] We assume for now that the marginal cost of producing water is constant for each source, that is, $C_k'(Z_k) = \alpha_k$. This means that the cost of producing water will be constant for each extra unit.

The water authority in this setting must determine the optimal production at each source and the allocation to the various farms (including zero from some sources to some farms and zero to some farms). We put forth two principles for determining the optimal allocation. The *least cost principle* states farms must obtain water from the (marginally) cheaper source. The cost of delivered water is the marginal cost at the source plus the unit transport cost. For the case of gravity irrigation where conveyance structures are already in place, this can be measured as the cost of producing one unit of *received* water for each farm, that is, $\dfrac{\alpha_k}{h(d_{ik})}$. Since marginal costs are constant for each source, the least cost source is exogenously identified by this criterion regardless of the quantities produced by each of the sources. Since there can be different cost functions across sources, the cheapest source may or may not be the one that is closest to the farm. For a set of nodes served by the same source, optimal allocation must also satisfy the *equimarginal principle*, that positive quantities received from the same source must have the same value of marginal products (as reckoned at the source).

With different costs for multiple sources, the water authority's problem now becomes:

$$\max_{Q,z} \sum_k \left[\sum_{i=1}^{n} \int_0^{Q_i^k} VMPS_i^k(\theta)d\theta - C_k(Z_k) \right] \tag{12.6}$$

subject to $Z_k = \sum_i Q_i^k$ and the non-negativity constraints $Q_i^k \geq 0$

The first order conditions of equation (12.6) require that, for each source:

$$VMPS_i^k(q_i^\star) = VMPS_j^k(q_j^\star) \tag{12.7}$$

for all farms i,j receiving positive water from source k.[7]

This means that for all agents receiving water from a particular source k, their *VMPS* should be equal. In other words, in the economic subnetwork generated from the potential network, the *VMPS* of the farms receiving water from the same source should be equal. Again, if this were not the case, then it is possible to reallocate water from farms with lower *VMPS* to the ones in the same economics subnetwork with higher *VMPS* and by doing so increase the total net benefits of the water network.

The first order conditions also yield:

$$VMPS_i^k(q_i^\star) = C_k'(Z_k) = \alpha_k \text{ for all } i \text{ farms served by source } k \tag{12.8}$$

Equation (12.8) just says that for a water allocation to be efficient, the *VMPS* of agents obtaining positive amounts of water from source k should equal the marginal cost of producing Z_k units of water.

Table 12.2a Least cost sources: low conveyance costs

From source 1:

Farm	h	Optimal q	Optimal Q	VMPS	Marginal cost of one unit of received water	Benefit
1	0.980	1.84	1.88	8.0	8.162	16.69
2	0.975	–	–	–	8.203	–
3	0.966	–	–	–	8.285	–
Total		1.84	1.88			16.69

From source 2:

Farm	h	Optimal q	Optimal Q	VMPS	Marginal cost of one unit of received water	Benefit
1	0.942	–	–	–	8.282	–
2	0.951	1.80	1.89	7.8	8.200	16.38
3	0.961	1.88	1.96	7.8	8.118	17.05
Total		3.68	3.85			33.43

Note: The function h is given by the formula $h(d_{ik}) = e^{-.02d_{ik}}$

Thus, for the allocation of water to be efficient, equations (12.7) and (12.8) should be satisfied for farms that receive water from source in a given economic subnetwork.

Let us take the example in Figure 12.2. Suppose that the first source S_1 costs more to produce an extra unit of water than the second source S_2. For instance, let the cost functions of S_1 and S_2 be $C_1(Z_1) = 8Z_1$ and $C_2(Z_2) = 7.8Z_2$, respectively. Assume again that $VMPS_i^k = (10 - q_i)h(d_{ik})$. Notice that for farms 2 and 3 the closest source in terms of distance is S_1 (1.25 versus 2.5 and 1.75 versus 2, respectively). However, the advantage of this proximity is washed out by the fact that producing water in this source is more expensive. This is shown in column 6 of Table 12.1, which computes the marginal cost of one unit of received water. For farms 2 and 3, the marginal cost is lower in source 2 than source 1, while for farm 1 the marginal cost is lower in source 1. Therefore, the *least cost principle* suggests that when we consider the different marginal cost of producing and transporting water, farm 1 will be connected to S_1 while farms 2 and 3 will be connected to S_2. The numbers in Figure 12.2, as well as Table 12.2a, summarize the optimal values of q, Q and Z.

From the arrows in Figure 12.2, we see that the economic water network S_1 sends water to farm 1 while S_2 supplies water to farms 2 and 3. This can also be seen in the positive entries of q in Table 12.2a for farm 1 from source 1 and farms 2 and 3 from source 2. The total net benefit is 5.08, which is higher than the net benefit that would be obtained if we source water exclusively from S_1 (4.78) or from S_2 (4.87). Notice that the *VMPS* for each of the farms is equal to the marginal cost of the source that they are connected to. Furthermore, the farms receive water from the source that provides the least cost, as seen in the sixth column, for instance, farm 2 is optimally served by S_2 since the marginal cost of one unit of received water from S_2 is lower (8.200 versus 8.203).

Table 12.2b Least cost sources: high conveyance costs

From source 1:

Farm	h	Optimal q	Optimal Q	VMPS	Marginal cost of one unit of received water	Benefit
1	0.961	1.67	1.74	8.0	8.326	15.53
2	0.951	1.59	1.67	8.0	8.410	14.63
3	0.932	–	–	–	8.580	–
Total		3.26	3.41			29.97

From source 2:

Farm	h	Optimal q	Optimal Q	VMPS	Marginal cost of one unit of received water	Benefit
1	0.887	–	–	–	8.794	–
2	0.905	–	–	–	8.620	–
3	0.923	1.55	1.68	7.8	8.450	14.30
Total		1.55	1.68			14.30
Total net benefit						3.87

To illustrate how the least cost source depends on the conveyance losses as well as production costs, consider the case wherein the distances in the links in Figure 12.2 were to double (so, for instance, the distance from source 2 to farm 2 is now 5 instead of 2.5). Table 12.2b summarizes the results with the new distances. Here we can see that farm 2 is now served by S_1 because the marginal cost of one unit of received water is lower (8.41 versus 8.62).

Endogenous matching of sources and demands

Now consider the case of increasing marginal cost for each of the sources. This can be justified, for example, if dam size is already beyond the point where scale economies have been exhausted. That is, building a dam beyond a certain height may entail an increase in average costs per expected quantity of water stored. In this case, the least cost source depends on the quantities produced by each of the sources and cannot be exogenously identified. Optimal matching must now be determined simultaneously with quantities produced and allocation among farms.

The water authority's problem in this section is similar to the last section. However, in that section where marginal cost is constant, farms will typically always receive water from a unique source unless we are in the trivial case where the cost advantage is exactly offset by the distance of the farm to the source.[8] In this section it will be possible for farms to receive water from multiple sources. Intuitively, the water authority will provide water from a source to a farm while it is still cheap to do so (that is, if marginal costs does not rise too much from the additional unit of water), and probably provide water from another source to that same farm if it is profitable to do so.

For the case where a farm receives water from multiple sources, the first order conditions of the water authority's problem imply that this equation must hold:

$$\frac{C'_k(Z_k)}{h(d_{ik})} = \frac{C'_m(Z_m)}{h(d_{im})} \tag{12.9}$$

for all sources k and m that farm i receives positive amounts of water. The condition tells us that farms receiving water from multiple sources must have the marginal costs of each source (weighted by the conveyance efficiency h of that source) equal to each other.

For instance, consider the same network in Figure 12.2 and suppose that the marginal costs are given by $C'_1(Z_1) = 2 + 5Z_1$ and $C'_2(Z_2) = 5 + 2Z_2$ for sources S_1 and S_2, respectively. Table 12.3 summarizes the results if we allocate water to the network under three different scenarios: Scenario 1 is the resulting allocation when farm 2 receives water from both S_1 and S_2; Scenario 2 is the resulting allocation when farm 2 receives water exclusively from S_1; and Scenario 3 is where farm 2 receives water exclusively from S_2.

Table 12.3 Summary of results of the network in Figure 12.2 and with increasing cost functions

Scenario 1: Farm 2 gets water from both sources

Farm	h from source 1	h from source 2	Optimal q (source 1)	Optimal Q (source 1)	Optimal q (source 2)	Optimal Q (source 2)	Benefit
1	0.980	0.942	1.017	1.037	–	–	9.65
2	0.975	0.951	0.316	0.324	0.66	0.69	9.24
3	0.966	0.961	–	–	1.06	1.10	10.05
Total			1.33	1.36	1.72	1.79	28.94
Total net benefit							9.40

Scenario 2: Farm 2 gets water exclusively from Source 1

Farm	h from source 1	h from source 2	Optimal q (source 1)	Optimal Q (source 1)	Optimal q (source 2)	Optimal Q (source 2)	Benefit
1	0.980	0.942	0.717	0.732	–	–	6.92
2	0.975	0.951	0.671	0.688	–	–	6.48
3	0.966	0.961	–	–	1.51	1.58	14.00
Total			1.39	1.42	1.51	1.58	27.40
Total net benefit							9.15

Scenario 3: Farm 2 gets water exclusively from Source 2

Farm	h from source 1	h from source 2	Optimal q (source 1)	Optimal Q (source 1)	Optimal q (source 2)	Optimal Q (source 2)	Benefit
1	0.980	0.942	1.283	1.309	–	–	12.01
2	0.975	0.951	–	–	0.84	0.88	8.06
3	0.966	0.961	–	–	0.93	0.97	8.89
Total			1.28	1.31	1.77	1.85	28.96
Total net benefit							9.34

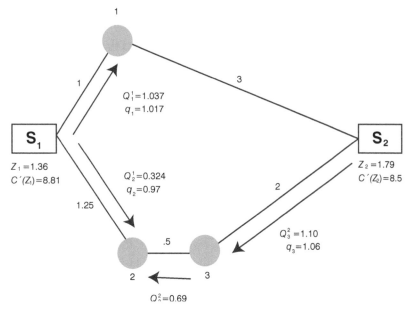

Figure 12.3 A network with two sources and three farms

Note that in Table 12.3, the allocation in Scenario 1 yields the highest net benefits for the water authority. Indeed, the water authority will balance sourcing water from one source or the other depending on the marginal costs of doing so. We can check that the condition in equation (12.9) is satisfied by plugging in the values of Z_1 and Z_2 (1.36 and 1.79, respectively) into the weighted marginal costs.

Figure 12.3 visually shows the information contained in Table 12.3, as well as the flow of water from sources to farms. Thus, in the economic water network, S_1 provides water to farm 1 and a small proportion to farm 2 while S_2 provides water to farm 3 and on to farm 2.

Two sources are in the same network if they optimally serve a common demand. Two demands are in the same network if they are optimally served by sources that are in the same network.

While in the constant marginal cost case it is straightforward to determine the least cost source from which a farm will obtain water, it is more complicated in the case of increasing marginal costs. One straightforward algorithm to solve this is to articulate all possible economic subnetworks of a given potential water network and to compute the allocation that will yield the maximum net benefit for each subnetwork. Of course as the number of farms and sources increases, finding the optimal design belongs to the class of "NP-hard" problems.[9] That is, computations using this algorithm become practically infeasible as the number of sources and farms increases.[10] The development of algorithms to determine the least cost source in larger networks is a matter of ongoing research in network economics.

Endogenous location of sources

Consider a water authority that faces substantial fixed costs that limit its ability to build multiple water sources. Thus, a related problem of efficient water distribution is where to place a source among potential locations for these sources. This section focuses the discussion on the problem of locating one source out of several locations and that the cost of building the infrastructure is the same across all locations. For example, suppose that only one of the two potential sources illustrated in Figure 12.2 can be developed, that is, either S_1 or S_2.[11]

In principle, we want to place our source in the location that maximizes the social net benefit. In doing so there are two things to consider. First, we have to consider the distance of each farm from the source because the farther away farms are from it the less water they receive, which in turn lowers the payoff to the farms. Second, we have to consider the cost of producing water. If marginal costs are sufficiently low such that it is optimal to send positive amounts of water to all farms, then we only have to consider the efficiency of transporting the water to these farms. However, when marginal costs are high, not all farms will be served and the water authority has to consider only the returns to the farms that are relatively nearer to the source. Indeed if the marginal cost of producing water is close to zero, the problem of locating one source in the water network is equivalent to the problem of finding the source with the highest efficiency, that is, the location which has the highest sum of the function h. The intuition is that since all farms will be served because of the (almost) zero marginal cost, the source that yields the highest sum of h will have the highest percentage of their water released reaching the farms. In the example in Figure 12.2, if marginal cost were zero then the location of the source will be S_1.

However, if marginal cost were high or rising, then some agents in the water network may not be served. The water authority therefore has to give priority to the payoff to the agents who are closer to the head. This will give the highest social net benefit among all possible locations for the source. As with the previous section on increasing marginal costs, a straightforward algorithm to determine the most efficient source location is to list all possible economic subnetworks of a given potential location and to compute the allocation that will yield the maximum net benefit for each of the subnetworks. But as the number of potential sources and number of farms increases, this "brute force" algorithm may be difficult to implement.

Cost sharing mechanisms and equity

Cost sharing mechanisms

Finally it is interesting to look at cases that examine desirable properties beyond efficiency. We have so far characterized allocations within the water network when we charge marginal cost. However, there may be several problems with marginal cost pricing. In this subsection, we discuss two potential equity problems: First, there may be head–tail inequities inasmuch as consumer surpluses for tail farmers are relatively small. Second, marginal cost pricing may be unnecessarily onerous if it generates a revenue surplus. Johansson et al. (2002) document many alternative mechanisms to marginal cost pricing that address other important dimensions of allocation, especially equity concerns. Here we consider fixed fee, equal pricing, and lump sum transfers.

In many countries, farmers are charged a fixed fee, which is below marginal costs. This mechanism suffers from the classic "head–tail" problem where farmers that are closer to the source (the "head" farmers) overdraw water and farmers that are far from the source (the "tail") under-draw water, relative to the efficient solution.[12] This type of inefficiency has a close parallel to the average cost sharing mechanism in the mechanism design literature and has been quantified in several network problems.[13]

Another candidate mechanism is *equal pricing* (EP), which maximizes the surplus of the agents subject to the condition that agents pay the same price per unit of received water and the water authority recovers some proportion of total cost. This mechanism generates a price for each farmer that divides the total cost of producing the water with the proportion of his received water to the total amount of received water by all of the farmers. That is, each farmer's total water bill is $\dfrac{q_i}{\sum_i q_i} C(Z)$. The main advantage of EP is that at the equilibrium the agents with similar utilities will demand the same amount of water in their farms and will be charged the same price. In particular, this implies that agents with similar utilities get the same net benefit (a key property in the fairness literature named *equal treatment of equals*). Using the same parameters as in Table 12.1, Table 12.4 shows the allocation of received water and how price per unit of received water and consumer surplus under the EP mechanism are equalized.

The main downside of EP is that it is inefficient relative to the case where we charge marginal cost. In Table 12.4, we can see that agents will receive the same amount of water at their farms. Therefore, agents who are farther away from the source are sent a larger amount of water. This is contrary to the efficient allocation, where agents closer to the source are sent larger amounts of water than agents farther away to the source. The allocation given by EP generates a loss in the consumer surplus equal to .18, which equals a loss of 28 percent of the net social benefit below the efficient solution.

One way of pursuing equity without sacrificing efficiency is the mechanism of marginal cost pricing with lump sum transfers. For example the head farmers in

Table 12.4 A mechanism that charges the served farmers the same price per unit of received water

Farm	h	q	Q	Price	Price per unit of received water	Benefit	Consumer surplus
1	0.98	0.97	0.99	9.03	9.27	9.26	0.23
2	0.96	0.97	1.01	9.03	9.27	9.26	0.23
3	0.94	0.00	0.00	–	–	–	–
4	0.91	0.00	0.00	–	–	–	–
5	0.89	0.00	0.00	–	–	–	–
6	0.85	0.00	0.00	–	–	–	–
7	0.79	0.00	0.00	–	–	–	–
Total		1.94	2.00			18.52	0.46

Table 12.1 could be charged lump sum taxes that are distributed to the tail farmers in accordance with a specific equity criterion. Since transfers are lump sum, they will not affect the optimal quantities of water demanded.

Trying to redistribute towards farmers with lower net benefits is problematic, however. Horizontal equity does not demand redistribution since, from an *ex-ante* point of view, farmers are typically not equal. Head farmers have the advantage of being close to the source, which confers a natural advantage over tail farmers. And even if the government is pursuing vertical equity, there is no necessary reason for building redistribution into each and every project. The government is likely to have more efficient mechanisms for addressing basic needs and redistributing income than burdening individual projects with those requirements (Kaplow and Shavell, 2002). This does not mean, however, that equity and efficiency are incompatible in irrigation pricing.

Another drawback of marginal cost pricing is that it may leave a revenue surplus above the appropriate level of cost recovery. First, marginal cost may be rising in the relevant range such that marginal cost pricing leaves a surplus above full costs. Second, under the principle of benefit taxation, direct-beneficiary farmers should only be responsible for the proportion of costs equal to the ratio of direct to total benefits. Indirect beneficiaries may include commercial interests who benefit from increased agricultural production and consumers who benefit from lower prices (Roumasset 1987, 1989). One mechanism for limiting charges to direct beneficiaries while maintaining the efficiency property of marginal cost pricing is the lump sum transfer approach described above. Here we consider a particular form of lump sum transfers, a block pricing scheme (BP mechanism) that leaves farmers with the same ratio of charges to benefits. There are several steps involved in BP: First, the water authority must identify the revenue target, that is, the percentage of total cost that should be recovered from direct beneficiaries. Second, the revenue requirement must be allocated among individual farmers. The BP mechanism charges every served farmer a first tier price at which they will be entitled to draw a certain amount of (sent) water, Q_i^e. For quantities exceeding the first block, prices are set equal to the constant marginal cost. The mechanism sets the quantity of the first block such that the ratio of costs to benefits is equal for each farmer and the revenue collected from direct beneficiaries is equal to the target revenue. Figure 12.4 illustrates. In Figure 12.4, the ith farmer pays the light-gray area in exchange for total benefits equal to the sum of the light-gray shaded and dark-gray shaded areas.

In order to compute for the entitlements and the charges under the BP mechanism, the following steps can be followed. First, we determine the indirect benefits at the optimal solution, IB^*, through a cost–benefit exercise. Given IB^* and the total benefits at the optimal solution, DB^*, we can compute the revenue target by using the formula

$$RT^* = \frac{DB^*}{DB^* + IB^*} \times MC \times Z^*$$, where MC is the marginal cost and Z^* is the total optimal

amount of sent water. The proportion of total benefits that will satisfy the revenue

requirement is therefore $\pi^* = \dfrac{MC \times Z^*}{DB^* + IB^*}$. To compute each farmer's entitlement, we

then solve for Q_i^e in: $\dfrac{MC(Q_i^* - Q_i^e)}{B_i} = \pi^*$ where Q_i^e is the entitlement for farmer i, Q_i^*

is the optimal amount of sent water to farm i, and B_i is the benefit for farm i (area under the demand curve to the left of Q_i^*).

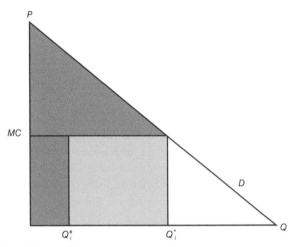

Figure 12.4 The block pricing mechanism

Figure 12.5 provides an illustration of the BP mechanism based on the parameters in Table 12.1. Suppose that indirect benefit is $IB^* = 5.52$. Given that $DB^* = 19.82$, $MC = 9$ and $Z^* = 2.13$, the revenue target is $RT^* = \dfrac{19.82}{19.82 + 5.52}(9 \times 21.3) = 15$. The proportion of total direct benefits that satisfies the revenue target given DB^*, IB^*, MC and Z^* is then $\pi^* = 0.76$, which is given in column 6 of Table 12.5. Given π^*, MC, Q_i^* in column 4 and the direct benefits B_i in column 5 of Table 12.5, we are able to obtain the entitlements and the corresponding charges for each of the served farmers by the formula $\dfrac{MC(Q_i^* - Q_i^e)}{B_i} = \pi^*$. The list of entitlements to each farmer is given in column 7 while the charges are given in column 8. Accordingly, the individualized entitlements achieve proportional benefit taxation for all farmers. In practice, the entitlements might be set according to farmer category to economize on administrative costs. Higher indirect benefits, lower MC, and higher water demands would all lead to higher farmer surpluses. Since the farmer subsidies implicit in the entitlements are inframarginal, they are lump sum transfers and do not distort the quantity decisions.

Table 12.5 Block pricing for a revenue target of 15

Farm	h	Optimal q (q*)	Optimal Q (Q*)	Benefit (B₍)	Ratio of charges to benefit (n*)	Entitlements (Q₍)	Charges [MC*(Q*-Q₍)]
1	0.98	0.82	0.83	7.85	0.76	0.17	5.94
2	0.96	0.63	0.66	6.13	0.76	0.14	4.64
3	0.94	0.44	0.47	4.34	0.76	0.11	3.28
4	0.91	0.15	0.17	1.51	0.76	0.04	1.14
5	0.89	0.00	0.00	0.00	–	–	–
6	0.85	0.00	0.00	0.00	–	–	–
7	0.79	0.00	0.00	0.00	–	–	–
Total		2.04	2.13	19.83			

Note: See text on the computation of columns 6 to 8. Marginal cost is equal to 9.

Compensation mechanisms

In the above setting where the source is known, farmers are inherently unequal by virtue of their locations relative to the source. In contrast, consider an endogenous facility-location problem wherein agents are uniformly distributed in a linear network and there are two potential equal-cost facilities at each extreme, only one of which can be developed (for example due to fixed costs). Picking the facility at one extreme will benefit agents who are closer to that facility and harm the agents who are closer to the other extreme of the network.[14] In this case, the BP mechanism would violate equal treatment of equals and some additional compensation may be appropriate.

To see this, suppose that we have three farms and two potential sources. Farm 1 is closest to source 1 and farm 3 is closest to source 2 while farm 2 is in the middle of farms 1 and 3. The distance between each farm is 1 kilometer, and the distance from source 1 to farm 1 and from source 3 to farm 3 is likewise 1 kilometer. Assume that $VMPS_i = (10 - q_i)h_{ik}$ for all farms and each source has the same marginal cost, that is, $C'_k(Z_k) = 9.5$ for both potential sources 1 and 2. Assume the h function is the same as that in the section dealing with a linear network. For the sake of transparency, assume also that there are no indirect benefits such that inframarginal block pricing is not called for. Table 12.6 shows the optimal amounts if source 1 and source 2 is chosen.

From the point of view of the water authority in this example, it does not matter whether to build source 1 or source 2 since it yields the same net social benefit. However, from an equity point of view, the choice whether to build source 1 or source 2 has implications on the welfare of farmer 1 and farmer 3. Indeed, if source 1 is built then farmer 3 gets no allocation of water and hence no consumer surplus. But if source 2 is built then farm 3 now gets a positive allocation and farm 1 receives nothing.

A simple compensation mechanism is to implement lump sum transfers that leave every agent with the same benefits regardless of which project is chosen. For instance,

Table 12.6 Optimum values in the symmetric example if source 1 or source 2 is chosen

From source 1:

Farm	h	Optimal q	Optimal Q	VMPS	Benefit	Consumer surplus
1	0.98	0.31	0.31	9.50	3.03	0.05
2	0.96	0.11	0.12	9.50	1.12	0.01
3	0.94	–	–	9.42	–	–
Total		0.42	0.43		4.15	0.06

From source 2:

Farm	h	Optimal q	Optimal Q	VMPS	Benefit	Consumer surplus
1	0.94	–	–	9.42	–	–
2	0.96	0.11	0.12	9.50	1.12	0.01
3	0.98	0.31	0.31	9.50	3.03	0.05
Total		0.42	0.43		4.15	0.06

if there were an equal chance of either source to be built in Table 12.6, then the lump sum tax that would take away .025 units of consumer surplus from farmer 1 if source 1 is built would be transferred to farmer 3. This simple scheme thus leaves all farmers indifferent between the development of source 1 or source 2. (Farmer 2 is already indifferent without transfer.) In this example, this compensation mechanism satisfies equal treatment of equals. This same principle can be generalized to non-symmetric distances, general network structures and different probabilities of projects being implemented by designing transfers that leave farmers indifferent between sources developed.

Concluding remarks and some open questions

This chapter has demonstrated how network economics can be a useful approach generalizing principles of efficient and equitable irrigation development.

For the case of given headworks and canals, we first determine optimal matching between sources and farmers by the principle of least cost in terms of water received. If farms optimally get water from the same source according to the least cost principle, then they are in the same economic subnetwork. The equimarginal principle then says that water should be allocated among farmers in the same economic subnetwork such that the values of marginal products reckoned at the source (*VMPS*) are equal. In case only one source can be developed, as in the section dealing with the endogenous location of sources, one applies the equimarginal principle to each potential source and chooses the source with the highest resultant welfare. Where volumetric pricing is feasible and farm productivities can be estimated, there is no need to compromise efficiency for equity inasmuch as lump sum transfer can achieve the equity objective in question without violating marginal cost pricing. Block pricing may be a convenient way to implement said transfers. We believe that future research is warranted to generalize these principles to the case of many sources and farmers with fully endogenous development of sources and canals.

The principles derived in this paper can be extended in several ways. First, a water basin may contain possible sources that could potentially be developed, for example, a lake, different points in a river, deep tubewells, shallow tubewells and rainfed reservoirs. Which of these should be developed cannot be determined by cost–benefit analysis of each individual source. Rather, the gain from adding a particular source depends on what other sources and transmission networks might be developed. Network economics can also be extended across inter-basin water transfers.[15] This involves more than just a comparison of two networks with and without a connection. As with sources within a basin, connecting sources has implications for the transmission network and water allocation. A second extension involves dynamics, especially involving conjunctive use. As detailed in another chapter (Jandoc et al. Chapter 11 in this Handbook), which source is used to deliver water to a particular farm changes over time. For example, as groundwater scarcity increases, then the economic network may also change. This has implications, in turn, for both allocation and optimal connectivity.

Applying and extending these principles may be facilitated by network economists working together with engineers and computer scientists, starting with a small number of endogenous sources and possible connections to farmers. Together, we believe they can advance a new generation of irrigation engineering.

Appendix A

In this appendix, we derive the first order conditions of the problem given in equation (12.6). Recall that the problem is:

$$\max_{Q,z}\left[\sum_k \sum_{i=1}^{n} \int_0^{Q_i^k} VMPS_i^k(\theta)d\theta - C_k(Z_k)\right]$$

subject to $Z_k = \sum_i Q_i^k$ and the non-negativity constraints $Q_i^k \geq 0$

The Lagrangian is given by:

$$L = \sum_k\left[\sum_{i=1}^{n}\int_0^{Q_i^k} VMPS_i^k(\theta)d\theta - C_k(Z_k)\right] - \lambda_k(Z_k - \sum_i Q_i^k)$$

Taking the partial derivatives of the Lagrangian over the variables of interest yields:

$$\frac{\partial L}{\partial Q_i^k} = VMPS_i^k - \lambda_k \leq 0, \quad Q_i^k \geq 0, \quad Q_i^k(VMPS_i^k - \lambda_k) = 0 \qquad (12.A1)$$

$$\frac{\partial L}{\partial Z_k} = C_k'(Z_k) + \lambda_k = 0 \qquad (12.A2)$$

If farm i is sent a positive amount of water from source k, that is, if $Q_i^k > 0$ then equation (12.A1) implies that:

$$VMPS_i^k = \lambda_k \qquad (12.A3)$$

Together with equation (12.A2), we have

$$VMPS_i^k = C_k'(Z_k) \qquad (12.A4)$$

which means that the *VMPS* of those who received positive amounts of water from source k should equal the marginal cost of source k.

If a farm receives water from multiple sources, then it must be true that equation (12.A4) must hold *for each of those sources*. By definition of *VMPS*, we can rewrite equation (12.A4) as

$$Pf'(q_i)h(d_{ik}) = C_k'(Z_k) \qquad (12.A5)$$

Since $Pf'(q_i)$ is the same for farm i, the first order conditions imply that for any two sources k and m that farm i obtains positive amounts water from, the equation

$$\frac{C_k'(Z_k)}{h(d_{ik})} = \frac{C_m'(Z_m)}{h(d_{im})} \qquad (12.A6)$$

must hold.

Notes

1 Charging farmers the wholesale marginal cost reckoned at the headworks means that tail farmers pay more for each unit of water received. Since optimality typically requires more water sent to tail farmers even though less water is received, this implies that tail farmers pay more for less. That is, full marginal cost pricing may require tail farmers to pay a greater total fee even though they receive less water than head farmers.
2 This may include the implicit rental cost of capital and operation costs.
3 We can mathematically represent a water network as a *graph* consisting of a set of nodes and links.

4 We consider a typical gravity-irrigation system with a conveyance infrastructure in place. That is, there is no cost of delivering water apart from the water lost en route to evaporation, seepage and percolation.

5 Note that the losses differ for the main and secondary canals, and we assume that there is a higher proportion of water lost per unit distance in the secondary canals, e.g. because of less canal lining.

6 There may be important dynamic considerations that we ignore, such as the depletion of the head level (the distance from some reference point to the top of the water table), which may increase extraction costs.

7 In Appendix A we provide a more detailed derivation of the first order Kuhn-Tucker conditions.

8 This happens when $\dfrac{\alpha_k}{h(d_{ik})} = \dfrac{\alpha_m}{h(d_{im})}$ for *all* sources from which farm *i* gets water.

9 See, e.g. Garey and Johnson (1979).

10 In practice, computer scientists often use approximation theory to obtain good, not perfect, solutions.

11 These source locations may be multiple upstream reservoirs, lakes and/or deep and shallow tubewells. This implicitly includes the choice between surface water and groundwater.

12 It should be noted that head farmers using more water than tail farmers is not an a priori indicator of inefficiency since this phenomenon is true even in the efficient solution.

13 See Moulin (2008) for a comparison of the average cost, marginal cost and serial cost. See Juarez (2008) for a comparison of the average cost relative to the random priority mechanism.

14 This harm might be due to the loss in value of the land and will also result in an unequal distribution of the benefits, since the agents closer to the facility will get a larger benefit

15 For a description in inter-basin transfers, see e.g. Ghassemi and White (2007).

References

Aadland, D. and V. Kolpin. 1998. "Shared Irrigation Costs: An Empirical and Axiomatic Analysis". *Mathematical Social Sciences* 35(2): 203–218.

Ambec, S. and L. Ehlers. 2008. "Sharing a River among Satiable Agents". *Games and Economic Behavior* 64(1): 35–50.

Ambec, S. and Y. Sprumont. 2002. "Sharing a River". *Journal of Economic Theory* 107(2): 453–462.

Bergantiños, G. and J. Vidal-Puga. 2007. "A Fair Rule in Minimum Cost Spanning Tree Problems". *Journal of Economic Theory* 137(1): 326–352.

Bird, C. 1976. "On Cost Allocation for a Spanning Tree: A Game Theoretic Approach". *Networks* 6(4): 335–350.

Chakravorty, U., E. Hochman, and D. Zilberman. 1995. "A Spatial Model of Optimal Water Conveyance". *Journal of Environmental Economics and Management* 29(1): 25–41.

Chakravorty, U. and J. Roumasset. 1991. "Efficient Spatial Allocation of Irrigation Water". *American Journal of Agricultural Economics* 73(1): 165–173.

Claus, A. and D. Kleitman. 1973. "Cost Allocation for a Spanning Tree". *Networks* 3(4): 289–304.

Dong, B., D. Ni, and Y. Wang. 2012. "Sharing a Polluted River Network". *Environmental and Resource Economics* 53(3): 367–387.

Food and Agricultural Organization (FAO). 2012. "World Agriculture Towards 2030/2050: The 2012 Revision". ESA Working Paper No. 12-03. Rome: FAO.

Garey, M. and D. Johnson. 1979. *Computers and Intractability: A Guide to the Theory of NP-Completeness*. San Francisco, CA: W.H. Freeman.

Ghassemi, F. and I. White. 2007. *Inter-Basin Water Transfer: Case Studies from Australia, United States, Canada, China and India*. New York: Cambridge University Press.

Johansson, R., Y. Tsur, T. Roe, R. Doukkali, and A. Dinar. 2002. "Pricing Irrigation Water: a Review of Theory and Practice". *Water Policy* 4(2): 173–199.

Juarez, R. 2008. "The Worst Absolute Surplus Loss in the Problem of Commons: Random Priority versus Average Cost". *Economic Theory* 34(1): 69–84.

Kaplow, L. and S. Shavell. 2002. *Fairness versus Welfare*. Cambridge, MA: Harvard University Press.

Márkus, J., M. Pintér, and A. Radványi. 2011. "The Shapley Value for Airport and Irrigation Games". MPRA Paper No. 30031. Budapest: Institute of Economics, Research Centre for Economic and Regional Studies,

Moulin, H. 2008. "The Price of Anarchy Serial, Average and Incremental Cost Sharing". *Economic Theory* 36(3): 379–405.

Ni, D. and Y. Wang. 2007. "Sharing a Polluted River". *Games and Economic Behavior* 60(1): 176–186.

Roumasset, J. 1989. "Decentralization and Local Public Goods: Getting the Incentives Right". *Philippine Review of Business and Economics* 1(26): 1–13.

Roumasset, J. 1987. "The Public Economics of Irrigation Management and Cost Recovery." Unpublished manuscript

Roumasset, J. and C. Wada. 2011. "Ordering Renewable Resources: Groundwater, Recycling, and Desalination". *The B.E. Journal of Economic Analysis and Policy* 11(1): 1–29.

Roumasset, J. and C. Wada. 2010. "Optimal and Sustainable Groundwater Extraction". *Sustainability* 2(8): 2676–2685.

WWAP (World Water Assessment Programme). 2012. *The United Nations World Water Development Report 4: Managing Water under Uncertainty and Risk*. Paris: UNESCO.

Vygen, J. 2005. "Approximation Algorithms for Facility Location Problems". Unpublished Lecture Notes. Available here: http://www.or.unibonn.de/cards/home/vygen/files/fl.pdf.

13

REAL-TIME INFORMATION AND CONSUMPTION

What can water demand programs learn from electricity demand programs?

John Lynham and Nori Tarui

Introduction

Water conservation has become an important policy issue. Though water management has been approached as an engineering problem (Olmstead and Stavins, 2007), theory, as well as an increasing body of evidence, suggests that economic approaches are crucial in understanding how conservation programs work and how they can be designed to be effective. This chapter identifies key lessons learnt from recent experiments on information-based demand-side management in both electricity and water markets. At present, there is a fascinating parallel between demand-side management programs in both sectors: the use of smart meters to provide consumers with real-time information on usage. The implementation of smart meter technology in the water sector has lagged behind the electricity sector. Although there is a strong push to introduce real-time information for water use,[1] we do not have a good sense of what the effects will be. But we are beginning to understand, based on a series of field experiments, how smart meters affect demand for electricity. A key contribution of this chapter is synthesizing what we know about the effects of providing real-time information on electricity use and how these might be relevant to water use.

We start by reviewing the conceptual frameworks that delineate why conservation is an efficiency issue (in the second section). Then, in the third section, we summarize the findings from the literature on how various price and non-price water conservation programs have performed. The fourth section focuses on a particular type of non-price campaign: information campaigns, in particular providing ancillary information (e.g. about conservation tips, neighbors' consumption, or externalities associated with water consumption) and better information (e.g. real-time information about consumption and prices). We highlight studies in the electricity sector, as this is where most of the work on information campaigns has been conducted. Based on this literature review, the final section identifies gaps in our knowledge that require future investigation.

To the extent that water overuse (or inefficient water allocation in general) is due to nonoptimal pricing, improving on how water is priced would be the first-best policy (see the second section). However, political constraints tend to limit the scope of water pricing reforms in many countries (third section). Recent developments in behavioral economics also indicate that information campaigns, taking the utility price structures as given, may improve efficiency of water use (second and fourth sections). For these reasons, this chapter focuses on the effectiveness of information campaigns on water conservation.

It is important to note that this chapter focuses on issues relevant to water conservation in the urban residential sector. Water conservation in the agricultural sector has its own distinct issues, and is beyond the scope of this chapter. What is discussed here may be relevant to the commercial sector, but most of the issues discussed are more acute in the residential setting—similar to the issue of the "energy efficiency gap," which is arguably more prevalent in the residential than in the commercial sector (Allcott and Greenstone, 2012). The review largely focuses on studies involving field experiments, which investigate various aspects of electricity conservation programs. Because water conservation and energy (electricity) conservation in the residential sector share similar characteristics, it is hoped that lessons learnt from one sector will be applicable to the other.[2] At the same time, there are differences in the objectives of conservation programs for electricity and water. For example, reducing peak load for production efficiency improvement is more important for electricity than for water. We will discuss what the differences between water and electricity conservation indicate in terms of different policy implications.

Conceptual frameworks

What makes water conservation a policy issue? This section summarizes the conceptual frameworks that explain why the levels of water conservation observed in residential sectors throughout many cities tend to be inefficient.

Inefficient pricing

Externalities associated with water use and water pricing

As demonstrated in other chapters of this book, a large body of literature has explored and illustrated what efficient water pricing rules would look like, by taking into account various aspects of water use. However, such optimal pricing is hardly adopted in practice. Water services are typically regulated by public utility commissions, with rates determined so that the water service provider's total costs are covered. In the residential setting, water and sewer charges may not be set to reflect their respective marginal costs.[3] Cross-subsidization across agricultural, commercial, industrial, and residential sectors (or across classes within a sector) is also prevalent across Organisation for Economic Co-operation and Development (OECD) countries as well as in developing countries (OECD, 1999; Dinar, 2000).

To the extent that water is priced below the efficient price (and given downward sloping demand), water use tends to be greater than the efficient level. In fact, studies indicate that water prices in North America tend to be consistently below the marginal social cost (Olmstead and Stavins, 2007). A first-best solution to this problem would be to raise the price to reflect the marginal social costs (including the marginal user costs) of water use.

Externalities associated with investment in water appliances

Developing new technologies that enhance water saving (i.e. that reduces the amount of water required for a given level of services) may be subject to spillovers, so that the investor's private return may be lower than the social return. Adopting new technology may also involve positive externalities (Jaffe and Stavins, 1994). Such externalities tend to induce suboptimal levels of innovation (and adoption) in water-conserving equipment, leading to inefficient water conservation. A first-best solution would be to close the gap between the private and social returns to investment. But analogous to what is discussed in the context of renewable energy policy (Fischer and Newell, 2008), subsidizing the installation of water-efficient equipment may not be the first-best solution to the issue of non-optimal technology innovation.

"Water efficiency gap"

The "energy efficiency gap" concept can be easily applied to the water context. The energy efficiency gap—also known as the energy paradox—has long been identified by researchers on energy policy (see Jaffe and Stavins, 1994, for one of the early comprehensive discussions on the concept). The energy efficiency gap refers to "the failure of consumers to make seemingly cost-effective investments in energy efficiency" (Gillingham and Palmer, 2013). The same notion applies as a rationale for policy interventions to influence water conservation in the residential sector. There are debates on the size of the energy efficiency gap. Engineering-based studies and observational studies find a large energy efficiency gap (e.g. Lovins and Lovins, 1991; McKinsey & Co., 2009). However, a consensus among recent studies, based on natural or controlled field experiments, indicates that the energy efficiency gap is smaller than engineering studies suggest (Allcott and Greenstone, 2012; Gillingham and Palmer, 2013). Gillingham and Palmer (2013) list several market failures as potential sources for the energy efficiency gap. Some of the most relevant to water use include the following:[4]

- Imperfect information: consumers may have imperfect information about the water savings from investing in water efficient products.
- Principal–agent issues in the context of residential rental properties: a landlord may pay for the water charge and choose water-use equipment (such as washing machines, toilets, and aerators on water faucets) while the tenant decides how much water to use.
- Credit constraints: limited access to credit may prevent consumers from making down payments to invest in cost-effective water efficiency improvements.

To the extent that the "gaps" listed above exist, pricing reforms alone will not be theoretically adequate for achieving efficient levels of conservation. First-best solutions for these market failures would include: educational or informational campaigns to increase consumers' awareness about water-efficiency investment; aligning principals' and agents' incentives by reforming contracts; and policies to improve the functioning of credit markets. When studying the cost-effectiveness and relative efficiency of each conservation program, the costs of program implementation must also be considered.

Credit constraints to households may be less binding for some water equipment (e.g. low-flush toilets, water-efficient washing machines) than for energy-efficiency improvements such as insulation of housing, installations of solar water heaters or solar photovoltaic panels, which tend to be more highly priced. However, the first two market failures may be extremely relevant. In particular, existing studies of some conservation programs shed light on the extent of market failures due to imperfect information and principal–agent problems.

Gillingham and Palmer (2013) also list "behavioral anomalies" that can explain the energy efficiency gap: non-standard preferences (such as hyperbolic preferences that result in time-inconsistent decisions regarding efficiency investments); nonstandard beliefs (systematically incorrect beliefs about the future, such as returns to investment in water efficient equipment and water prices); and non-standard decision-making. Non-standard decision-making refers to limited attention, framing, and suboptimal heuristics used for choices. Limited attention might apply if consumers do not study their water bills carefully to figure out how much they pay for water for various purposes. If suboptimal heuristics are prevalent in consumers' water-use decisions, then a pricing reform that is meant to enhance efficiency of water use may not achieve its goal. As an example, consider the fact that current residential water pricing tends to involve nonlinear pricing (e.g. two-part tariffs and tiered pricing with increasing or decreasing block rates), similar to the way in which electric rates are structured. Given such nonlinear pricing, Ito (2014) finds that consumers appear to respond to average, instead of marginal, prices of electricity.

Standard theory of water pricing tends to assume away non-standard preferences such as the ones described above. The theory also tends to assume that preferences are selfish: a water user's utility depends on the user's own water consumption, but not on others' consumption or on negative externalities imposed on others due to excessive water use. However, a large body of experimental evidence suggests that human beings behave as if their preferences depend on how their consumption compares to others. Other-regarding behavior and social preferences have garnered significant attention in game theory and experimental economics (Fehr and Schmidt, 1999; Bolton and Ockenfels, 2000; Charness and Rabin, 2002). In the context of water use, consumers may care about whether their consumption levels exceed their neighbors or how their consumption impacts the local environment. Policies in the presence of non-standard preferences would include improved feedbacks to consumers on the prices that they face and the amount of water they use, possibly in comparison with how much their neighbors use. As Gillingham and Palmer (2013) note, however, behavioral anomalies complicate economic welfare analysis of policies to improve the efficiency of water conservation.[5]

How water demand programs work: evidence

The previous section has outlined how water use tends to be highly inefficient. This section summarizes the findings from the literature on how various price and non-price water conservation programs have performed in addressing this inefficiency. In terms of non-price approaches, we focus on technology adoption and mandatory water use restrictions, leaving the discussion of information campaigns to the fourth section.

Price approaches

The price elasticity of demand for water tells us how much of a price change would be needed to achieve a given conservation target. A large body of literature has explored how sensitive water demand is to price changes (Espey et al. (1997) is often cited as a comprehensive review on the topic). The prevalence of nonlinear rate structures in the residential sector (e.g. tiered rate structures with increasing or decreasing block pricing), present a challenge to researchers in estimating the elasticity of water demand correctly. As for electricity consumption, the prevalence of two-part tariffs has also prompted researchers to study whether consumers respond to marginal or average prices (e.g. Ito, 2014). Olmstead and Stavins (2007) note that, though the estimated elasticity of residential water demand is greater than zero, it tends to be low (short-run elasticity averages about 0.3 to 0.4; Espey et al. (1997)). Thus, on one hand price change could induce water conservation; on the other hand, low elasticity may indicate that a rather large price change would be necessary for achieving a large-scale conservation target.

With a thorough discussion of recent evidence, Olmstead and Stavins (2007) summarize what economic research has discovered regarding the effectiveness of various conservation policies. Perhaps representing a consensus view of most water economists, they argue that price-based approaches to water conservation are more cost-effective than non-price approaches. However, pricing reforms explicitly designed for the purpose of water conservation are rarely observed. There have been some experiments in the context of time-of-use and block electricity pricing (e.g. Caves and Christensen, 1980; Caves et al., 1984; Herriges and King, 1994; Sexton et al., 1989).[6] However, similar experiments or pricing reforms on water (in the residential sector in particular) have been scarcely observed. Dinar's (2000) edited volume contains several case studies of water pricing reforms over agricultural, industrial and residential sectors (many in developing countries). For example, Musgrave (2000) reports on the transition from a flat fee to a two-part tariff with a volumetric charge in the Hunter District in New South Wales, Australia. The total water use, as well as average household use, dropped significantly upon the reform in 1982. However, the focus of Dinar (2000) is on the political economy factors that prevent the implementation of theoretically efficient pricing reforms in many countries. Dinar and Saleth (2005) succinctly summarize the difficulties in improving the efficiency of water pricing:

> Use of prices for rationing scarce water use is almost non-existent. Examples where pricing for water rationing has been attempted include: Israel, United Kingdom, and Broadview water district in California. ... A common problem in such cases is that the intent is to ration water but the design of the rate structures tends to be a compromise of political pressure by interest groups and thus, falls short of its intent.
>
> (Dinar and Saleth, 2005, p.3)

The authors move on to argue that the fact that efficient water pricing schemes are rare provides sufficient evidence for the persistence of a vast gap between the development of pricing theory and its practical application—a point the next section will revisit.

Non-price approaches

In some cases, regulations require adoption of water-efficient equipment or encourage voluntary adoption with subsidies or rebates on the purchase of efficient equipment. Adoption of low-flush toilets is an example. Note that these approaches are not the first-best solutions to address inefficient water pricing or technology spillovers. They may address consumers' imperfect information on water efficiency and distortions due to credit constraints.

With new technology, however, there are several reasons why an actual water saving might be lower than what engineering estimates suggest. One reason is behavioral changes: consumers may use water-efficient equipment longer or more frequently. Early models of low-flush toilets may have necessitated consumers to conduct "double flushing" (Olmstead and Stavins, 2007). However, a study by Bennear et al. (2013) finds no significant evidence of double flushing with high-efficiency toilets (HET) installed through a voluntary rebate program after 2008 in North Carolina. Davis (2008) reports that, after receiving high-efficiency washers in a field experiment, households increased clothes washing on average by 5.6 percent.

Bennear and colleagues' (2013) study identifies another concern about the rebate program: its "additionality." The authors study the performance of a utility's HET rebate program in North Carolina. Based on the water billing data for voluntary program participants and non-participants, they find that the water savings due to the HET installations were significant and comparable to what an engineering estimate suggests. However, based on a survey of participants, they learnt that about 63 percent of households would have invested in more efficient toilets without the rebates. Based on this limited additionality, the authors raise questions about the cost-effectiveness of the rebate program. Principal–agent issues may also explain suboptimal adoption of water-efficient technologies. By using the Residential Energy Consumption Survey, Davis (2012) shows that, controlling for household income and other household characteristics, renters are significantly less likely to have energy-efficient refrigerators, clothes washers and dishwashers.[7]

Few experimental studies on water-use restrictions exist. There are several accounts on the performance of water-use restrictions in a number of cities in the United States.[8] Based on residential water billing data from Santa Barbara, Renwick and Green (2000) study the relative impacts of various demand-side management policies on residential water use in Santa Barbara, California. They find that more stringent mandatory policies, such as use restrictions and water rationing, reduced aggregate water demand more than voluntary measures, such as public information campaigns and retrofit subsidies. Their estimates suggest that, while moderate (5–15 percent) reductions in aggregate residential water demand can be achieved through modest price increases and voluntary approaches including information campaigns, stringent mandatory restrictions or large water price increases would be necessary for a large-scale (i.e. more than 15 percent) reduction in aggregate demand. Coleman's (2009) finding, based on three-year data from Salt Lake

City in Utah, was that water consumption decreases by no more than 7 percent as a result of a public information campaign, controlling for other factors.

A few studies apply randomized-control or natural experiments to investigate the efficacy of mandatory use restrictions. Among them, two working papers study the impact of drastic energy rationing in Brazil (Gerard, 2013 and Costa, 2013). Due to exceptionally low rainfall, a part of Brazil (dependent on hydro-electric power) experienced a significant energy shortage in 2001. Residential customers in these regions were assigned individual quotas, with fines for exceeding the quotas for large consumers and bonuses for consuming below the quotas for small consumers.[9] During the rationing, the average monthly consumption of the rationed households dropped by 46 kWh/month (28 percent) relative to the non-rationed households. The two papers both find that the reduction in energy consumption persisted long after the rationing was removed, up to 2011. Analysis of household level microdata shows that the temporary rationing had a lasting impact on people's utilization of existing appliances, rather than increasing long run adoption of energy-efficient technology. Even three years after the rationing, there is evidence that rationed households use fewer freezers, and maintain the thermostats of showers with electric heating at a lower level, relative to non-rationed households. This strongly suggests that households adopted new consumption behaviors during the rationing, shifting to a new stable steady state with lower electricity use. In summary, both studies find that a price-based approach would have needed a substantial price increase in order to achieve the same level of conservation.

These studies' applicability to water conservation (or energy conservation in general) may be limited. Perhaps more than electricity, water is considered as satisfying basic human needs (Dinar, 2000). Draconian rationing of water may raise a humanitarian concern. Renwick and Green (2000) note that strict rationing policies, where households are allowed a fixed quantity of water and then are cut off from service, are rarely observed in the United States for residential water usage due to sanitary and other considerations. We now consider one of the most common non-price factors used to influence energy and water: information.

Information campaigns

The recent literature on energy and water conservation has made a large amount of progress in assessing the effectiveness of informational campaigns, which are becoming the most popular mode of non-price demand-side management. We start with ancillary information campaigns, which have been used to manage both electricity and water use, before discussing real-time information campaigns, which have been used predominantly in the electricity sector, with one or two exceptions.

Ancillary information

Residential water customers typically receive water bills once a month, which contain fairly aggregated monthly information about the customer's own water usage. Several conservation programs provide customers with additional information in order to induce water saving. Such programs are very popular in the context of electricity. Examples include the services provided by OPOWER, a private company that partners with electric utilities to implement energy efficiency programs through information provision to customers.

The theoretical underpinnings of the efficacy of information-based conservation programs include imperfect information for consumers as well as behavioral anomalies and social preferences. List and Price (2013) identify Schultz et al. (2007) as the most influential among the social psychology literature that examines the use of social-norm marketing, feedback, and tailored information campaigns to promote environmental conservation. In their experimental study, a half of randomly chosen participants received feedbacks that involve descriptive normative messages detailing average neighborhood usage of electricity while the other half received the same feedbacks plus an injunctive message (a happy emoticon for those with lower-than-average consumption and a sad emoticon for those above). While those households whose consumption exceeded the average neighborhood usage reduced energy consumption, those with consumption below average showed an increase in consumption—demonstrating what is known as the "boomerang effect." They also report that those given injunctive messages showed smaller boomerang effects. The first finding indicates that consumers do have social preferences while the second finding implies other-regarding preferences.

Schultz et al. (2007) study the effect of a one-time feedback on energy consumption (effects realized in a few weeks after the feedback was provided). Ayres et al. (2013) investigate the effect of similar information provision (that involves both descriptive norms as well as injunctive norms), conducted by OPOWER, where subjects received either monthly or quarterly feedback reports on natural gas and electricity consumption in two large-scale, natural field experiments. The reduction in energy consumption averaged 1.2 percent in an experiment for 7 months, while it averaged 2.1 percent in the other for 12 months. These findings echo what Allcott (2011) finds with data from 17 OPOWER projects across the United States (consisting of utility bills from more than 580,000 households). Compared to the control group with no feedbacks (besides the standard monthly bills), those that received monthly feedbacks reduced electricity consumption by about 2 percent, and the effect was sustained for two years after the start of the programs.

In the context of water use, Ferraro et al. (2011) and Ferraro and Price (2013) report the treatment effects of a similar randomized-control experiment in a county in Georgia. About 100,000 households are enrolled in a control group and three treatment groups that received a one-time feedback on their water consumption behavior: (a) those who received "Technical Advice" (a tip sheet listing ways to conserve water effectively); (b) those who received Technical Advice and a message based on "Weak Social Norm" (that involves a message appealing to pro-social preferences highlighting the importance of conserving water); and (c) those who received messages based on Weak Social Norm as well as "Strong Social Norm," which involves a social comparison that contrasts the household's use in the previous year with the median use in the county. While the treatment effect (in terms of changes in water consumption after the intervention) was insignificant or small with (a), it was the largest with (c), the social comparison treatment. However, perhaps due to the nature of the intervention (the feedback was given only once), the effects did not persist over the two-year period after intervention.

Real-time information

The studies reviewed above investigate the treatment effects of providing technical advice and information on peer usage. A separate set of studies has examined the effects of more frequent information on energy use to consumers (e.g. weekly, daily, hourly, or real-time). Faruqui, Sergici and Sharif (2010) reviewed a number of pilot programs worldwide that focus on energy-conservation impacts of in-home displays (IHDs) as well as alternative electricity rate structures. Among the programs that tested the effect of IHDs on electricity consumption (where the test period ranged from three months to two and a half years), the average reduction in energy use was 7 percent, and the largest reduction observed was 13 percent. A review article by Fischer (2008) finds that energy conservation effects are larger when feedbacks are given frequently. Houde et al. (2013) is a recent study based on a relatively large randomized-control experiment involving real-time feedbacks to the treatment-group households. Households in the treatment were given access to the own energy usage data via installed electricity monitoring devices and the "Google PowerMeter" application on their computers. The treatment-group households had access to PowerMeter for the duration of the experiment (eight months) while the control-group households had access to PowerMeter for only the last five months. The authors estimated the average treatment effect to be about 5.7 percent for the duration of the experiment. They also find that electricity consumption reductions due to feedbacks occur primarily during peak household activity periods, and that the effects do not persist over time: the effects become insignificant starting the third week after the treatment begins. These findings lead the authors to suggest that energy conservation due to real-time feedbacks might be primarily due to *temporary* changes in habitual behaviors.

Jessoe and Rapson's (2014) study combines randomized treatments of real-time feedbacks and price increases: 437 households are randomly assigned to a control group with no intervention, the "price-only" treatment group that faced pre-announced price increases, and the "price+IHD" treatment group that had access to in-home displays in addition to the price treatment. In-home displays show real-time information about the households' energy use. Among the treated households, those with in-home displays are significantly more responsive (in adjusting their energy consumption) to price changes than those without. The consumption decreases observed with price-only households (relative to the control-group households) were not statistically significant.

As reviewed earlier, studies on the elasticity of residential electricity demand tend to find low elasticity values. Jessoe and Rapson's (2014) finding suggests that the reason may be partially due to consumers' imperfect information about pricing. This finding is also consistent with Gaudin's (2006) observational studies on the effects of billing design on residential water demand: it finds that the residential price elasticity may increase when price information is posted on water bills.

Lynham et al. (2013) explore the causal mechanisms through which real-time information affects energy consumption by conducting a randomized-control trial with residential households. Their experiment attempts to disentangle two competing mechanisms: (i) learning about energy use via real-time feedback is sufficient to sustain energy conservation, or (ii) having a constant reminder of energy use is necessary for energy conservation. They have two main results. First, they find a statistically significant treatment effect from receiving real-time information. Second, they find that learning plays a more prominent role than saliency in driving energy conservation.

Kenney et al. (2008) report on the effects of real-time feedbacks, as well as pricing policies and water-use restrictions, on water conservation in Aurora, Colorado. Interestingly, they find the treatment effects of having access to real-time feedbacks via "water smart readers" (WSR) to be positive, i.e. an increase in water consumption. The authors state that this result holds even when residents' self selection into smart meter adoption is controlled for. The authors attribute the positive effect to the possibility that the WSR enabled consumers to learn about their water consumption relative to the thresholds under block pricing. They observe that the frequency with which consumers with WSRs entered into the most punitive pricing tier (the third block) diminished. Without the WSR, consumers face uncertainty on how far their consumption is from a higher block with higher rates, causing them to err on the side of caution by consuming less than they would have otherwise preferred. This finding provides an important lesson on the efficacy of real-time information as a conservation tool. On one hand, as the authors note, improved information enables consumers to make utility-maximizing decisions based on a broader information set. On the other hand, improved information may not necessarily imply lower consumption.

What works for water demand-side management versus electricity demand-side management

As mentioned above, there are differences in the nature of residential water demand and supply on one hand, and residential electricity demand and supply on the other. These differences lead to different objectives of conservation programs for electricity and water. They also imply that the information campaigns that might work for electricity demand-side management may not necessarily work for water demand-side management.

The usefulness of real-time feedbacks may be more limited with water than with electricity as peak-load management is not a significant issue in water conservation. However, water demand does exhibit seasonality (Renwick and Green, 2000), where frequent feedbacks on prices and consumption may play a role. The above studies thus point to the possibility that water conservation is enhanced with feedbacks to consumers that are more frequent than on a monthly basis.

The promise of real-time information as a demand-side management instrument depends on how cost-effective it would be compared to other conservation programs including those involving social comparison. There are several factors to be considered. The size of the treatment effects, as well as the costs of program implementation (i.e. the cost of providing monthly energy reports versus the costs of installing smart meters for real-time feedbacks), would matter a lot. Another benefit aside from feedbacks to consumers is the possibility of cutting down load when it is too high at peak times. The latter may be a benefit because shaving peak load (or shifting load from peak periods to off-peak periods) tends to reduce power generation costs. Such benefits of peak shaving may not be a problem with water use. Smart readers may help, however, if there is a seasonal shortage in water and the board of water supply wants to signal that to consumers. However, we believe these benefits should be considered in isolation as it is possible to provide one without the other.

Research opportunities

We reviewed the theoretical frameworks on water conservation and demand-side management programs as well as empirical studies that either test the theory or seek to estimate treatment effects of particular interventions. Here we summarize what we believe to be important gaps in the literature where further research may be useful.

How much conservation is necessary?

Studies focusing on the treatment effects of water and energy conservation programs do not discuss how much conservation is necessary or efficient. Investigating such questions is usually beyond the scope of such studies. However, it would be useful to have in perspective how much conservation is needed when assessing the welfare impacts of alternative conservation programs. For example, the studies reviewed above suggest that informational intervention that involves social comparison may have long-term conservation effects of about 2 percent. In contrast, use restrictions could achieve a larger reduction in water consumption. In order to compare which approaches are more efficient, we need to know not only the costs of implementing the programs under consideration but the optimal level of water conservation. General-equilibrium modeling of water and water policies might be further utilized for that purpose.

Short run versus long run

Whether the effects of water conservation programs persist in the long run is an important policy issue that influences the cost-effectiveness of policies. Mandatory or voluntary technology programs, such as those on low-flush toilets, appear to have a long-term conservation effect (Bennear et al., 2013). Regarding information programs, the literature indicates that a one-time intervention (such as Schultz et al. (2007) in the context of electricity use and Ferraro and Price (2013) on water use) tends to have a short-lived treatment effect. Monthly feedbacks that involve social comparison appear to induce persistent effects (Allcott and Rogers, forthcoming). Although Allcott and Rogers find that there is a pattern of "action and backsliding," this pattern dampens over time and the reduction in consumption is persistent. This suggests that consumers have purchased a new stock of physical capital or changed their consumption habits. From a policy point of view, conservative assumptions about long-run persistence may lead to underestimates of the cost-effectiveness of information programs.

Gneezy et al. (2011) summarize the experimental literature on habit formation, with a focus on smoking and exercise. In terms of smoking, studies that do measure long-term effects don't find that habits persist. Participants in Volpp et al. (2006) were smokers at the Philadelphia Veterans Affairs Medical Center who were randomized into incentive and non-incentive treatments. The study combined incentives to participate in a five-class smoking-cessation program with incentives for smoking cessation. The incentive group was offered US$20 for each class attended and US$100 if they quit smoking for 30 days post-program completion. As expected, the incentivized participants were more likely to complete the classes (26 versus 12 percent) and to quit in the short run (16 versus 5 percent). However, after six months, quit rates between the treatments were not significantly different (6 versus 5 percent). Within the realm of exercise, Charness and Gneezy (2009) conducted two field experiments in which university students were offered incentives to attend the university's gym. The authors were able to observe attendance before, during, and after the intervention. The main result of these experiments is that requiring people to visit the gym at least eight times, in order to be paid, significantly improved attendance rates during and, more importantly, after the intervention. The improvement in gym attendance was entirely driven by the change for those people who had not previously been regular attendees. In general, habit formation remains an area of fertile research and further investigation is necessary for us to truly understand what factors sustain treatment effects (and why).

Heterogeneity in treatment effects

Studies have found differences in how consumers respond to a conservation program. When receiving information about the neighbors' consumption, for example, studies have identified "boomerang effects": those above the neighbors' consumption level tend to conserve more while the opposite is true when the self consumption is below the neighbors' levels (Schultz et al., 2007; Ayres et al., 2013). Costa and Kahn (2013) examine the effect of peer information on consumption patterns for pro-environmental consumers compared to consumers who don't have strong environmental preferences. They find that while environmentalists reduced consumption versus the control group, non-environmentalists actually increased their consumption. This is clear evidence that treatment effects can be completely different for different types of people.

Ferraro and Miranda (2013) do not find heterogeneous responses to purely technical information on water use or to traditional water conservation messages that combine technical information and moral suasion. However, norm-based messages that combine technical information, moral suasion and social comparisons exhibit strong heterogeneity: households that are wealthier, owner-occupied and use more water are more responsive. These are also the groups that tend to be least responsive to financial incentives in their study. Interestingly, they find no evidence of a boomerang effect. Ito et al. (2013) also touch on heterogeneous treatment effects based on a field experiment on dynamic residential electricity pricing and social pressure. They find that, compared to low-income consumers, high-income consumers are more responsive to social pressure. Conversely, the response to dynamic pricing is less for high-income consumers.

These studies reveal that the multiple dimensions on which households differ matter for how they respond to conservation programs. To what extent policy makers can make use of such heterogeneity in their program design has not been explored empirically. These studies also suggest a research opportunity for investigating the mechanism behind treatment effects, which we turn to next.

The mechanisms of conservation: connecting theory to experimental design

As reviewed in this chapter, the literature has made significant progress in estimating *how much* energy conservation is achieved by providing households with information feedbacks. However, few of them tell us *why* such feedbacks change energy-use behavior.

In other words, what are the *causal mechanisms* through which conservation programs change consumer behavior? Although many experiments have found treatment effects from providing information, we don't really know why or how these effects happen. As an example, consider the following statement from Allcott (2011) on the well-known OPOWER program:

> Because the entire Treatment group received the social comparisons, historical consumption information, and energy efficiency tips, it will not be possible to determine what aspect of the Reports drives that treatment effect. ... I can evaluate the overall efficacy of the program motivated by behavioral science, but I cannot test more refined hypotheses about the channels through which the program works. Although popular media outlets have concluded it is the peer comparison feedback in particular that reduces households' electricity usage, the treatment arms of the pilot program itself provide no evidence that this is actually true.

Studies such as Ayres et al. (2013) are informative as to "how" residents achieve conservation—on the intensive margin or extensive margin: households in their sample responded mainly through behavioral changes as opposed to investment in energy-efficiency appliances. Jessoe and Rapson (2014) suggest learning is behind their treatment effects, but this argument is based on survey results.

For example, in the case of real-time feedbacks, is it because they update consumers' beliefs about how much power different appliances use? Is it because smart meters reduce uncertainty about energy use and electricity charges that consumers face? Is it because smart meters make energy use more salient as residents are constantly reminded of their energy consumption? While the first two hypotheses imply the presence of a "learning effect," the last hypothesis refers to the "saliency effect" of real-time information. Lynham et al. (2013) is one of the few attempts to try and explore the mechanisms behind treatment effects. Although they have a small sample size, their results indicate that learning is more important than saliency. But they also observe that both effects decay over time.

Explaining the popularity of non-price approaches

As Olmstead (2013) observes, water conservation programs tend to rely on non-price water conservation mandates, incentives, and other policies as opposed to price-based mechanisms. Why do we see such a prevalence of non-price approaches? The discussion may be akin to the choice of pollution control instruments: why are command-and-control approaches so much more prevalent than incentive-based approaches such as cap-and-trade and taxes?

Olmstead (2013) also argues that there is no model in the literature that explains how or why water utilities choose one or more non-price water conservation policies. Given the perceived need for water conservation policies, predicting the types of policies that are likely to be adopted seems highly important. A promising line of research includes an approach by Jacobsen et al. (2013), which sheds light on the circumstances under which non-price approaches may perform better than price approaches. They find that, in the presence of heterogeneous "green" preferences, technology standards are almost always preferable to price instruments.

Incentives for utilities

Ayres et al. (2013) note that, although some utilities are beginning to provide informational feedback programs, often utilities do not have adequate incentives to reduce energy on their own. It is well known that utilities under traditional rate-of-return regulation lack incentives for demand-side management, which tends to reduce their profits (Brennan, 2010). Electric utilities' incentives for resource conservation may be different from those of conserving water supply. However, to the extent that the disincentive for energy conservation might apply to water service providers, further research on what institutional arrangements align the utilities' incentives for conservation (in addition to the focus of most demand-side management research on consumers' incentives for conservation) will be useful.

Conclusion

Real-time water metering has been hailed by some as the next revolution in demand-side management of water use. This chapter reviewed what we know about residential water use and attempts to reduce consumption though both price and non-price channels. Demand for water shares a lot of similarities with demand for electricity. In particular, demand for both goods may be more responsive to non-price than price incentives. This chapter identified key lessons learnt from recent experiments on information and demand-side electricity management. We explored the effects of providing better information (e.g. about consumption and prices) and ancillary information (e.g. about conservation tips, neighbors' consumption, or externalities associated with water consumption). The effects in the electricity domain have been small and short-lived. The only published study on real-time information and water demand actually found an increase in water usage. There is clearly much that remains unsolved. There is a lot that we assume we know about demand-side management but have not yet formally tested.

Notes

1 See "How Smart Metering Can Solve the Water Crisis" by Jaymi Heimbuch. http://science. howstuffworks.com/environmental/conservation/issues/how-smart-metering-can-solve-the-water-crisis.htm
2 As we shall see below, however, water and energy conservation programs share many design characteristics.
3 For example, in Honolulu, Hawaii, the sewer service charge is based on the assumption that about 20 percent of the water used by a household goes to watering yards or plants, washing cars and other water uses that do not enter the sewer system (Department of Environmental Services, City and County of Honolulu http://www.honolulu.gov/envwwm/wwm_faq.html accessed on August 25, 2014).
4 Gillingham and Palmer (2013) also classify inefficient pricing, discussed above, as a regulatory failure.
5 Similarly, based on the finding that consumers respond to average prices instead of marginal prices, Ito (2014) also argues that such suboptimizing behavior makes nonlinear pricing

unsuccessful in achieving its policy goal of energy conservation and critically changes the welfare implications of nonlinear pricing.

6 Sexton et al. (1987, 1989) find evidence that electricity customers do respond to time-of-use pricing. The authors find that, with real-time feedbacks on energy use provided to treatment-group households, significant peak shifting occurred under time-of-use pricing. They also find, however, that the average demand remained the same.

7 Allcott and Greenstone (2012) argue that, given the limited magnitude of the efficiency gap in Davis (2012) as well as other unobserved household characteristics that are not controlled for, the study does not provide an "iron-clad" evidence of (significant) principal–agent issues.

8 In a review article on water conservation programs in the United States, Olmstead and Stavins (2009) cite three studies including Renwick and Green (2000).

9 Therefore, strictly speaking, they were not purely mandatory restrictions.

References

Allcott, H. (2011). "Social Norms and Energy Conservation." *Journal of Public Economics* 95(9–10): 1082–1095.

Allcott, H. and Rogers, T. (forthcoming) "The Short-Run and Long-Run Effects of Behavioral Interventions: Experimental Evidence from Energy Conservation." *American Economic Review.*

Allcott, H. and Greenstone, M. (2012). "Is There an Energy Efficiency Gap?" *Journal of Economic Perspectives* 26(1): 3–28.

Ayres, I., Raseman, S. and Shih, A. (2013) "Evidence from Two Large Field Experiments that Peer Comparison Feedback Can Reduce Residential Energy Usage". *Journal of Law, Economics and Organization* 29 (5): 992-1022.

Bennear, L.S., Lee, J.M., and Taylor, L.O. (2013). "Municipal Rebate Programs for Environmental Retrofits: An Evaluation of Additionality and Cost-Effectiveness." *Journal of Policy Analysis and Management* 32(2): 350–372.

Bolton, G. and Ockenfels, A. (2000). "ERC: A Theory of Equity, Reciprocity, and Competition." *American Economic Review* 90(1): 166–193.

Brennan, T.J. 2010. "Decoupling in Electric Utilities." *Journal of Regulatory Economics* 38(1): 49–69.

Caves, D.W. and Christensen, L.R. (1980). "Residential Substitution of Off-Peak for Peak Electricity Usage under Time-of-Use Pricing." *Energy Journal* 1(2): 85–142.

Caves, D.W., Christensen, L.R., and Herriges, J.A. (1984). "Consistency of Residential Customer Response in Time-of-Use Electricity Pricing Experiments." *Journal of Econometrics* 26(1): 179–203.

Charness, G. and Rabin, M. (2002). "Understanding Social Preferences with Simple Tests." *Quarterly Journal of Economics* 117(3): 817–869.

Charness, G. and Gneezy, U. (2009). "Incentives to Exercise". *Econometrica,* 77(3), 909–931.

Coleman, E.A. (2009). "A Comparison of Demand-Side Water Management Strategies Using Disaggregate Data." *Public Works Management and Policy* 13(3): 215–223.

Costa, D.L. and Kahn, M.E. (2013). "Energy Conservation 'Nudges' and Environmentalist Ideology: Evidence from a Randomized Residential Electricity Field Experiment." *Journal of the European Economic Association* 11(3),: 680–702.

Costa, F.J.M. (2013). "Can Rationing Affect Long Run Behavior? Evidence from Brazil". Working Paper.

Davis, L.W. (2008). "Durable Goods and Residential Demand for Energy and Water: Evidence from a Field Trial." *RAND Journal of Economics* 39(2): 530–546.

Davis, L.W. (2012). "Evaluating the Slow Adoption of Energy Efficient Investments: Are Renters Less Likely to Have Energy Efficient Appliances?" In Fullerton, D. and Wolfram, C. eds. *The Design and Implementation of U.S. Climate Policy.* Chicago, IL: University of Chicago Press. (Also available as NBER Working Paper No. 16114, 2010.)

Dinar, A. ed. (2000). *The Political Economy of Water Pricing Reform.* New York: Oxford University Press.

Dinar, A. and Saleth, R.M. (2005). "Issues in water pricing reforms: from getting correct prices to setting appropriate institutions." In Folmer, H. and Tietemberg, T. eds. *International Yearbook of Environmental and Resource Economics,* Cheltenham: Edward Elgar.

Espey, M., Espey, J., and Shaw, W.D. (1997). "Price Elasticity of Residential Demand for Water: A Meta-Analysis." *Water Resources Research* 33:1369–1374.

Faruqui, A., Sergici, S., & Sharif, A. (2010). "The Impact of Informational Feedback on Energy Consumption—A Survey of the Experimental Evidence". *Energy*, 35(4), 1598–1608.

Fehr, E. and Schmidt, K. (1999). "A Theory of Fairness, Competition, and Cooperation." *Quarterly Journal of Economics* 114 (3): 817–868.

Ferraro, P.J., Miranda, J.J., and Price, M.K. (2011). "The Persistence of Treatment Effects with Norm-Based Policy Instruments: Evidence from a Randomized Environmental Policy Experiment." *American Economic Review: Papers & Proceedings* 101(3): 318–322.

Ferraro, P.J. and Miranda, J.J. (2013). "Heterogeneous Treatment Effects and Causal Mechanisms in Non-Pecuniary, Information-Based Environmental Policies: Evidence from a Large-Scale Field Experiment." *Resource and Energy Economics* 35: 356–379.

Ferraro, P.J. and Price, M.K. (2013). "Using Non-Pecuniary Strategies to Influence Behavior: Evidence from a Large-Scale Field Experiment." *The Review of Economics and Statistics* 95(1): 64–73.

Fischer, C. (2008). "Feedback on Household Electricity Consumption: A Tool for Saving Energy?" *Energy Efficiency* 1: 79–104.

Fischer, C. and Newell, R.G. (2008). "Environmental and Technology Policies for Climate Mitigation." *Journal of Environmental Economics and Management* 55(2): 142–162.

Gaudin, S. (2006). "Effect of Price Information on Residential Water Demand." *Applied Economics* 38(4): 383–393.

Gerard, F. (2013). "The Impact and Persistence of Ambitious Energy Conservation Programs: Evidence from the 2001 Brazilian Electricity Crisis". Discussion Paper 13–06. New York: Resources for the Future

Gillingham, K. and Palmer, K. (2013). "Bridging the Energy Efficiency Gap: Insights for Policy from Economic Theory and Empirical Analysis". Resources for the Future Discussion Paper 13-02.

Gneezy, U., Meier, S., and Rey-Biel, P. (2011). "When and Why Incentives (Don't) Work to Modify Behavior." *The Journal of Economic Perspectives* 25(4): 191–209

Herriges, J.A. and King, K.K. (1994). "Residential Demand for Electricity under Inverted Block Rates: Evidence from a Controlled Experiment." *Journal of Business & Economic Statistics* 12(4): 419–430.

Houde, S., Todd, A., Sudarshan, K., Flora, J.A., and Armel, K.C. (2013) "Real-Time Feedback and Electricity Consumption: A Field Experiment Assessing the Potential for Savings and Persistence." *Energy Journal* 34(1): 87–102.

Ito, K. (2014) "Do Consumers Respond to Marginal or Average Price? Evidence from Nonlinear Electricity Pricing." *American Economic Review*, 104(2): 537–563.

Ito, K., Ida, T., and Tanaka, M. (2013). "Using Dynamic Electricity Pricing to Address Energy Crises: Evidence from Randomized Field Experiments". Working Paper. Kyoto: Kyoto University

Jacobsen, M., LaRiviere, J., and Price, M. (2013). *Public Goods Provision in the Presence of Heterogeneous Green Preferences*. Paper presented at the American Economic Association Annual Meeting.

Jaffe, A.B., and Stavins, R.N. (1994). The Energy-Efficiency Gap: What Does It Mean? *Energy Policy* 22(10): 804–810.

Jessoe, K., and Rapson, D. (2014). "Knowledge is (Less) Power: Experimental Evidence from Residential Energy Use". *The American Economic Review*, 104(4), 1417–1438.

Kenney, D.S., Goemans, C., Klein, R., Lowrey, J., and Reidy, K. (2008). Residential Water Demand Management: Lessons from Aurora, Colorado. *Journal of the American Water Resources Association* 44(1): 192–207.

List, J.A. and Price, M.K. (2013). *Using Field Experiments in Environmental and Resource Economics* (No. w19289). Cambridge, MA: National Bureau of Economic Research.

Lovins, Amory B. and Hunter Lovins, L. (1991). "Least-Cost Climatic Stabilization". *Annual Review of Energy and the Environment* 16: 433–435.

Lynham, J., Nitta, K., Saijo, T. and Tarui, N. (2013) "Why Does Real-Time Information Reduce Energy Consumption?" Working Paper, University of Hawaii at Manoa.

McKinsey & Co. (2009). "Unlocking Energy Efficiency in the U.S. Economy." http://www.mckinsey.com/clientservice/electricpowernaturalgas/downloads/US_energy_efficiency_full_report.pdf.

Musgrave, W. (2000). "The Political Economy of Water Price Reform in Australia". In Dinar, A. ed. *The Political Economy of Water Pricing Reform*. New York: Oxford University Press.

Olmstead, S.M. (2013). "Climate Change Adaptation and Water Resource Management: A Review of the Literature." *Energy Economics* Available online 19 September 2013. http://dx.doi.org/10.1016/j.eneco.2013.09.005

Olmstead, S.M. and Stavins, R.N. (2007) "Managing Water Demand: Price vs. Non-Price Conservation Programs". Pioneer Institute White Paper, No.39. Boston, MA: Pioneer Institute.

Olmstead, S.M. and Stavins, R.N. (2009). "Comparing Price and Nonprice Approaches to Urban Water Conservation." *Water Resources Research* 45(4): 1944–1973. http://onlinelibrary.wiley.com/doi/10.1029/2008WR007227/abstract

Organisation for Economic Co-operation and Development (1999). *The Price of Water: Trends in OECD Countries. Paris: OECD*

Renwick, M.E. and Green, R.D. (2000). "Do Residential Water Demand Side Management Policies Measure Up? An Analysis of Eight California Water Agencies." *Journal of Environmental Economics and Management* 40(1): 37–55.

Schultz, P.W., Nolan, J.M., Cialdini, R.B., Goldstein, N.J., and Griskevicius, V. (2007). "The Constructive, Destructive, and Reconstructive Power of Social Norms". *Psychological Science* 18: 429–434.

Sexton, R.J., Johnson, N.B., and Konakayama, A. (1987). "Consumer Response to Continuous-Display Electricity-Use Monitors in a Time-of-Use Pricing Experiment." *Journal of Consumer Research* 14(1): 55–62.

Sexton, R., Sexton, T., Kling, C., and Wann, J. (1989). "The Conservation and Welfare Effects of Information in a Time-of-Day Pricing Experiment." *Land Economics* 65: 272–279.

Volpp, K.G., Levy, A.G., Asch, D.A., Berlin, J.A., Murphy, J.J., Gomez, A., ... and Lerman, C. (2006). "A Randomized Controlled Trial of Financial Incentives for Smoking Cessation." *Cancer Epidemiology Biomarkers & Prevention* 15(1): 12–18.

14

WATER SCARCITY AND WATER INSTITUTIONS

Ariel Dinar and Yacov Tsur

Introduction

Water scarcity can be succinctly summarized by three easy-to-remember numbers: 1,700, 1,000 and 500 m^3 per person per year. A region is experiencing water stress when annual natural water supplies available on a sustainable fashion drop below 1,700 m^3. When natural annual water supplies drop below 1,000 m^3, the population faces water scarcity, and below 500 m^3 "absolute scarcity" (Falkenmark et al. 1989). Annual renewable water supplies, while fluctuating from year to year, are on average constant, with a possible long-term trend due to climate change. In contrast, population grows constantly in most regions. As a result, per capita water availability declines over time, turning more regions water stressed as time goes by. Table 14.1 (based on World Bank data) reports the population that live in water stressed, water scarce and absolute scarcity conditions in 2012 and 2050 (the latter is based on World Bank population predictions). The m^3 per person per year figures refer to internal renewable resources (internal river flows and groundwater from rainfall) divided by the World Bank's population estimates. Due to spatial and temporal variability, a more refined (e.g., regional and seasonal) data will reveal larger populations under water shortage conditions.[1]

Table 14.1 Worldwide water shortages

		2011	2050
Water stress	Population (million)	2,842.29	6,209.14
(less than 1,700 m^3 per person per year)	Number of countries	65	80
Water scarce	Population (million)	905.56	51.63
(less than 1,000 m^3 per person per year)	Number of countries	43	28
Absolute scarcity	Population (million)	5,595.66	2,956.83
(less than 500 m^3 per person per year)	Number of countries	65	57

Source: Adapted from Falkenmark et al. (1989).

Worldwide irrigation water consumes the bulk of the available renewable fresh water resources (over 70 percent). Irrigated agriculture is practiced on about 18 percent of total cultivable land (267 million hectare in 1997, of which 75 percent are in developing countries), and it produces 44 percent of agricultural output (Alexandratos and Bruinsma, 2012). Although the rate of expansion of the irrigated area has been diminishing, it is expected to continue to expand by an additional 200 million hectares by 2050 (Alexandratos and Bruinsma, 2012). Consequently, the irrigation sector, which already consumes a large share of global water, will continue to increase its demand for water in the foreseeable future.

Meeting food demand of a growing population at the existing structure of water use requires increasing water withdrawal. But the cost of water supply rises each time a new dam is built and concerns over the adverse environmental and social effects of large water projects are mounting. As a result prospects for increasing natural water supply are at best limited and in many regions nonexistent. The course of action left open, then, is to do more with the available water supplies, i.e., to increase the efficiency of water use, and to augment the supply of natural water with manufactured water, by treating and reusing water used by households and by desalinating seawater. Achieving this goal requires taking account of the full cost of water, which beyond the engineering cost of water conveyance also includes the alternative cost associated with the different uses of water at present and in the future. It is important that water users consider the true cost of water when deciding on water demand and allocation between various activities.

This chapter surveys various policy measures that have been used to deal with the allocation of scarce water resources and the institutions that have been developed to implement these policies. Since the evolution of water policies and water institutions depends on a variety of endemic conditions characterizing a country or region, including hydrology, climate, culture, economics and history, a large diversity of water institutions and policies has emerged and no attempt is made to provide an exhaustive survey. Instead, we present the analysis in the context of a stylized water economy which contains aspects common to many real-world situations. The stylized water economy, presented in the next section, provides the context and framework to discuss, in the following section, various water allocation policies and water institutions that have been developed in different countries and regions. The final section concludes the chapter with a brief summary and discussion.

The water economy

A water economy consists of (i) the physical resource base (precipitation, rivers, lakes, aquifers), (ii) consumers and users (irrigators, households, industry, environment), (iii) suppliers and the associated infrastructure (extraction-conveyance-treatment infrastructure), and (iv) regulatory and institutional infrastructure (water laws and property rights, prices and quotas, water institutions). We begin with a description of these components, following the schematic framework of Tsur (2009).

Water resources

In addition to precipitation, water can be derived from M sources (rivers, lakes, reservoirs, aquifers) whose stocks at time t, denoted $Q_t = (Q_t^1, Q_t^2, ..., Q_t^M)$, evolve in time according to

$$Q_{t+1}^m = Q_t^m + R^m(Q_t) + x_t^m - g_t^m, m = 1, 2, ..., M,$$

where $R^m(\cdot)$ represents deterministic recharge, x_t^m is stochastic recharge and g_t^m is the rate of extraction from source m. Recharge at time t emanates from current precipitation and from subsurface (lateral) flows. The latter depends on current and past precipitation. Precipitation may vary spatially across the water basin. Accordingly, we divide the basin into $N \geq 1$ subregions and denote by w_t the N-dimensional vector whose elements w_t^n are the precipitation in subregion n during period t, $n=1,2,...,N$. The w_t, $t=1,2...$, are independently and identically distributed with (N-dimensional) distribution F_w defined over a nonnegative support.

Current and past precipitations generate the M-dimensional stochastic recharge vector $x_t = (x_t^1, x_t^2, ..., x_t^M)'$ according to

$$x_t = \Gamma w_t + \Omega x_{t-1}$$

where Γ and Ω are, respectively, $M \times N$ and $M \times M$ (known) matrices. The mth row of Γ represents the immediate effect of precipitation on stock m'th recharge, while the mth row of Ω represents the (diminishing) effects of past precipitation. In view of the above relation, the water stocks evolution can be rendered as

$$Q_{t+1} = Q_t + R(Q_t) + \Gamma w_t + \Omega x_{t-1} - g_t,$$

where $R(Q) = (R^1(Q), R^2(Q), ..., R^M(Q))'$ and $g_t = (g_t^1, g_t^2, ..., g_t^M)'$. The extraction quotas g_t are restricted to lie in a set of feasible extractions.

In addition to the natural water sources, two types of produced sources may also be available: desalinated water (of brackish or seawater) and recycled (treated sewage) water. We refer to desalination as source $M+h$, $h=1,2,...,H$, where H is the number of desalination plants. Recycled water has two distinctive features that separate it from the other sources. First, exogenous (health and environmental) regulations often require treating sewage water, disregarding whether it will later be reused. Second, the same regulations often forbid mixing treated effluent with potable water, implying that reusing the treated water requires separate conveyance and distribution systems. These properties affect the allocation of recycled water (see Tsur, 2009, and references therein).

Consumers and users

The basin contains S private sectors (urban, agriculture, industry) and a few public sectors, including parks, estuaries and wilderness areas. We consider a single public sector, called the environment (e.g., instream water), indexed $S+1$.[2] Consumers are scattered spatially in L locations (districts, regions, municipalities). The inverse water demand of sector $s=1,2,...,S$, in location $l=1,2,...,L$, is denoted $D^{sl}(\times)$: when the water price ($\$ \times m^{-3}$, say) is $D^{sl}(q)$, sector s in location l demands the water quantity q. We assume stationary water demands.

Agricultural (irrigation) demand

The number of agricultural subsectors depends on the level of aggregation and may contain, for example, orchards, vegetables, fiber (cotton), cereals, other field crops and livestock. Agricultural sector s in location l has J activities (crops), indexed $j=1,2,...,J$. Let $y_j(q)$ denote crop j's water-yield value function, not including the water cost.[3] The corresponding inverse demand for irrigation water is $y_j{}'(.)$.[4] Typically, $y_j(.)$ is increasing and strictly concave, so that $y_j{}'(.)$ is decreasing and its inverse exists. The water demand of agriculture sector s in location l is $q_{sl}(p_w)=\sum_j y_j{}'^{-1}(p_w)$ and the corresponding inverse demand is $D^{sl}(.)=q_{sl}^{-1}(.)$. The diminishing marginal productivity of water implies that $D^{sl}(.)$ is decreasing

Industrial demand

Industrial sectors use water as an input of production. As above, the number of industrial sectors depends on the level of aggregation and the sectors are defined according to how water is used in the production process. The inverse water demand of industrial sector s in location l, $D^{sl}(.)$, is derived in much the same way as for agricultural water demand (see Renzetti 2002a).

Residential demand

We assume that needed for subsistence is provided at a nominal cost and consider demand for water above this quantity. The utility of household i depends on the per-capita consumption of water ($\tilde{q}$) and other goods ($\tilde{z}$). The (individual household's members) demands for $\tilde{q}$ and $\tilde{z}$ are the outcome of

$$v_i\left(p_w,p_z,y_i,n_i\right)=max_{(\tilde{q},\tilde{z})}u_i\left(\tilde{q},\tilde{z}\right)s.t.\ p_w\tilde{q}+p_z\tilde{z}\leq y_i,$$

where y_i is the household's income, n_i is the household's size (number of members) and p_w and p_z are the prices of $\tilde{q}$ and $\tilde{z}$, respectively. With $\tilde{q}_i(p_w,p_z,y_i,n_i)$ denoting household i's (per capita) water demand, the residential water demand in location l is (retaining only the water price argument)

$$q_{sl}\left(p_w\right)=\sum_{i\in \text{location } l} n_i\tilde{q}_i(p_w,p_z,y_i,n_i)$$

and the corresponding inverse water demand is $D^{sl}(.) = q_{sl}^{-1}(.)$. Residential water includes (in addition to household's dwelling) water consumed in service, public and commercial institutions (see Baumann et al., 1998; and Renzetti 2002b).

Environmental water

Environmental sectors include public urban parks and instream water in wilderness areas and estuaries. They differ from the sectors discussed above due to their public good features. We briefly outline how to incorporate environmental water, assuming for simplicity a single environmental sector indicated as sector $E = S+1$. Let q^{El} represent allocation of environmental water in location l. Household i's demand for q^{El} is measured in terms of the household's willingness to pay (WTP) to preserve q^{El}

against the alternative $q^{El} = 0$ and unchanged environmental water allocations in all other locations $\boldsymbol{q}_{-l}^{E} \equiv \left(q^{E1}, q^{E2}, \ldots, q^{El-1}, q^{El+1}, \ldots, q^{EL} \right)$. Suppose that the utility household i derives from $\boldsymbol{q}^{E} \equiv (\boldsymbol{q}_{-l}^{E}, q^{El})$ is represented by the term $u_i^{E}(\boldsymbol{q}^{E})$, which is added to $v_i \left(p_w, p_z, y_i, n_i \right)$. Household i's WTP for q^{El} when environmental water allocation is $\boldsymbol{q}^{E}$, denoted $WTP_i^{l}(\boldsymbol{q}^{E})$ and defined by

$$v_i \left(p_w, p_z, y_i - WTP_i^{l}(\boldsymbol{q}^{E}), n_i \right) = v_i \left(p_w, p_z, y_i, n_i \right) + u_i^{E} \left(\boldsymbol{q}_{-l}^{E}, 0 \right),$$

represents household i's demand for environmental water. There is a large (and growing) literature dealing with the estimation of such WTPs (see Freeman 2003; Bockstael and McConnell 2007; and references they cite).

Consumers (users) surplus

The gross surplus (not including the water cost) of sector s in location l from consuming the water quantity q is

$$B^{sl}(q) = \int_0^q D^{sl}(\theta) d\theta, \, s = 1, 2, \ldots, S, l = 1, 2, \ldots, L.$$

Since $D^{sl}(.)$ is positive and decreasing, $B^{sl}(.)$ is increasing and strictly concave. The surplus generated by q^{El} is the sum of the $WTP_i^{l}(\boldsymbol{q}^{E})$ over all households i in the economy, $B^{El}\left(\boldsymbol{q}^{E}\right) = \sum_i WTP_i^{l}(\boldsymbol{q}^{E}), l = 1, 2, \ldots, L$, and the surplus generated by $\boldsymbol{q}^{E}$ is

$$B^{E}\left(\boldsymbol{q}^{E}\right) = \sum_{l=1}^{L} B^{El}(\boldsymbol{q}^{E}).$$

Cost of water supply

Water supply entails extraction-production, conveyance, treatment and distribution. Each activity requires capital, labor, energy and material inputs. The capital cost constitutes the bulk of the fixed cost, while the costs of the other inputs make up the variable cost.

Capital cost

The activities associated with water supply include extractions, treatment (domestic sector), conveyance and distribution, sewage collection (domestic and industrial sectors), recycling and desalination. The capital stock of each activity is measured in terms of the full cost of installing the infrastructure (pipes, pumps, canals etc.) necessary to carry out the activity. Let K represent the vector of capital stocks. An element of K is the capital stock of a particular activity and determines the capacity of the associated supply activity, i.e., the maximal quantity of water that can be supplied during a year, but otherwise has no effect on the water supply rate. We denote these capacity functions by $F(K)$. Examples of three capital stocks and their associated capacity functions are given in Table 14.2. A complete list can be found in Tsur (2009).

The annual cost of capital is the interest and depreciation on the (current-value) capital stock, which constitutes the bulk of the fixed cost of water supply. For example, with r and d representing the interest and depreciation rates, respectively, the annual capital cost associated with extraction from source m is $(r+d)_e^{m}$.

Table 14.2 Capital stocks involved in the water sector

Capital notation	Capacity constraint	Activity
K_e^m	$F_o^m(\cdot)$	Extraction from source m
K_{tr}^l	$F_{tr}^l(\cdot)$	Treatment at location l
K_c^{ml}	$F_c^{ml}(\cdot)$	Conveyance from m to l

Variable cost

The variable costs of supply are due to energy, labor and material inputs and vary across activities and sectors. For example, supplying q m^3 per year from source m to residential use in location l entails the variable cost due to extraction plus conveyance plus treatment plus distribution plus sewage treatment. Irrigation water entails no sewage collection and may not require treatment (depending on the source), while desalinated water supplied to residential use requires treatment (desalination), conveyance, distribution and sewage collection/treatment.

If mixing recycled water with potable water is not allowed, recycled water requires conveyance and distribution systems of its own, which are included in the recycled capital, with the associated variable cost (see Tsur 2009 for details).

Water policy

At the beginning of year t, after observing the precipitation w_t, hence also the recharge x_t, the water state $Z_t = (Q_t, x_t)$ is observed. Given Z_t, the policy decisions for year t entail: (i) extraction quotas g_t^m, $m = 1,2,...,M$, for the M naturally replenished sources, and production of desalinated and recycled water; (ii) allocation of the extracted (from the M natural sources) and produced (from the desalination and recycling sources) water among the end users; and (iii) investment in the capital infrastructure that determines the capacity of the various supply activities. Characterizing the optimal water policy, although solving for the optimal policy in any given circumstance could turn out to be highly demanding in terms of computation, is in principle straightforward (see discussion in Tsur 2009). The main issue is how to implement a desirable policy in a practical manner. This implementation task is carried out by a plethora of institutions that have been evolving to address the legal, physical, cultural and hydrological characteristics specific to the region under study. In the next section we discuss commonly found water institutions available to implement these policies, without committing to a particular situation. The discussion in the next section is organized along the content of the various components of the stylized model in this section. We start with review of institutions in a global context, move to institutions in certain countries, and wrap up with institutions in certain sectors or institutions addressing allocation of certain types of water (surface, groundwater and recycled water).

Water institutions

The discussion so far has provided a schematic description of a water economy. The use of market-based mechanisms to allocate water resources among the different sectors and users is hampered by a number of market failures, including externalities related to common pool resource extraction, uncertainty associated with water supply and demand, and political economy considerations of water stakeholders in the location (region, state, country) under study. In addition, prevailing water institutions (property rights, water laws) in a water economy may not be conducive to market-based water allocation mechanisms, exacerbating prospects for efficient water allocation. These difficulties are even more apparent with increased water scarcity due to population growth and impacts of climate change on water supply (Dinar and Jammalamadaka 2013). In this section we introduce regulatory interventions (institutions) that have been attempted by states and local water management agencies for improving the management of their water resources.

We group institutions into nation-wide institutions versus regional/local/sectoral (e.g., state wide water reform versus river basin decentralization, irrigation sector institutions, groundwater institutions); demand-side versus supply-side institutions (e.g., pricing, rights, trading versus investments); and structural versus non-structural institutions (e.g., infrastructure versus administrative permits).[5] We discuss various cases of institutional interventions at state, basin/region, and sector levels as they have been documented in countries facing various degrees of water scarcity. The effectiveness of the institutions will be analysed along the lines of the findings from the analytical framework proposed in the earlier part of the chapter.

Water institutions in a global context

Documentation of water institutional performance at the country level does exist for many countries as case studies that are hard to compare and derive conclusions from. Only a few studies applied a comparative framework that allows one to draw conclusions across countries (Saleth and Dinar 2004, 2005, 2006; Dinar and Saleth 2005).

From the perspectives of the institutional and political economy theory of institutional reforms (e.g., North 1990), the reform process observed in the sample countries also provides some interesting evidence for some of the theoretical and empirical results reported in the literature on institutional economics in general and water institutions in particular. Saleth and Dinar (2004) have proposed an institutional transaction cost theory for explaining water institutional reforms by using the general formulation of Coase (1960) and North (1990). According to this theory, various socio-economic, political and resource-related factors are considered to affect the relative success of institutions. These factors can be either endogenous (e.g., water scarcity and financial weakness) or exogenous to the water sector (e.g., macro-economic crisis, political reforms and donor pressure).

One approach that allows one to compare institutional performance across countries uses country water institutions health index (WIHI), which were then applied to 43 countries across the globe (Dinar and Saleth 2005). The WIHI uses a set of 16 institutional components that are aggregated into one index that allows a comparison across countries. This approach was extended both spatially (by adding more countries) and temporally (by adding time periods) for the assessment (Araral and Yu 2013). Using the WIHI values and the area graphs of their components helps in ranking water economies over space. The study by Dinar and Saleth (2005) also included an analysis that found a very close relationship between WIHI and major economic, social and state governance indicators, which suggests that successful water institutions are related to economy-wide performance. Furthermore, Araral and Yu (2013) indicate that certain water institution indicators vary with a country's level of economic development, suggesting a possible Kuznets curve in water institutions performance.

Comparison of country-level institutional reforms

A detailed analysis of water institutions' performance for several countries was conducted, using the framework developed by Saleth and Dinar (2004) and reported in McKay (2005), Hearne and Donoso (2005), Doukkali (2005), Heyns (2005), Backeberg (2005) and Samad (2005) for Australia, Chile, Morocco, Namibia, South Africa and Sri Lanka, respectively. Some of the comparative analysis is summarized below.

Key features of the resource base

Understanding the forces behind and the nature of the institutions applied in the water sector in a country can be greatly dependent on knowing the physical features of the water sector as well as the political economy in the country that led to changes from previous to present institutions (Table 14.3).

We would expect to see less successful institutional performance in countries that have non-uniform conditions of any of the physical characteristics mentioned in the table, as well as other characteristics that can be added.

The previous institutional arrangements in the country

As path dependency is a major factor in explaining a great deal of the performance of present institutions, we will compare (Table 14.4) the various pre-reform situations in the seven countries surveyed.

Table 14.3 Some water sector characteristics

Characteristic	Australia	Chile	Morocco	Namibia	South Africa	Sri Lanka	Israel
Uniform weather and physical conditions	No	No	No	Yes	No	No	No
Dominated sectoral use of water	Irrigated agriculture	Hydro	Irrigated agriculture	Irrigated agriculture and livestock	Irrigated agriculture	Irrigated agriculture	Irrigated agriculture
Majority of supply from	Surface water	Surface water	Surface water	Surface water	Surface water	Surface water	Ground-water
Water supply coverage	High	High	Medium	Low	Medium	Medium	High

Source: Adapted from Saleth and Dinar (2005) for all countries except Israel, which is based on authors' elaboration.

Table 14.4 Pre-reform characterization

Country/ Parameter	Australia	Chile	Morocco	Namibia	South Africa	Sri Lanka	Israel
Colonial ruling	Yes	Yes	Yes	Yes	Yes	Yes	Yes
State role in water sector	Strong	Weak	Strong	Strong	Strong	Strong	Very Strong
Role of water law	Non-existent	Existed	Existed to some extent	Non-existent	Non-existent	Non-Existent	Existed
Inter-sectoral conflicts	Irrigation – Urban	Irrigation – Hydro	Irrigation – Hydro – Urban	Livestock – Urban	Irrigation – Hydro	Not Known	Irrigation – Urban

Source: Adapted from Saleth and Dinar (2005) for all countries except Israel, which is based on authors' elaboration.

Tables 14.3 and 14.4 and data in Saleth and Dinar (2004) suggest that path dependency matters in water sector performance and successful institutions. We can expect different institutions and institutional performance level if the water sector is managed by the public sector or by the private sector. We expect more demand-side management with private sector involvement and simpler institutions used when the water sector serves fewer sectors. Existing legal arrangements and their nature (which are more difficult to address) are probably the most important factor in explaining the institutional setup and its level of performance.

The regulatory framework for water resources management

Salman and Bradlow (2006) review and compare the water resources management regulatory frameworks of 16 (developing and developed) countries[6] from which it is possible to draw recommendations for efficient management of water resources. Some of these recommendations are summarized below. Legal framework of water resources at the state levels should:

- streamline the institutions' scope to address aspects such as water management, protection of quality and prevention of pollution, licensing, abstraction and ownership;
- be comprehensive, but framed in a simple way and consistent with government objectives and agency abilities of implementation, and the means to confer rights to use on individuals;
- allow for an easy implementation of policies that reflect public interests while protecting interests of individuals;
- water tariffs should be aligned with general development policies and retain flexibility to allow adjustment to changes in environment and global situations.

Given these recommendations, a comparative framework was suggested, which consists of the following parameters: statutory framework; underlying principles and priorities; regulation of water uses (ownership rights, allocation, transfer of rights, termination of rights); protection of water; regulation of water infrastructure; institutional arrangements (river basin authorities, water user associations, advisory committees); financial arrangements; enforcement of regulations; and dispute resolution mechanism.

While the discussion in this section refers to state-wide institutions and institutional reforms in the water sector, it is useful also to look at specific institutions applied to the water sector in various countries and also at more stylized examples that compare across various institutions with similar policy objectives. The next section will focus on examples related to water demand management.

Special features in country water law and regulations

While Salman and Bradlow (2006) reviewed the regulatory frameworks of many countries, there are a handful of documented cases which shed light on specific aspects of water law and their interaction with water institutions.

Arizona Groundwater Management Act and its implementation

Arizona, a prominent water-scarce state in southwestern USA, relies heavily on groundwater for reliable water supply. The prior appropriation water rights doctrine is the basis for allocation of water among users (see Colby and Jacobs 2007). Due to lack of flexibility associated with historical water rights, in years of high water scarcity this doctrine proves extremely inefficient and unjust not only among senior and junior right holders, but also with regards to environmental aspects of the river, leading to drying up of lower stretches of many rivers. The prior appropriation doctrine (with a minor modification) is used also to govern groundwater allocation in western USA. Both the principles of the doctrine, and the use of each water source (surface flows and groundwater) separately from the other, ignores the physical linkages between the river and aquifers and leads to harm to the water basin and the society that relies on its water (Glennon 2007). As our understanding of the links between groundwater and river flows and as claims for the limited water sources grow, that linkage becomes important and has been reflected in changes to the water law.

Arizona has passed the 1980 Arizona Groundwater Management Act (AGWM), which imposes severe limits on drilling groundwater wells and in certain cases requires developers to demonstrate assured water supply, or to show alternative water supplies that would replenish the extracted (or mined) groundwater they need for their development activity (Glennon 2007). The AGWM also serves as a law governing storage and recovery of water, using a system of permits, accounts and infrastructure (Megdal 2007). The storage and recovery system and its institutions take care of the default in the law, recognized by Glennon (2007), by taking a basin-wide approach and recognition of the location-importance of storage and recovery sites. Since its implementation, this water storage and recovery institution has been proven quite successful in terms of water stored, although many outstanding issues still exist. Between 1996 and 2004 a total of nearly 454 acre-foot (1 acre-foot = 1233.5 m^3) of water have been stored, demonstrating a steep increase from 5,000 acre-foot in 1996 to 120,000 acre-foot in 2004.

Recent water law reforms in Australia

Water law reforms in Australia comprise four phases, each corresponding to the needs and changing conditions of climate and economic situations (McKay 2008). We will summarize the evolution of the Australian water law and its support of water institutions reforms that took place in Australia (based on McKay 2008). The English Common Law used in early stages of the state (Phase 1, since 1786) applied different rules to treat surface water and groundwater. It empowered land owners to appropriate water they could reach from the land they own, even when this use entails detrimental effects on neighbors. As water has been used more remotely and canals had to be constructed to deliver water, the law was not providing answers to many situations, and starting in 1886 (Phase 2) the law vested "Use Power and Control Power" to the states. The Common Law rules were replaced by a system of licensing and allocation schemes, again, separately for groundwater and surface water. Each state in the Commonwealth was allowed to establish its own system, leaving the basin as a whole at risk. In 1994 the Commonwealth government agreed on the "Council of Australian Governments" (CoAG) water reform that lasted until 2004 (Phase 3). This law introduced several new features (regulations and institutions), including: pricing reforms with the objective of full cost recovery; requirement for each jurisdiction to introduce a comprehensive water allocation system; water and land are no more linked, so that water can be traded; water services to be separated from water resource management; urban utilities to introduce a two-part tariff; states to introduce arrangements for water trade; and new irrigation project investments to be justified on the basis of economic and environmental sustainability. While the framework in Phase 3 was advanced and improved water allocation efficiency, it still faced difficulties in addressing the severe and prolonged drought in Australia in the early 2000s. Major reforms culminated in 2004 (Phase 4) in the "National Water Initiative," followed in 2007 by the "Commonwealth Water Bill." Both legal frameworks represent a major reform in the water sector of Australia with important core elements such as: clear water access entitlements; efficient water market structure; water metering; and very sophisticated and efficient water pricing policies.

Surface water demand management institutions

Institutions that regulate water use from the demand-management point of view include pricing, taxing and various water trading schemes. These regulatory measures operate via signals regarding the economic value of water, and their objective is to make the allocation of a given quantity of water more efficient. In the following we provide several examples from a number of countries to demonstrate various levels of success associated with such institutions.

Demand management in Israel

With a chronic water shortage, exacerbated by the rapid population growth, Israel has developed an impressive water conveyance infrastructure that connects all water sources with all demand nodes in the country. However, with the growth in demand (due mainly to population growth), cost of capital increased and scarcity breaking records occurred. Demand management institutions have been gaining prominence in implemented water policies (see Kislev 2011). A notable aspect of Israel's water allocation is the shift of the agricultural sector from natural to recycled and marginal (saline and flood) water: the use of fresh/potable water in irrigation has been reduced from 892.3 million m³ (MCM) in 1995 to 413.7 MCM in 2011, while the use of recycled/marginal water has increased from 392 MCM in 1996 to 628 MCM in 2011.[7] This shift represents a reduction of 54 percent in the use of natural water and an increase of 60 percent in the use of recycled/marginal water over a period of 15 years—and the reallocation process is still ongoing.

Pricing

Full cost recovery pricing has recently been implemented in Israel, both for the agricultural (irrigation) and urban (households and industries) sectors (see Kislev 2011; Katz 2013). Water is priced volumetrically according to an increased block-rate pricing (see the 2011's irrigation and domestic block rates in Kislev 2011, p. 43 and pp. 63–65, respectively). Over the period 2007–2009, Israel's Water Authority increased the water tariffs for domestic and industrial users. Available consumption data[8] reveal that the reduction in domestic water consumption following the increases in tariffs brought a reduction in use of over 10 percent, about 100 MCM annually. This volume is equivalent to the annual production of a large scale desalination plant.

The purpose of this demand management measure is to send the proper signal regarding water scarcity to the users and to fully cover the cost of infrastructure investment. A representative household monthly water tariff in 2012, for example (Katz 2013, p. 151), includes two blocks with the lower block at a basic consumption level of up to 4 m³/per capita and price of 9 Shekel/m³, and the next block at 14 Shekel/m³ for the remainder of the consumption.[9] For the agricultural sector (Katz 2013, p. 153) the tariff structure includes three main blocks, corresponding to up to 50 percent, the additional 30 percent and the additional 20 percent of the annual quota, with additional rates for deviations above the annual quota.[10]

Extraction levies

Differential levies are imposed on wholesalers with extraction rights. The levies are designed to introduce higher basin efficiency and internalization of environmental cost in the basin. Extraction levies are imposed on all extractions, including Mekorot's—the national water company which supplies nearly 70 percent of all water in the country (Kislev 2011; Katz 2013).

Water rights, water markets/transfers

Following the five-year drought that ended in 2008, the Government of Israel introduced a legal framework for bilateral transfer of unused quota by a group of farmers to satisfy above quota demands of other groups (DiSegni 2013). The water transfer mechanism is subject to several constraints such as limit to bilateral transfer in the same region and subject to hydrological constraints; transfers are not permanent, and restricted to 30 percent of the annual quota; original quota + transfer should be lower than a cap, based on the 1989 quota; and, all transfers are subject to clearance by the Water

Authority. Under these restrictions, actual transfers (DiSegni 2013, p. 142) were quite small between 2008 and 2011, quoting between 192 and 479 transactions that cover only 1.15–2.64 percent of the total water allocation.

Water pricing in Spain

Spain faces severe water scarcity in many of its regions. In the past 20 years, Spain has undertaken a radical water reform that included various water pricing and water transfer initiatives stemming from the debate leading to the 2001 Consolidated Water Act (Garrido and Calatrave 2010). The new institutional setup introduces a comprehensive pricing system that comprises extraction levies at the source, an effluent control levy at the downstream, a regulation levy for the administrative cost of regulating water use, and a tariff for water use to cover infrastructure of water projects. It is important to note that the levels of the levies changes with location and sector. For example, farmers are exempt from the levy in Catalonia. Garrido and Calatrave (2010) documented several versions of two-part tariff pricing of irrigation water. The first part is always set on a per hectare basis, while the second part of the tariff can be based on number of water applications, duration of the turn in the canal, volume of water (volumetric tariff), or combinations of these.

Institutions for groundwater management

We turn now to discuss several examples of institutions that have been used for groundwater management.

Institutions' experience in Spain

Lopez-Gunn (2010) highlights the likely difference between collective management institutions and command and control institutions, using the case of the Eastern La Mancha and Western La Mancha aquifers. Lopez-Gunn (2010) classifies the institutional setups of groundwater management in Spain, and Esteban and Albiac (2011) and Esteban and Dinar (2013a, 2013b) provide comparative results of the effectiveness of these institutions. Esteban and Albiac (2011) find that the contrast in management outcomes between the Western and the Eastern La Mancha aquifers is related to the different types of policy instruments and institutions implemented for each aquifer. The results of these policies underline the importance of nurturing the stakeholders' collective action under the appropriate institutional setting. Esteban and Dinar (2013a) estimated the effectiveness of sequencing and packaging various groundwater regulation institutions, focusing on two instruments: water quota and uniform water tax. They demonstrate how packaging and sequencing sets of interventions, with possible triggers to initiate their time of implementation, may be more effective in achieving a sustainable groundwater management, in terms of smaller water table drop and higher users' welfare, than a single institution, when environmental externalities exist.

The policy instruments are applied to the Western La Mancha aquifer in Southeast Spain, a major aquifer that is managed by a command and control approach. In a different study, Esteban and Dinar (2013b) quantify the value of cooperation in the

Eastern La Mancha Aquifer. They find that cooperation allows users to internalize the damages caused by their activities and reduce excessive extractions. A cooperative game theory model is applied. The model empirically shows how the uncontrolled extractions in each sub-aquifer affects neighboring groundwater users but also cause severe impacts to the interlinked ecosystem. The results in the case of the Eastern La Mancha aquifer illustrate how both extraction and environmental externalities interact in affecting the likelihood of cooperation among the users. The paper estimates the value of cooperation and its stability with and without the environmental externality.

Groundwater institutions in southern California

Groundwater basin management in the Southern California Metropolitan Water District (MWD) includes many groundwater basins, with fierce competition over their water by many users. There are various types of management or governing structures of groundwater within the service area of MWD. One can find five specific types of management structures (MWD 2007):

1 Formally adjudicated with respect to production, water levels or downstream flow within the basin;
2 Managed by an agency created and given authority by State statute;
3 Managed pursuant to an adopted groundwater management plan developed in accordance with the State water code provisions;
4 Managed informally by city ordinance or by consensus among some or all of the producers; and
5 Not governed, managed or adjudicated.

More than 90 percent of the groundwater resources within the Metropolitan service area are adjudicated or formally managed pursuant to statute or adopted groundwater management plan. Basins with a court judgment are included in the "adjudicated" category.

METHODS OF SUSTAINABLE BASIN OPERATIONS

Long-term recharge and production or discharge of groundwater are the objectives of sustainable groundwater basin management. The Watermaster function is an institution that regulates the basin water. It is a board comprised of representatives of the basin producers, and makes decisions regarding setting the annual operational yield. It focuses on strict accounting pursuant to the established rules. Two examples of various management operation styles are provided below (MWD 2007):

- *Fixed maximum pumping.* Some basins are managed or adjudicated to maintain a fixed maximum amount of groundwater pumping from year to year (e.g., the Central Basin, located in the Los Angeles County, California). In that case the aquifer has been adjudicated with a fixed pumping allocation above the safe yield of the basin, which requires supplemental recharge with imported and recycled water to support the fixed pumping rights of the groundwater producers.

- *Variable maximum pumping.* Other basins are managed or adjudicated to allow for variations in groundwater pumping but still maintain sustainable operations (e.g., the Main San Gabriel Basin). The governance structure of this basin provides for setting an annual operational safe yield with associated adjustments in pumping rights that are not subject to payment for replacement water. Producers may pump in excess of the annual pumping right set through this process, but this excess production is subject to payment for imported supplemental water to recharge the basin.

Closing remarks on water institutions

The section on water institutions aimed at providing examples of various water institutional interventions in various countries that face various levels of water scarcity. We reviewed country attempts to reform their water law, introduce market based allocation, manage groundwater in a sustainable manner, and develop pricing schemes that produce the scarcity signal to the potential users. The examples we introduced may not be representative of the trends of water institutional reforms around the world, but they provide the range of water scarcity faced by the various countries with an attempt to address it via implementation of a variety of water institutions.

Concluding comments

The chapter provides a framework to address issues of water allocation under scarce situations and derivs basic principles to guide water policy makers. These basic principles are summarized below.

We were interested in also reflecting on real-world situations. Any real-world situation presents a myriad of factors that limit the set of feasible policy tools and requires departure from the basic principles derived in the first part of the chapter. Attempts at addressing real-world situations are provided in the second part of the chapter. Various surveyed attempts are presented, under which institutions were designed and implemented in several countries facing different political and legal, as well as physical, constraints.

Several lessons can be learned from both the analytical and from the survey part of the chapter. Neither the analytical part nor the survey part of the chapter addressed equity considerations associated with the various institutional arrangements analysed and surveyed. Additional considerations may include accommodation of basic human needs, costs of implementation of the institutions, shocks to the physical system of water supply, and political economy considerations of institutional reforms.

For example, securing water for basic human needs in the institutional frameworks may not be straightforward. They may be hard to monitor and implement, and finally lead to their malfunctioning and collapse. The Chilean case study is endowed with a combination of quotas and prices to address a secured quantity of water for basic needs of the poor in Santiago, while still keeping the efficiency principle and scarcity signal that is sent by the pricing system in place.

Implementation costs vary with the specific institution selected. For example, while volumetric water pricing may be efficient in that it allocates water and maximizes social welfare, still, if implementation costs are considered, a significant social cost could be lost due to major difficulties, technical, legal, and social, associated with the implementation of volumetric pricing schemes. The same is true with regards to the effect of political economy impacts. Examples include the case of volumetric prices reform in Morocco, where water metering devices have been dismantled and the pricing reform was called back (see the description of Morocco water economy in Tsur et al. 2004, pp. 175–197).

And finally, we have learned that institutions have to be able to adjust to shocks to the physical water system as well as to changes in water demand due to changes in water availability and population over time. For example, the Spanish case study suggests that in managing groundwater resources, changes in demand and supply may affect the stability and effectiveness of the institution. Such concerns have been addressed in the case of the Metropolitan Water District of Southern California by the adjudication institution, however, with very high annual transaction costs, where demand for and supply of water change and don't match the adjudicated quotas.

Notes

1 For example, while China as a whole does not suffer from water stress conditions (2,093 and 1.723 m^3 per person per year in 2011 and 2050, respectively), large parts of northern China, home to hundreds of millions of inhabitants, suffer from water scarcity or absolute scarcity throughout the year or seasonally.
2 Water allocated to the environment has features of a public good, hence the analysis of this sector differs from that of the S private sectors.
3 These functions are defined as follows: Let $\tilde{y}_j\left(q,b,z\right)$ denote crop j's production function, where q is water input, b is a vector of fixed inputs (e.g., land and family labor) and z is a vector of purchased inputs (labor, fertilizers, pesticides, machinery) with price vector r. Then, $y_j(q) = \max_z \{p_j\, \tilde{y}_j\left(q,b,z\right) - rz\,\}$ s.t. $b \le \bar{b}$, where the output price p_j, the fixed inputs constraint $\bar{b}$ and the input prices r are suppressed as arguments.
4 To see this, note that when the price of water is p_w, profit is $y_j(q)-p_w q$ and the water input that maximizes profit satisfies $y_j\phi(q) = p_w$. Thus, the water demand at that price is $y_j\phi^{-1}(p_w)$.
5 This classification is not mutually exhaustive and exclusive and entails some duplications and overlaps.
6 The study includes the European Union and the following individual countries: Armenia, Brazil, Cameroon, China, Costa Rica, France, Germany, Kazakhstan, Mexico, Morocco, Nepal, Senegal, South Africa, Vietnam and Yemen.
7 See allocation data in Israel Water Authority's website at: http://www.water.gov.il/Hebrew/ProfessionalInfoAndData/Allocation-Consumption-and-production/20112/1996-2011.pdf
8 These data can be found in Israel's Water Authority website at: http://www.water.gov.il/Hebrew/ProfessionalInfoAndData/Allocation-Consumption-and-production/20112/1996-2011.pdf
9 At the time this chapter was written, one Israeli Shekel was worth 0.28 US dollars.
10 For 2013 rates see http://www.water.gov.il/Hebrew/Rates/DocLib1/prices-1.1.13.pdf.

References

Alexandratos, N. and Bruinsma, J. 2012, World agriculture towards 2030/2050: the 2012 revision. ESA Working paper No. 12-03. Rome: FAO.

Araral, E. and Yu, D. 2013. Comparative water law, policies, and administration in Asia: Evidence from 17 countries. *Water Resources Research*, 49, doi:10.1002/wrcr.20414.

Backeberg, G.R. 2005. Water institutional reform in South Africa. *Water Policy*, 7(1):107–124.

Baumann, D.D, Boland, J.J. and Hanemann, W.M. 1998. *Urban Water Demand Management and Planning*. New York: McGraw-Hill.

Bockstael, N.E. and McConnell, K. 2007. *Environmental and Resource Valuation with Revealed Preferences*. New York: Springer

Coase, Ronald H. 1960. The problem of social cost, *Journal of Law and Economics*, 3(1): 1–44.

Colby, B.G. and Jacobs, K.L. 2007. *Arizona Water Policy*. Washington, DC: Resources for the Future.

Dinar, A. and Saleth, R.M. 2005. Can water institution be cured: A water institutions health index. *Water Science and Technology: Water Supply Journal*, 5(6): 17–40.

Dinar, A. and Kumar Jammalamadaka, U. 2013. Adaptation of irrigated agriculture to adversity and variability under conditions of drought and likely climate change: Interaction between water institutions and social norms. *International Journal of Water Governance*, 1: 41–64.

DiSegni, D.M. 2013. Market-based regulations on water users. In Becker, N. Ed., *Water Policy in Israel*, Berlin: Springer, pp.137–146.

Doukkali, M.R. 2005. Water institutional reform in Morocco. *Water Policy*, 7(1): 71–88.

Esteban, E. and Albiac, J. 2011. Groundwater and ecosystems damages: Questioning the Gisser-Sánchez effect. *Ecological Economics*, 70(11): 2062–2069

Esteban, E. and Dinar, A. 2013a. Modeling sustainable groundwater management: Packaging and sequencing of policy interventions. *Journal of Environmental Management*, 119: 93–102.

Esteban, E. and Dinar, A. 2013b. Cooperative management of groundwater resources in the presence of environmental externalities. *Environmental and Resource Economics*, 54: 443–469.

Falkenmark, M., Lundquist, J. and Widstrand, C. 1989. Macro-scale water scarcity requires micro-scale approaches: aspects of vulnerability in semi-arid development. *Nat. Resour. Forum*, 13: 258–267.

Freeman, A.M. 2003. *The Measurement of Environmental and Resource Value: Theory and Methods*. Washington, DC: RFF Press. 2nd ed.

Garrido, A. and Calatrave, J. 2010. Trends in water pricing and markets. In Garrido, A. and Llamas, R. Eds., *Water Policy in Spain*. London: CRC Press, pp. 131–144.

Gleick, P.H. 2000. *The World's Water 2000-2001: The Biennial Report on Freshwater Resources*. Washington, DC: Island Press.

Glennon, R. 2007. The disconnect between water law and hydrology. In Colby, B.G. and Jacobs, K.L. Eds., *Arizona Water Policy, Management Innovations in an Urbanizing, Arid Region*, Washington, DC: RFF Press, pp. 106–120.

Hearne, R.R. and Donoso, G. 2005. Water institutional reform in Chile. *Water Policy*, 7(1): 53–70.

Heyns, P. 2005. Water institutional reform in Namibia. *Water Policy*, 7(1): 89–106.

Katz, D. 2013. Policies for water demand management in Israel. In Becker, N. Ed., *Water Policy in Israel,* Berlin: Springer, pp.147–164.

Kislev, Y. 2011. The Water Economy of Israel. Policy Paper No. 2011.15. Taub Center for Social Policy Studies in Israel.

Lopez-Gunn, E. 2010. Making groundwater institutionally visible. In Garrido, A. and Llamas, R., Eds., *Water Policy in Spain*. London: CRC Press, pp. 165–174.

McKay, J. 2005. Water institutional reform in Australia. *Water Policy*, 7(1): 35–52.

McKay, J. 2008. The legal framework of Australian water: progressing from common law rights to sustainable shares. In Crase, L. Ed., *Water Policy in Australia-The Impact of Change and Uncertainty*, Washington, DC: RFF Press, pp. 44–60.

Megdal, S.B. 2007. Arizona's recharge and recovery programs. In Colby, B.G. and Jacobs, K.L. Eds., *Arizona Water Policy, Management Innovations in an Urbanizing, Arid Region*, Washington, DC: RFF Press, pp. 188–203.

MWD (Metropolitan Water District of Southern California). 2007. *Groundwater Assessment Study: A Status Report on the Use of Groundwater in the Service Area of the Metropolitan Water District of Southern California*. Report No. 1308, September. Los Angeles, CA: MWD.

North, D.C. 1990. *Institutions, Institutional Change, and Economic Performance*. Cambridge, MA: Cambridge University Press.

Renzetti, S. 2002a. *The Economics of Industrial Water Use*. Dordrecht: Kluwer Acad.

Renzetti, S. 2002b. *The Economics of Water Demands*. Dordrecht: Kluwer Acad.

Saleth, R.M. and Dinar, A. 2004. *The Institutional Economics of Water: A Cross-Country Analysis of Institutions and Performance*. Cheltenham: Edward Elgar.

Saleth, R.M. and Dinar, A. 2005. Water institutional reforms: theory and practice. *Water Policy*, 7(1): 1–19.

Saleth, R.M. and Dinar, A. 2006. Water institutional reforms in developing countries: insights, evidence and case studies. In Lopez R. and Toman M. A. Eds., *Economic Development and Environmental Sustainability*, London: Oxford, pp. 273–306.

Salman, M.A.S and Bradlow, D.D. 2006. *Regulatory Frameworks for Water Resources Management: A Comparative Study*. Washington, DC: The World Bank.

Samad, M. 2005. Water institutional reform in Sri Lanka. *Water Policy*, 7(1):125–140.

Tsur, Y. 2009. On the economics of water allocation and pricing. *Annual Review of Resource Economics*, 1(1): 513–536.

Tsur, Y., Roe, T., Doukkali, R. and Dinar, A. 2004. *Pricing Irrigation Water: Principles and Cases from Developing Countries*. Washington, DC: RFF.

15

MANAGING CLIMATE RISKS THROUGH WATER TRADING

Bonnie Colby, George Frisvold, and Matthew Mealy

Introduction

This chapter examines the potential of voluntary water trading to reallocate water supply risks associated with climate change effects on regional water supplies, focusing upon the Lower Colorado River Basin of the southwestern United States. In many regions worldwide, increased variability in precipitation, along with shifts from snow to rain are key impacts of climate change. These changes present new challenges for storing and delivering reliable water supplies. We examine the usefulness and potential pitfalls of reliance on seasonal climate forecasts for designing water trading arrangements to buffer an urban area against supply reliability risk.

Voluntary trading of water is a promising strategy to improve supply reliability under climate change for regions with multiple water-using sectors. Trades are motivated by differing marginal values for water across locations and water uses, as well as differences in costs associated with supply shortfalls. In many regions, trades to improve supply reliability involve moving water out of agricultural use because irrigation accounts for a large portion of water use worldwide. In regions with a mix of irrigated crops, the opportunity cost of ceasing to irrigate lower profitability crops may be small compared to the value of that water made available during shortage to urban, industrial and environmental uses or for higher value crop irrigation. In the western United States, the high seniority of long-established agricultural water entitlements makes those entitlements more "drought proof" and thus more valuable to other users who face high costs when their supplies are curtailed. Intermittent and temporary water transfers that occur only in dry years have advantages over permanent water trades. Dry-year only transfers cannot be activated to support water use in normal years as that would make them unavailable as a buffer in future dry periods. Intermittent transfers also are less financially costly than permanent acquisitions and tend to generate less concern and conflict over third party impacts of reducing irrigated acreage.

With climate change, water supplies in many regions are projected to become more variable and subject to more extreme flood and drought events (Hartmann and

Wendler 2005; Garrick and Jacobs 2006; Rajagopalan et al. 2009; Kenney et al. 2010). Water managers struggle with an increasingly unreliable water supply. Water and risk-sharing agreements specifically designed to improve supply reliability are becoming an important climate adaptation strategy. Urban water providers often are willing to pay a premium for access to secure water supplies because of high costs of supply curtailment, including household water rationing and bans on watering landscape. Low streamflows, associated with more variable hydrology, also expose municipalities to difficulties in complying with regulations requiring minimum flows for water quality or endangered fish protection.

The balance between limited and variable water supplies and growing demands in the Lower Colorado River Basin means that costs of drought-induced supply variability have the potential to be widespread and acute, affecting many different sectors of the economy. Voluntary transactions with agricultural districts holding senior entitlements are a key mechanism for urban water providers to adapt to the more variable water supplies projected for the Lower Colorado River Basin. Such arrangements will generally involve temporary movement of water out of agricultural use, because agriculture accounts for a substantial proportion of water use in the Basin. Low financial returns to irrigating various field crops, and the relative seniority of some agricultural water allocations, make transfers out of agriculture attractive to those urban and industrial users which would face high costs if their supplies were curtailed. Energy producers also may find themselves seeking supplies that are more reliable. Hydropower generation declines when streamflow and reservoir storage levels decrease, and higher temperatures increase both water and energy demand. In addition, public agency and non-governmental organization (NGO) programs seek reliable water to protect and restore water-dependent habitat, and are potential participants in supply reliability transactions.

The economic consequences of increased water supply variability in the Basin are harsher today than they would have been decades ago. Supply-side approaches, such as reservoir construction, have become less economically and environmentally viable while population growth and environmental considerations place additional demand on shrinking water supplies.

The research summarized here examines the use of climate forecasts to inform urban water provider decisions to participate in supply reliability transactions. We develop an econometric model of agricultural water use for agricultural areas diverting their water from the lower Colorado River. We then incorporate seasonal climate forecasts (Climate Prediction Center) for temperature and precipitation into that model in order to predict agricultural water use under a variety of summer temperature and rainfall conditions. These models allow us to consider the economic value associated with a seasonal forecast for urban water supply managers. Junior urban water supplies are modeled as dependent upon the water supplies remaining after senior agricultural uses. City water managers, therefore, must decide whether to engage in water leases to augment their summer season supplies based on forecasts of the quantity of water that agriculture will consume.

We conclude from our simulation that improved forecast skill for temperature and precipitation at more advanced lead times has high economic value in the context of water trading for supply reliability. Improved forecasts could enable urban water managers to secure supplemental supplies during shortages with more advance notice and at a lower unit price, providing substantial financial savings to urban water providers and their customers over the peak use summer season.

Literature review

We briefly review studies on the benefits of improved supply reliability and then on the value and application of seasonal climate forecasts to water resources management, primarily in the western United States.

Various studies examine the potential costs of climate-related water shortages and the value of mechanisms to improve supply reliability. Jenkins et al. (2004) optimization model of California's water supply system shows economic benefits of improvements to system operation and water allocations, including voluntary transfers, as having a statewide expected value of US$1.3 billion/year. Much of this is due to reduced costs to agricultural and urban users of accommodating environmental requirements. For the Rio Grande Basin of the southwestern U.S., Ward and Pulido-Vasquez (2012) find that it is hydrologically and economically feasible to manage the basin's water supplies sustainably. This can be achieved at a cost in the range of 6–11 percent of the basin's average annual total economic value of water considered over a 20-year time horizon. Heinz et al. (2007) consider the value of modeling water demand and supply and averting shortages in the context of the European Water Framework Directive. Integrated economic-engineering-hydrologic modeling is presented to improve cost-effectiveness in water management measures and to assist in public participation in decision processes.

Jenkins et al. (2003) estimate economic losses associated with reduced urban water supplies for projected 2020 population levels. Their economic loss functions indicate an average annual cost to end users of urban water scarcity in California in 2020 of US$1.6 billion per year. Olmstead and Stavins (2009) examine the costs in urban areas of achieving reductions in water use as a response to scarcity. They find that although water prices have rarely been used to allocate scarce supplies, using prices to manage water demand is more cost-effective than implementing non-price conservation programs. As with all policy choices, political considerations are central and explain the tendency toward non-price policy tools. In the face of political barriers to raising water prices, facilitating market-based transactions (water sales and leases) can usefully signal the opportunity cost of water to entitlement holders and ease the effects of drought on water users that are willing to pay for more reliable supplies (Colby and Frisvold 2011; Colby et al. 2011). Overall, the literature on costs associated with urban water shortages suggests that urban water managers can face high political and economic costs and seek measures to avert shortages when possible.

Designing and operating water markets will also entail transactions costs such as the fixed costs establishing the system and additional costs of expanding the system if drought causes shortages to become more severe (Booker et al. 2005). Developing institutions to protect third parties from the negative consequences of water trades will also involve additional costs. Concern over third party impacts are often the basis for existing institutional limits on trading.

Early economic research on climate forecasts focused on how forecasts are created and their potential applicability. Pagano and Hartmann (2001) examine Arizona water supply during the 1997–98 El Niño season, interviewing agencies about their experiences with and usage of climate forecasts. The authors conclude there is need for stronger relationships between forecasters and water managers, and that policies that allow more flexibility will permit water managers to make better use of forecasts. Pagano and Garen (2004) find that although climate forecasts do produce benefits to water managers (including farmers, city planners, and energy managers), benefits are limited by low forecast skill.

Hamlet and Lettenmaier (2000) focus on water supply issues in the Pacific Northwest. Using six different forecast scenarios, climate forecasts were used to model streamflows. The authors demonstrate the ability to forecast winter El Nino Souther Oscilliation (ENSO) with a lead time of roughly six months, also achieving accurate summer streamflow forecasts. Yao and Georgakakos (2001) analyze the response of Folsom Lake operations to historic climate variability and conclude that more reliable inflow forecasts immensely benefit reservoir performance.

Meza and Hansen (2008) review *ex ante* assessments analyzing the economic value of seasonal climate forecasts for agriculture. Uncertainty associated with climate forecasts forces decision makers to interpret certain impacts of a forecast, which can either be correct or incorrect and can lead to incorrect planning systems. The authors analyze a series of papers regarding climate change and agriculture and combine all of this data to form overall conclusions about the value of forecasting. They conclude that seasonal climate forecasts have positive, but modest, benefits to the agriculture sector. The authors suggest focusing upon improving forecast skill in those regions with the highest climate variability and those with high agricultural value, and broadening measures of forecast value to include environmental benefits and development.

Hamlet and Huppert (2002) analyze the Columbia River and the economic value of long lead forecasts for hydropower. They take six-month lead forecasts and use them to predict energy release cycles and demands. The authors find high streamflows lead to spot market energy sales, as well as increases in non-firm energy production and conclude that the use of long-lead forecasts result in an increase in hydropower revenue.

The consequences of misinterpretation of seasonal forecasts have not been widely examined in the existing literature. Hulme et al. (2009) examine a case study in Victoria, Australia, in which decision makers used a climate forecast to develop a water model for the next 15 years. In this example, the climate changed dramatically, the model developed did not anticipate the changes experienced and the area faced critical supply shortages. Hulme suggests that climate forecasts be used to predict what events will not happen, or to predict the worst that can happen rather than what is likely to happen.

Towler et al. (2010) develop a generalized approach that uses seasonal forecasts to predict the likelihood of exceeding a prescribed water quality limit. They generate four streamflow scenarios for the Bull Run River in the Pacific Northwest United States based on a seasonal probabilistic precipitation forecast, and then utilize a local logistic regression to estimate the likelihood of turbidity exceedance given the seasonal climate forecast. The authors find that, while the turbidity exceedance forecasts are useful in predicting turbidity threshold exceedance, they offered only slight advantages over using past turbidity occurrence data. This is due, most likely, to the coarse spatial resolution of seasonal climate forecasts.

Vano et al. (2010) examine the impact of projected climate change leading to earlier snowmelt and reduced summer flows for the reservoir systems of Everett, Seattle, and

---

Tacoma, Washington, on the ability of these systems to meet current and future summer demand. The authors consider the effects of climate change on water supply reliability via reservoir system models. They find that, assuming fixed demand, supply reliability in all three reservoirs will remain above 98 percent through the 2020s, though summer and fall storage will begin to shrink. However, with increased demand, reliability falls dramatically, especially after the 2020s, with corresponding negative effects on hydropower generation and flood control. The authors note that this analysis assumes current operating policies, however, and that some mitigation of climate effects could be achieved by adapting reservoir-operating policies.

Georgakakos et al. (2012) evaluate the performance of an adaptive decision model in managing the Northern California Sacramento/San Joaquin River reservoir system under historical and predicted future climate conditions. The authors find that, compared to the current policy, the adaptive policy: 1) maintains reservoir levels and reduces spillage more effectively, 2) better balances water deliveries between wet and dry years, 3) improves energy generation, and 4) meets environmental/ecological targets. The adaptive policy significantly outperforms the current policy in the future period. Georgakakos et al. conclude that adaptive management "can effectively counteract the adverse impacts of climatic change and maintain historical system performance," while management practices based on historical climate responses will perform poorly in the face of climate change.

Towler et al. (2013) discuss the application of climate-based seasonal streamflow forecasts to river management of an unregulated watershed in southwest Montana. First, the authors define the risk tolerance for the watershed—the minimum acceptable flow threshold and the probability of its exceedance—as 60 cubic feet per second (cfs) and estimate the number of days in which that threshold is exceeded. Then, using probabilistic precipitation forecasts in conjunction with a modeling approach that portrays relevant low-flow characteristics in terms of climate predictors, the authors produce season-ahead forecasts of the risk profile of the watershed for a given season. The authors discuss potential applications of these tools for climate risk management, including proactive fish conservation efforts. The authors conclude that their models provide a viable approach to relate climate to streamflow characteristics. They note that their risk-based framework could be utilized to further advances in climate prediction and adaptation.

Loch et al. (2013) find that increased water transfer activity will be a key part of adaptation to climate change impacts for all water use sectors in South Australia. This report considers economic, social and environmental impacts of water reallocation through markets, as well as needed institutional innovations. The authors concluded that water market transactions are essential in improving national capacity to deal with water scarcity under climate change.

Sankarasubramanian et al. (2009a) note that while probabilistic, seasonal forecasts are becoming increasingly available, water managers have been slow to use them. Water managers face a risk in altering water system operations using a probabilistic forecast because a supply shortfall that follows from following existing operating rules is "protected" by official policies. The authors propose a participatory water allocation process that can effectively use probabilistic forecasts. Users express their demand for water in terms of water quantity and timing needed at a particular reliability level, an associated willingness to pay, and acceptable compensation in the event of contract nonperformance. The water manager assesses feasible allocations using the probabilistic forecast to address these criteria across users, resulting in a set of short-term contracts over the operating period for which the forecast was issued. The authors develop such a model using data for a Brazilian urban water system and explore the roles of forecast skill and reservoir storage in this framework.

Sankarasubramanian et al. (2009a) examine the role of seasonal forecasts in setting up water supply contingency measures during extreme years and demonstrate the importance of using updated climate forecasts throughout the most crucial water supply season. They focus upon seasonal and intra-seasonal water allocation for a multipurpose reservoir system in the Philippines. The performance of the system is compared under three scenarios: forecasts issued at the beginning of the season, monthly updated forecasts during the season, and use of historic weather values. Operation of the system using monthly updated forecasts represents a significant improvement during below-normal water supply years compared to reliance on past weather data or the start-of-season forecast. The authors conclude that significant opportunities exist to improve water allocation by using updated seasonal climate forecasts.

In research conducted to date, there has been relatively little emphasis on links between use of forecasts, forecast accuracy and the economic benefits that reliable forecasts may yield through providing better information for water management decisions. This linkage between forecasts and economic benefits is explored in this chapter, focusing on the economic value of forecasts in water leasing decisions by urban water managers seeking to improve supply reliability.

Model of agricultural water use in the Lower Colorado River Basin

We develop a model of water use by irrigation districts which divert the Colorado River for agricultural purposes in western Arizona and southern California under long term water delivery contracts with the US Bureau of Reclamation (Reclamation). We estimate the model based on 14 years of data, and then apply the model to simulate agricultural water use under several seasonal climate forecasts. The large water entitlements represented in this model are senior to the water supplies available to large cities in the region and some districts have been involved in various agreements to improve urban supply reliability. Using Reclamation data on water use, we model water diverted as a function of acres planted, crop shares, precipitation and temperature. Water use and crop mix data is available going back to 1995.

The key dependent variable analyzed is monthly water diversions by agricultural contractors, as reported by Reclamation, measured in acre-feet. The explanatory variables were chosen based on factors hypothesized to affect diversions. The largest factor is clearly the number of acres planted. Another set of variables reflect the share of different crop groups: cotton, grains, corn, forage, tree and fruits/vegetables (fruit and vegetable are grouped together). This data comes from Reclamation annual reports on the lower Colorado River (Bureau of Reclamation, 2010).

One of the districts, Imperial Irrigation District, charges its member growers a modest marginal cost per acre-foot of water used and we include this in our model. No other district charges a marginal cost per unit of water used.

Table 15.1 Descriptive statistics

Variable name	N	Minimum	Mean	Maximum
Ln (water diversions)	1,512	2.89	8.97	12.84
Precipitation	1,512	0	0.211	5.37
Ln (water price)	1,512	0	0.189	2.71
Ln (temp)	1,512	3.85	4.26	4.57
Ln (acres)	1,512	7.16	9.84	13.23
Share cotton	1,512	0	0.13	0.52
Share grain	1,512	0	0.12	0.39
Share forage	1,512	0.05	0.38	0.85
Share corn	1,512	0	0.19	0.2
Share tree	1,512	0	0.09	0.74
Share fruits and vegetables	1,512	0	0.22	0.58

Source: authors.

Monthly average temperature and total monthly precipitation are used to capture the effect that variation in regional climate change may have on the amount of water being consumed. These values were taken for Arizona districts from the Arizona Meteorological Network (AzMet) and for California districts from the California Irrigation Management Information System (CIMIS).

Specific transformations have been applied to the data. To enable us to focus on percentage changes rather than absolute values, we apply a logarithmic transformation to monthly diversions, total acreage, temperature and water price. The rainfall variable is not log-transformed because some months have zero rainfall. The summary statistics for our data set can be found in Table 15.1, while a list of the variables, definitions and expected signs can be found in Table 15.2.

In order to correctly specify the model, specific statistical tests were performed. The Breusch-Pagan Lagrange multiplier test statistic was used to test for contemporaneous correlation, examining whether errors between separate panels are correlated (Greene 2008). The test for autocorrelation examines whether there is a time correlation between different monthly or yearly periods using a Durbin Watson test statistic (Greene 2008). To test for heteroscedasticity, we used a Lagrange multiplier test distributed as a Chi-squared (Greene 2008). In order to test for all three possible sources of misspecification, the statistical software program "SHAZAM" was used, enabling us to test for all three errors at the same time. Table 15.3 shows the results and corresponding test statistics.

All three sources of misspecification are present in the data. Using a procedure that accounts for contemporaneous correlation, autocorrelation and heteroscedasticity, we arrive at our final model. The parameter estimates and standard errors are shown in Table 15.4. Monthly diversions increase by 0.64 percent for every 1 percent increase in irrigated acreage. They also increase by about 0.363 percent for every 1 percent increase in temperature. Diversions decline with rainfall and the share of acreage allocated to grain and fruits and vegetables. Diversions increase with the share of acreage planted to tree crops. A 10 percent increase in the price of water per acre-foot contributes to a 0.72

Table 15.2 Agricultural water use model variables

Variable	Definition	Units of measure	Expected sign
Ln (water diversions)	Total amount of water diversions used by irrigation district	Acre-feet	Dependent variable
Precipitation	Total amount of precipitation in a given month	Inches	–
Ln (price)	Natural log of price per acre-foot of water paid by district farmers	Dollars	–
Ln (temp)	Natural log of average monthly temperature	Degrees fahrenheit	+
Ln (acres)	Natural log of total acres planted in a specific district	Acres	+
Share cotton	The acres planted for cotton divided by the total acres planted	% acres	–
Share grain	Acres planted for grains divided by total of acres planted	% acres	–
Share forage	Acres planted for forage divided by the total acres planted	% acres	–
Share corn	Acres planted for corn divided by the total acres planted	% acres	–
Share tree	Acres planted for tree crops divided by the total acres planted	% acres	+
Share fruits and vegetables	Acres planted for fruits and vegetables divided by total acres planted	% acres	–

Source: authors.

Table 15.3 Test statistics

Econometric problem tested for	Test implemented	Test statistic	Correction needed?
Contemporaneous correlation	Breusch-Pagan	$Chi^2 = 1004.5$ (91 d.f.)	Yes
Autocorrelation	Durbin Watson	Durbin Watson = 1.1859	Yes
Heteroscedasticity	Lagrange multiplier	$Chi^2 = 1505.4$ (13 d.f.)	Yes

Source: authors.

percent reduction in diversions. Overall, the model has high explanatory power, with Buse R^2 of 0.902.

The regression directly provides estimates of the predicted value of the log of diversions. Calculating the predicted value of actual diversions (as opposed to their log transformation) requires additional adjustment. If μ is the predicted value of the log of diversions then the expected value of diversions, $E(y)$, is

$$E(y) = exp(\mu + \sigma^2/2)$$

Table 15.4 Econometric results: monthly water use, Lower Colorado Basin River agriculture, June–August

Variable	Estimated coefficient	Standard error	Significance p–value
ln (diversions) lagged 12 months	0.450	0.035	0.000
ln (per acre–foot water price)	−0.072	0.020	0.000
ln (total crop acres)	0.643	0.038	0.000
Share of acres planted to cotton	0.280	0.320	0.383
Share of acres planted to grains	−1.475	0.579	0.011
Share of acres planted to forage	−0.035	0.189	0.853
Share of acres planted to corn	−1.000	0.624	0.109
Share of acres planted to tree and nut crops	1.347	0.238	0.000
Share of acres planted to fruits and vegetables	−0.270	0.067	0.000
ln (monthly mean temperature)	0.363	0.124	0.003
Monthly precipitation	−0.187	0.030	0.000
Constant	−2.701	0.569	0.000

Source: authors.

Notes: Dependent variable: log of monthly irrigation water diversions
Buse $R^2 = 0.9020$; Buse Raw Moment $R^2 = 0.9957$
Variance of estimate σ^2: 0.13767

where σ^2 is the variance of the regression estimate.

The model conforms to expectations about relationships between irrigation water use and the independent variables. Higher temperature increases agricultural use, precipitation decreases water use and the water price variable is highly significant with a negative effect on water use. A higher share of crops with high water requirements increases use and a higher share of crops with lower water requirements reduces use. This agricultural water use model provides a tool for simulating an urban water manager's water acquisition decision based on seasonal climate forecasts.

Simulation—urban water leasing decisions

Using the model we developed on agricultural water use and applying summer season climate forecasts from 1998–2003, we simulate the decision of a hypothetical urban water manager. The urban water manager uses climate forecasts to anticipate how much water agriculture will consume, in order to determine whether additional water needs to be leased for urban use. Our hypothetical urban water manager must consider whether agriculture is likely to use a larger or smaller amount of water than average in the summer season. Urban water supplies are modeled as being dependent on the residual (leftover) water available after agricultural use. This hypothetical situation represents a common condition in the western United States, in which agriculture holds the most senior water rights and urban supplies are affected by how much water agriculture consumes. If climate forecasts and the water diversion model predict that agriculture is expected to use more than an average amount of water, then the urban water manager will lease supplemental water to satisfy the upcoming June, July and August water delivery season.

Table 15.5 Urban water manager planning methods simulated

Planning method	Temperature and precipitation information used
Perfect foresight	Actual next summer temperature and precipitation known 13 months in advance, serves as baseline for comparison
1-month lead time	Summer temperature and precipitation values from forecast with 1-month lead time
7-month lead time	Summer temperature and precipitation values from forecast with 7-month lead time
13-month lead time	Summer temperature and precipitation values from forecast with 13-month lead time

Source: authors.

The urban water manager estimates the amount of water that agriculture will need in the upcoming June–August season using seasonal forecasts, and computes the difference between the predicted use and the average annual agricultural use. (The average amount of water used by agriculture is calculated by averaging the total amount actually used over the summer seasons 1998–2003.) We focus on summer because water demand is highest in the southwest for both agricultural and urban use in the summer. Despite large reservoirs on the lower Colorado River, experience with extended drought over the last decade indicates that junior entitlement holders (primarily urban areas) are subject to cutbacks.

Our simulation analyzes the differences in anticipated amounts of water consumed by farms based on different forecast lead times. Using the agricultural water use levels predicted from our model, we simulate how much water an urban water manager will need to lease and the acquisition costs, with costs depending on how far in advance the water leases are arranged.

A baseline planning method of perfect foresight is used as a control for comparative purposes. In the perfect foresight baseline, we use actual temperature and precipitation values for each year to estimate how much water is used by agriculture. These agriculture water use values serve as a baseline (perfect foresight) from which to assess the effects of using seasonal climate forecasts.

To examine the effect of using seasonal forecasts, we use the actual Climate Prediction Center seasonal climate forecasts for the relevant years with a 1-, 7-, and 13-month lead time. These forecasts are made available on public websites at the lead times specified and are available to decision makers. Forecasts used in this analysis are for Climate Divisions 94 and 96, which contain the agricultural areas diverting from the mainstem of the lower Colorado River. Table 15.5 summarizes the forecast lead times and perfect foresight assumption used in our simulations.

The parameter estimates from our agriculture water use model allow us to predict the amount of water used for agriculture based on temperature and precipitation values provided by forecasts with different lead times. In order to focus upon the role of seasonal forecasts, we use actual observed monthly values for acres planted, price and crop shares so that these are the same regardless of planning method. The only variables that change across simulations are temperature and precipitation. Since our base model analyzes the log of water use, we also re-transform our dependent variable back to the untransformed water use variable. With this transformation, we arrive at a calculation of estimated water diversion by agricultural water users in the summer season.

We use four levels of water acquisition costs (per acre-foot of water leased) to examine the financial consequences of different lead times for the urban water planner. The acquisition cost per unit corresponds to the lead time in which the urban water manager has access to information about the level of agricultural water use in the upcoming summer season and makes a decision about acquiring additional water supplies. The lease price figures we use here are illustrative of lease arrangements observed in the region, but must be considered hypothetical due to lack of robust transaction price data for leases negotiated at various lead times.

The lowest lease price, at US$50 per acre-foot, reflects a water manager acquiring water through a voluntary lease transaction arranged 12 or more months in advance. The second price, US$70 per acre-foot, represents the cost when a decision maker arranges a water lease more than 7 months in advance, but less than 12 months in advance. The third price, US$120 per acre-foot, reflects the cost when a decision maker arranges a water lease more than 1 month in advance, but less than 7 months in advance. These figures are based on recent pilot water leasing agreements between the Bureau of Reclamation and Arizona irrigation districts (Bureau of Reclamation 2008, 2009). The highest potential price paid is US$210 per acre-foot. This price reflects an urgent situation in which the water lease is negotiated within one month or less of urban water supply need. O'Donnell (2010) found this price reflective of such water leases in California. A summary of the decisions an urban water manager may make based on different planning methods and lead times, and the economic ramifications of those decisions, are presented in Table 15.6.

The 13-month, 7-month and 1-month lead time forecasts all predict that agriculture will use more water than average and therefore water managers believe they must negotiate water leases for the upcoming summer season. Comparing these columns to perfect foresight, we see that some water acquisitions prove to be unnecessary given actual summer precipitation and temperature and lower-than-forecast irrigation water use. Using the forecast information leads to an overinvestment in water acquisitions quantified as the difference between perfect foresight lease costs and lease costs associated with relying upon seasonal forecasts with 13-month, 7-month or 1-month lead times (Table 15.6). This "overinvestment" ranges up to US$24 million dollars for a single summer season's water lease costs.

Conclusions

Our simulation model allows us to estimate a monetary "stake" associated with using seasonal forecasts for estimating agricultural water use. Irrigated agriculture accounts for over 75 percent of water use in the Lower Colorado River Basin. Accurate forecasts of agricultural use with lead times of seven months or more can provide significant benefits to urban water managers who may need to lease supplemental water supplies.

Each of the seasonal forecast lead times (1, 7 and 13 months) provide the urban water manager with overestimations of water use by agriculture. Consequently, water leasing decisions made using forecast information result in the urban supply manager purchasing water that eventually proves to be unnecessary once actual agricultural

Table 15.6 Simulation results: summer (June–August) water use and predicted water use, 1998–2003

	Average water use summer 1998–2003	1-month lead	7-month lead	13-month lead	Perfect foresight 13-month lead
Difference (average minus predicted) acre-feet	1,941,460	141,649	124,406	141,270	114,321
Cost/acre-foot		US$210/acre-foot	US$120/acre foot	US$50/acre-foot	US$50/acre-foot
Lease cost		US$29,746,290	US$14,928,720	US$7,063,500	US$5,716,050
Cost difference (versus perfect foresight)		US$24,030,240	US$9,212,670	US$1,347,450	

Source: authors.

summer use is observed. However, if the city can store this water for future use the acquisitions may still have value. Moreover, urban water managers may view leases of water that turn out to be unnecessary for the upcoming summer as a worthwhile form of insurance against the opposite situation—deciding not to lease supplemental water based on forecasts and then being caught short in the summer. This latter predicament would be costly as the urban water provider likely would be forced to pay a very high price for short turnaround water leases, or would incur significant political and financial costs due to cutting back the quantities customarily supplied to their customers.

We do not find that there is a consistent preferred lead time in terms of forecast used by an urban water manager looking ahead to estimate summer agricultural use. This result confirms the findings of Livezey (2008), who found that lead time currently is not a significant factor affecting the skill of forecasts. Our model does indicate that improved temperature and precipitation forecasts at 7- and 13-month lead times would have significant value in water supply planning. More accurate forecasts would permit better estimation of agricultural water use during peak use seasons when competition for water is highest, and this would allow water lease arrangements to be made further in advance and more cost-effectively.

There are many possible extensions of this work in understanding the economic value of seasonal forecasts for water management. While we focused upon the role of forecast lead time, follow up analyses could employ a similar model to examine the incremental effects of changes in forecast accuracy at various lead times. Our findings depend upon variations in the cost of leasing water at various lead times, variations indicated by water lease arrangements in the southwestern U.S. The cost per unit of water leased could also vary with the volume of water placed under a lease contract and various changes in leasing cost structure could be explored using a simulation model of the type developed here.

To summarize, temporary and intermittent water transfers have key advantages over permanent acquisitions to improve supply reliability in the face of climate change. In years of high and average water supply, farmers use their water for irrigation. They earn revenues by leasing their water and refraining from irrigating in dry years. Such dry-year arrangements provide cities and environmental restoration programs the option of leasing water only during periods when supply augmentation is needed. Temporary transfers allow for season-to-season decisions about leasing supplemental water and avoid some of the economic and political costs of permanently acquiring water rights. With temporary transfers, ownership of the water entitlement remains with the agricultural user and this can mitigate political opposition and third-party impacts of cropland fallowing.

Addressing climate-related increases in regional water supply variability requires careful weighing of probabilities and costs of shortfalls. Pre-arranged temporary transfers to secure urban water supplies are likely to be more cost-effective than leases arranged on short notice as a supply shortfall looms. Seasonal climate forecasts can be a valuable tool for urban water managers weighing tradeoffs associated with leasing water from agricultural districts.

References

Booker, J. F., Michelsen, A.M., and Ward, F.A. (2005) Economic impact of alternative policy responses to prolonged and severe drought in the Rio Grande Basin. *Water Resources Research*, vol 41, no 2. W02026, doi:10.1029/2004WR003486

Bureau of Reclamation (2008) Agreement Between the United States Bureau of Reclamation and the Yuma Mesa Irrigation and Drainage District to Implement a Demonstration Program for System Conservation of Colorado River Water, Agreement No. 08-XX-30-W0534 (Oct. 2008)

Bureau of Reclamation (2009) Agreement Between the United States Bureau of Reclamation and the Yuma Mesa Irrigation and Drainage District to Implement a Demonstration Program for System Conservation of Colorado River Water, Agreement No. 08-XX-30-W0522 (Dec. 2009)

Bureau of Reclamation (2010) *Colorado River Accounting and Water Use Report: Arizona, California, and Nevada*. Bureau of Reclamation Lower Colorado Region Boulder Canyon Operations Office

Christensen, N. and Wood, A. (2004) The Effects of Climate Change on the Hydrology and Water Resources of the Colorado River Basin. *Journal of Climatic Change*, vol 62, no 1–3, pp 337–363

Climate Prediction Center (2010) ENSO Cycle: Recent evolution, Current status and Predictions

Colby, B. and Frisvold, G. (2011) *Addaption and Resilience: The Economics of Climate-Water-Energy, Challenges in the Arid Southwest*. Washington, DC: Resources for the Future Press.

Colby, B., Jones, L. and Pittenger, K. (2011) Economic Tools for Climate Adaptation: Water Transaction Price Negotiations, Chapter 4 in B. Colby and G. Frisvold (eds), *Adaptation and Resilience: The Economics of Climate-Water-Energy Challenges in the Arid Southwest*, Washington, DC: Resource for the Future Press.

Garrick, D. and Jacobs, K. (2006) Water management on the Colorado River: from surplus to shortage in five years. *Southwest Hydrology*, vol 5, no 3, pp 8–9.

Georgakakos, A.P. et al. (2012) Value of Adaptive Water Resources Management in Northern California under Climatic Variability and Change: Reservoir Management. *Journal of Hydrology*, vol 412–413, pp 34–46

Gong, G., Wang, L., Condon, L., Shearman, L. and Lall, U. (2010) A Simple Framework for Incorporating Seasonal Streamflow Forecasts into Existing Water Resource Management Practices. *Journal of the American Water Resources Association*, vol 46, no 3, pp 574–585

Greene, W. (2008) *Econometric Analysis*, 6th Edition. Upper Saddle River, NJ: Pearson Education.

Hamlet, A. and Lettenmaier, D. (2000) Long-Range Climate Forecasting and Its Use for Water Management in the Pacific Northwest Region of North America. *Journal of Hydro-informatics*, vol 2, no 3, pp 163–182

Hamlet, A. and Huppert, D. (2002) Economic Value of Long-Lead Streamflow Forecasts for Columbia River Hydropower. *The Journal of Water Resource Planning and Management*, vol 128, pp 91–101

Hartmann, B. and Wendler, G. (2005) The significance of the 1976 Pacific climate shift in the climatology of Alaska. *Journal of Climate*, vol 18, pp 4824–4839

Heinz, I., Pulido-Velazquez, M., Lund, J.R., and Andreu, J. (2007) Hydro-economic Modeling in River Basin Management: Implications and Applications for the European Water Framework Directive. *Water Resources Management*, vol 21, no 7, pp 1103–1125

Hulme, R., Pielke, R. and Dessai, S. (2009) Keeping prediction in perspective. *Nature Reports Climate Change*, vol 3, pp 126–127

Jenkins, M., Lund, J.R. and Howitt, R.(2003) Using Economic Loss Functions to Value Urban Water Scarcity in California. *Journal of the American Water Works Association*, vol 95, no 2, pp 58–70

Jenkins, M. et al. (2004). Optimization of California's Water Supply System: Results and Insights. *Journal of Water Resources Planning and Management (Impact Factor: 1.71)*, vol 130, no 4, pp. 271–280

Kenney, D., Ray, A., Harding, B., Pulwarty, R. and Udall, B. (2010) Rethinking vulnerability on the Colorado River. *Journal of Contemporary Water Research & Education*, vol 144, no 1, pp 5–10.

Livezey, R. (2008) The First Decade of Long Lead U.S Seasonal Forecast. *American Meteorological Society*, vol 89, pp 843–854

Loch, A. et al. (2013) The role of water markets in climate change adaptation. National Climate Change Adaptation Research Facility, Gold Coast

Meza, F. and Hansen, J. (2008) Economic Value of Seasonal Climate Forecasts for Agriculture: Review of Ex-Ante Assessments and Recommendations for Future Research. *The Journal of Applied Meteorology and Climatology*, vol 47, pp 1269–1286

O'Donnell, M. (2010) Innovative Water Supply Reliability Arrangements. Master's thesis. The University of Arizona School of Agricultural and Resource Economics

Olmstead, S. and Stavins, R. (2009) Comparing Price and Nonprice Approaches to Urban Water. *Conservation Water Resources Research*, vol 45, no 4

Pagano, T. and Garen, D. (2004b) Use of Climate information in the Western United States. *ASCE Conf. Proc.* 118, 377

Pagano, T. and Hartmann, H. (2001) Using Climate Forecasts for Water Management. *Journal of the American Water Resources Association*, vol 37, pp 1139–1153

Rajagopalan, B. et al. (2009) Water supply risk on the Colorado River: Can management mitigate? *Water Resources Research*, vol 45, no 8, DOI: 10.1029/2008WR007652

Sankarasubramanian, A., Lall, U, Souza Filho, F.A. and Sharma, A. (2009a) Improved Water Allocation Utilizing Probabilistic Climate Forecasts: Short-Term Water Contracts in a Risk Management Framework. *Water Resources Research*, vol 45, pp 1–18

Sankarasubramanian, A., Lall, U., Devineni, N. and Espinueva, S. (2009b) The Role of Monthly Updated Climate Forecasts in Improving Intraseasonal Water Allocation. *Journal of Applied Meteorology and Climatology*, vol 48, pp 1464–1482

Towler, E., Rajagopalan, B., Summers, R.S. and Yates, D. (2010) An Approach for Probabilistic Forecasting of Seasonal Turbidity Threshold Exceedance. *Water Resources Research*, vol 46, pp 1–10

Towler, E., Roberts, R., Rajagopalan, B., and Sojda, R.S. (2013) Incorporating Probabilistic Seasonal Climate Forecasts into River Management Using a Risk-Based Framework. *Water Resources Research*, vol 49, pp 1–12

Vano, J.A. et al. (2010) Climate Change Impacts on Water Management in the Puget Sound Region, Washington, USA. *Climatic Change*, vol 102, no 1–2, pp 261–286

Ward, F. and Pulido-Velazquez, M. (2012) Economic Costs of Sustaining Water Supplies: Findings from the Rio Grande. *Water Resources Management*, vol 26, no 10, pp 2883–2909

Yao, H. and Georgakakos, K. (2001) Assessment of Folsom Lake Response to Historical and Potential Future Climate Scenarios. *Journal of Hydrology*, vol 249, pp 148–175

Data websites

Azmet.com. The Arizona Meteorological Network, March 2009. http://ag.arizona.edu/AZMET/az-data.htm

Bureau of Reclamation, March 2010. http://www.usbr.gov

Cimis.com. California Irrigation Management Information Systems, April 2009. http://www.cimis.water.ca.gov/cimis/welcome.jsp

Climate Prediction Center, August 2009. http://www.cpc.noaa.gov

Imperial Irrigation Website, January 2010. http://www.iid.com

PART IV

Water markets and institutions around the world

16

A CALIFORNIA POSTCARD

Lessons for a maturing water market

Ellen Hanak

Introduction

California has the largest population, economy, and agricultural sector of the 50 states within the United States of America. It has achieved this status despite a highly variable climate and a sharp geographic mismatch between the locations of water supplies and demands: most precipitation falls in the state's northern and eastern mountain ranges, and most population and farming centres are in coastal plains and inland valleys to the south and west. During the early to mid-twentieth century, local, state, and federal agencies constructed a vast, interconnected water infrastructure network to store water for seasonal irrigation and dry years and to deliver water to distant demand centres. In recent decades, the management emphasis has shifted toward using water resources more efficiently, reflecting new environmental sensibilities and constraints on resource availability that preclude meeting new demands solely through continued expansion of water supply infrastructure.

Water marketing—the temporary, long-term, or permanent transfer of rights to use water in exchange for compensation—is a key part of this new approach. The prices negotiated for these transfers provide useful information to all parties about the economic value of water, creating incentives to conserve water, to invest in local infrastructure to reduce conveyance losses from evaporation and leakage, and to coordinate infrastructure uses state-wide. In this way, the market helps California's overall water use become more economically efficient. Short-term transfers (within a given year) are especially useful for coping with droughts. Long-term and permanent transfers facilitate longer-term shifts in economic activity and the associated changes in the pattern of water demands. Given the physical, financial, and environmental limits on expanding overall water supplies in California—and the prospect of supply reductions caused by a warming climate—the water market will become an increasingly valuable tool for supporting a healthy economy, along with other tools that improve the economic efficiency of water use and water infrastructure (Joyce et al. 2009; Hanak et al. 2011; Medellín-Azuara et al. 2012; Hanak and Lund 2012).

California's water market enjoys some celebrity in national and international water management circles. One of its early manifestations was a large, state-run "Drought Water Bank"—a successful emergency brokerage operation introduced in 1991, several years into a major drought (Howitt 1994a; Israel and Lund 1995). California was also one of the first to experiment on a large scale with environmental water trading, notably through the use of an "Environmental Water Account" in which the state purchased water to reduce the impacts of water diversions on endangered native fish species in the early to mid-2000s (Brandt 2001; Burke et al. 2004; Hollinshead and Lund 2006). As usual, celebrity also comes with some bad press. In conversations about water marketing, California's market is sometimes compared unfavorably (especially by economists) with markets that operate at arm's length, such as electronic water trading in Australia's Murray-Darling Basin.[1] Instead, water trading in California is a somewhat regimented business, with a set of rules that can seem unfathomable to anyone not steeped in the complexities of the state's system of water rights, and inimical to the efficient allocation of resources.

The state's water market has nevertheless evolved over the past 25 years to become an essential, if imperfect, instrument of modern water management, and one that will become ever more important as the state accommodates anticipated population growth and grapples with the effects of climate change. This chapter provides an overview of the evolution of this market. It begins with a short primer on water rights in California and how they relate to water trading. It then describes the evolving policy context for California's water market, presents trends in market growth using data collected by the author, and discusses lessons from the California experience for fostering the responsible development of this water management tool.[2]

Water rights, "wet water," and the market

California's Water Code provides two basic guidelines on who can participate in the water market: Sellers must have the rights to use the water throughout the term of the proposed transfer, and the water they sell must be "wet"—i.e., physical water, not merely unused "paper" rights (as described below). From a practical standpoint, sellers and buyers must also be able to get the water from the source to the destination, making suitable infrastructure a key ingredient.

Water-use rights

In California, surface water rights have generally been held for many decades under the state's "first in time, first in right" legal system—similar to the appropriative water rights system operating in most parts of the American west.[3] Most of these rights have been allocated on the basis of seniority. One main premise of the market is that senior rights-holders (who have more reliable—and hence more valuable—supplies) often have relatively low-value uses for at least some of their water. The market provides incentives for water-rights-holders with more ample supplies and relatively lower-value uses to transfer some water to parties with less ample supplies and higher-value uses.

Most surface water rights are held by local public agencies: special-purpose public agencies (or "special districts") and some municipalities. Legally, some of these agencies actually hold long-term "contract entitlements" rather than "rights" to surface water; in these cases, the local parties have contracts with federal or state agencies that run large water projects and hold the associated water rights. The ultimate rights-holder for the federally-owned Central Valley Project (CVP) is the U.S. Bureau of Reclamation (USBR). The California Department of Water Resources (DWR) plays a similar role for the state-owned State Water Project (SWP).

Groundwater is also an important source of water in California—constituting about a third of state-wide use on average, and more in dry years and in some regions.[4] In contrast to surface water, there are few places where groundwater rights are "quantified" (i.e., where users have rights to withdraw a specific quantity of water). Thus, the right to pump groundwater is generally available to all private individuals overlying the aquifer, as well as municipalities that have staked a claim to groundwater use as appropriators. Adjudicated basins—where groundwater rights *are* quantified—are located principally in urbanized areas of Southern California; trading among rights-holders within these basins is common.[5] Specialized local agencies that regulate access to aquifers are also found in some urbanized areas; these districts charge pumping fees and manage recharge programs with the proceeds. Groundwater management elsewhere in the state remains largely voluntary, without pumping restrictions or pumping fees (Nelson 2011; Association of California Water Agencies 2011). As described below, local restrictions have limited the extent to which groundwater can be sold in many places where basins are not fully managed.

Water transfers in California usually are formal agreements to trade between public agencies holding water rights or contract entitlements, with many of these trades made on behalf of individual farmers.[6] Because most water is still used for irrigation in California, most water is leased or sold by farmers or irrigation districts, who market water to other farmers with scarce supplies and higher value crops, to growing cities, and to environmental programs.[7] In some water districts, individual farmers have specific amounts of surface water (or "allocations") assigned to them and are therefore in a position to sell or lease this water, whereas in others the district will make this determination and compensate farmers who agree to participate in the transaction. In rural areas, some water districts' governing boards are elected by a weighted vote of property owners, while in others, a one-person-one-vote rule applies. It is often thought that districts with the property-weighted voting rules (which more heavily represent local farmers) are more likely to sell or lease water to other parties.[8]

Wet water

"Wet water" is the term commonly used in contrast to "paper water"—water rights held on paper for which actual water is not available. Under the appropriative water rights doctrine governing most of California's surface water, the "use it or lose it" requirement dictates that rights lapse for any water not used for five consecutive years (Water Code § 1241); this restriction is designed to prevent hoarding and speculation. It is generally acknowledged that there are many more claims on surface water than the physical water typically available in the system.[9] This discrepancy arises from a combination of inactive claims that are still on the books and use rights that are only available in high flow years.[10] In addition, some water is used more than once within the same season, because most active claimants return some of the water they divert after using it. Such "return flows" (e.g., from irrigation drainage and wastewater treatment plants) are then available for reuse by others.[11]

Water-rights-holders must generally demonstrate that the water they propose to lease or sell is indeed "wet"—i.e., water they would have used otherwise in that season or legally stored for later use. Without this safeguard, the seller could end up transferring "paper" water that someone else is already legally using, causing harm to that user (or in legal parlance, "injury"). This policy stands in stark contrast to the early operation of Australia's water market, where participants were allowed to transfer any water they held title to, including so-called "sleeper" rights that were not in use (Quiggin 2006; Hansen et al. 2013).

There are four potential sources of wet water:

- excess water stored in surface reservoirs to which the seller has rights,
- other excess amounts of surface water that the seller has the right to use, but does not need and cannot store,
- "conserved" surface water that the seller saves by reducing his or her own use, and
- groundwater.

The first two sources of excess surface water are not widely available. Only a limited number of rights-holders have surpluses available in surface storage, and the use-it-or-lose-it principle limits conditions under which other excess surface supplies are actually considered wet. CVP and SWP contractors are the only ones who have been able to regularly sell unstored excess surface supplies, because their rights to use water are determined by contract rather than the appropriative doctrine. Otherwise, rights-holders may sell the excess surface water generated in very wet years, when they are likely to have less need for irrigation water. These are times when overall market demand is more limited as well.

The third source of wet water, conservation—or reduced "net" water use—is a more generally available option. Two principal ways to achieve conservation are land fallowing and switching to crops that use less water. In some cases, conservation savings can also be achieved through investments to improve the efficiency of the conveyance and use systems (e.g., canal lining, installation of drip irrigation, and water recycling), although such investments may be discounted when they reduce the amount of water returned to the system.[12]

Groundwater, the fourth potential source of wet water, can be transferred directly or can be used on-site in lieu of surface water transferred to another party. The latter practice is known as "groundwater substitution" or "groundwater exchange." Groundwater-related transfers are subject to less oversight from the state than surface water transfers because the state's Water Code does not apply to most groundwater.[13] In many counties, the ability to transfer groundwater, either directly or through groundwater substitution, depends on local ordinances, which have sought to prevent harm to local users by limiting groundwater exports (described further below). Because many groundwater basins have a hydrologic connection to surface waters, groundwater-related transfers can also be limited by the "wet water" principle. In particular, groundwater pumping can reduce surface flows in adjacent streams. For this reason, DWR has developed guidelines for Sacramento Valley groundwater transfers, restricting the location of wells that can be used for transfers and setting pumping ratios that deduct for the loss of surface water that occurs with pumping.[14]

Although many aspects of water management in California are highly sophisticated, the system is also known for information gaps that seem surprising for a modern economy. For instance, state-level reporting of some surface water use is incomplete, and groundwater use is only systematically reported in some managed basins. These information gaps may limit the extent of trading. As described below, most surface water trading now occurs between parties that are part of large federal and state projects, for which monitoring is more extensive. Other parties wishing to gain approval to trade water outside of their service areas must demonstrate their use patterns and the consistency of those patterns with their authorized water rights—an up-front investment cost that may be hindering the development of trades in some areas. Groundwater trades are mostly governed by local authorities, and the practice is most prevalent in areas with transparent monitoring of withdrawals.

Infrastructure

Of course, water marketing cannot happen without a hydrologic connection between sellers and buyers (Israel and Lund 1995). Fortunately, California's extensive water infrastructure network has enabled a comparably extensive ability to run a state-wide market. Numerous large water projects developed in the early to mid-twentieth century—including the CVP, the SWP, and investments to harvest water from the Colorado River and various local rivers—have forged hydrologic connections among most population and farming centres. Where the connections between transacting parties are indirect, the market works through a series of exchanges, where one or more intermediary agencies take the seller's water and make an equivalent amount available to the buyer. Exchanges are also important in cases where the terms of water rights do not readily allow direct transfers, for instance between members of different water projects.

As described below, the market has recently experienced new frictions because of infrastructure constraints at a key conveyance hub in the Sacramento-San Joaquin Delta, limiting both north-to-south and east-to-west transfers.

The approval process

Most transfers require approval from one or more state or federal authorities, but the process varies with the nature of the water right and the source of water. The State Water Resources Control Board (SWRCB) must approve transfers (i.e., changes in purpose, place of use, or point of diversion) involving any surface water rights established since 1914 (the year the state's modern Water Code came into effect). Transfers of surface water among contractors within the CVP and SWP generally do not require SWRCB approval because they do not involve a change in the purpose, place of use, or point of diversion assigned to the overall water right, but the projects themselves must authorize these transfers. Transfers of groundwater and surface water held in pre-1914 appropriative rights do not require SWRCB approval because the board does not have direct regulatory jurisdiction over these types of water rights. However, transfers of these rights do require public notice and review under the California Environmental Quality Act (CEQA), and also the National Environmental Protection Act (NEPA) when federal water rights are involved in the transfer.[15] Such transfers also come under state or federal jurisdiction if government-owned conveyance facilities are involved, which is likely to be the case in many areas of the state. In particular, transfers conveyed through the Sacramento-San Joaquin Delta typically need to use the state-owned pumping plant, requiring DWR approval. And, as noted above, groundwater-related transfers from many rural counties require a county permit demonstrating no injury to local groundwater users.

The evolving policy context for water trading

For much of the past several decades, state and federal policies have supported the development of water marketing through a suite of actions, including legal revisions to facilitate marketing, brokering of deals among transacting parties, and direct purchases of water. More recently, various state and federal authorities have also placed new restrictions on water transfers. At the same time, local resistance to transfers has arisen in some source regions over concerns that transfers could harm local economies—as evidenced by the rise of county ordinances restricting groundwater exports and other objections.

State and federal support initiatives

State initiatives to support the market began in 1977, a year of severe drought. Two reports commissioned at that time, one by the governor and one by the legislature, strongly endorsed water marketing as an element in the state's strategy for handling its future water needs (Governor's Commission to Review California's Water Rights Law 1978; Phelps et al. 1978). The governor's commission also advocated various changes in the Water Code to facilitate transfers, notably provisions to ensure the security of water rights for parties leasing water to others and to ensure access to the use of conveyance facilities owned by other parties. Although many of the recommendations were accomplished in the years that followed, the 1980s saw little uptake in market activity.

In the early 1990s, several events significantly changed the trading climate. First, natural conditions provided the occasion for a large-scale experiment in water trading when a multiyear drought prompted the state to initiate an emergency water bank in 1991. The following year, in response to findings that the federally run CVP was harming the indigenous wildlife of the San Francisco Bay-Delta Estuary, Congress passed the Central Valley Project Improvement Act (CVPIA). The Act mandated that nearly 1,000 million cubic meters of project water (about 11 percent of total project yield) be returned to in-stream uses to regenerate salmon runs, and that another 500 million cubic meters be allocated to wildlife refuges. The CVPIA also contained provisions to facilitate water marketing and introduced a mechanism for the project to purchase additional water for environmental purposes.

In 1994, the SWP contractors concluded negotiations for the Monterey Agreement, a revision of project operating rules that facilitated water marketing by the contractors. It authorized the first permanent transfer of contract entitlements from some agricultural contractors to urban contractors and established other measures to make it easier for contractors to transfer water to one another.

Two other significant activities were undertaken by the state and federal governments in the late 1990s and in 2000. Under instructions from the U.S. Secretary of the Interior in 1996 and 1997, California began to devise a plan to reduce its use of Colorado River water to the contractually allocated amount of 5,400 million cubic meters over a 15-year period. This "4.4 Plan" (so-named for the volume in local measurement units—4.4 million acre-feet) created strong incentives for water transfers between agricultural and urban users of Colorado River water within California, leading to a suite of long-term agreements that were finalized by 2003. In 2000, state and federal authorities launched the Environmental Water Account (EWA), a program of environmental water purchases under CALFED—a multiagency state and federal program with the goal of restoring health to native fishes in the San Francisco Bay-Delta Estuary while securing water supplies to agricultural and urban users. Although the CALFED program was superseded by new governance arrangements under legislation passed in late 2009, the EWA continues to operate on a diminished scale.

During the 2000s, the state also ran smaller drought water-purchase programs in two dry years (2001 and 2009), and the federal government sought to further ease transfer restrictions for CVP contractors and others wishing to use CVP-owned transfer facilities. Federal-state cooperation also increased, as the CVP and SWP were granted temporary "joint place of use" south of the Delta during the drought of the late 2000s. Under this arrangement, water allocations for the two projects were treated as though they derived from the same water right, making it possible for contractors from the two projects to transfer water to each other without seeking SWRCB approval for each transfer—a time-consuming step in the normal approval process.

Local concerns in source regions

In tandem with state and federal efforts to expand the market in the 1990s, concerns over the prospect of damage to local economies led to an increase in county-level groundwater ordinances in the state's rural areas (Hanak 2003; Hanak and Dyckman 2003). These ordinances all restrict direct groundwater exports; most also restrict groundwater substitution transfers, and some also aim to restrict the practice of storing groundwater by non-local parties (Figure 16.1). The absence of state-level no-injury protections for groundwater derives from incomplete state groundwater regulation; county ordinances have been deemed legal because this type of injury needs to be prevented to avoid harm to other legal water users.[16] (In economic terms, transfers that reduce other users' access to water generate negative physical externalities and must be mitigated to avoid economically inefficient transfers.) However, rather than encourage a process of review and proper mitigation of transfers, the ordinances appear to have worked principally to discourage groundwater-related transfers altogether from these counties.[17]

Some groups are also concerned about the potential for fallowing-based transfers to cause local economic harm. Transfers compensate farmers who fallow fields for the crop revenues they forego; but local businesses, workers, and governments do not automatically receive compensation for any economic harm they might suffer when the crop activity declines. Fallowing-related damages (or "pecuniary externalities") are not proscribed under state law, which generally views them as a natural consequence of a shifting economy, much like the opening or closing of a manufacturing plant may

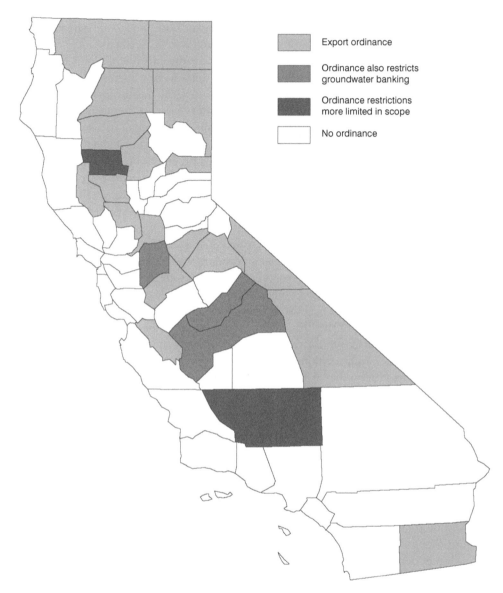

Figure 16.1 County ordinances restrict groundwater export from many rural counties (source: updated from Hanak, 2003).
Note: All but one of these ordinances were in place by 2002. Stanislaus County adopted an ordinance in 2013.

affect neighbouring businesses and property values for better or worse. Studies of actual and projected fallowing operations suggest that the aggregate local losses are likely to be limited, because farmers tend to fallow their least profitable fields.[18] State law does require public hearings if a local agency wishes to transfer water available through fallowing and the volume exceeds 20 percent of the agency's water supplies (Water Code § 1745.05). When transfers use a public entity's conveyance facilities, that entity is also required to ensure that they do not cause significant economic harm (Water Code §

1810). Transacting parties have voluntarily developed community mitigation programs for two large, long-term fallowing-based transfers of Colorado River water; these may serve as useful models for similar transfers in the future.[19]

New state and federal restrictions

In the late 2000s, some state and federal policies also increased restrictions on marketing. Over time, DWR has progressively increased its scrutiny of transfers between points north and south of the Delta over concerns of potential injury to the SWP. The SWP holds relatively junior water rights to Sacramento Valley water, which it ships to most of its contractors through pumps in the south Delta. Concerns that transfers are impinging on these flows have led to increased restrictions on acceptable locations and ratios for groundwater substitution transfers and tighter restrictions on acceptable fallowing arrangements.[20]

Since 2007, the ability to move water through the Delta has also become more limited as a result of concern for endangered fish species, whose populations plummeted in the preceding years. Although the new pumping restrictions have not specifically targeted water marketing, they limit the physical infrastructure available for transfers, which have lower priority than CVP and SWP when it comes to using the project-owned pumps in the south Delta.

Finally, new environmental restrictions have been imposed on transfers for reasons other than protection of in-stream flows, the traditional focus of environmental no-injury protections. Fallowing of fields to make water available for transfers has been severely restricted in rice-growing areas (primarily in the Sacramento Valley) to prevent harm to habitat of the endangered giant garter snake, a listed species that now depends on artificial wetlands created by irrigation water.[21] Groundwater-related transfers have also become subject to Clean Air Act restrictions against the use of diesel pumps—a regulation from which farmers are normally exempt when they pump water for their own use. As discussed below, it is likely that these assorted restrictions have dampened market activity in recent years.

Trends and phases in market development

Water marketing has grown significantly in California over the past three decades. Figure 16.2 shows actual flows through 2011 under short- and long-term lease contracts (black and dark grey bars), estimated flows under permanent sale contracts (light grey bars), and the additional volumes committed under long-term and permanent contracts that were not transferred in those years (white bars). Background shade indicates years characterized as dry or critically dry. Annual trades in the early 1980s averaged around 125 million cubic meters. The market took off during a multiyear drought in the late 1980s and early 1990s, spurred by direct state purchases and the development of an emergency drought water bank. The market continued to grow when the rains returned; and by the early 2000s, the annual volume of water committed for sale or lease was around 2,500 million cubic meters, with roughly 1,600 million cubic meters actually moving between parties in any given year. These volumes increased slightly by the end of the decade, and trades now represent about 5 percent of all water used by businesses and residents in the state.[22]

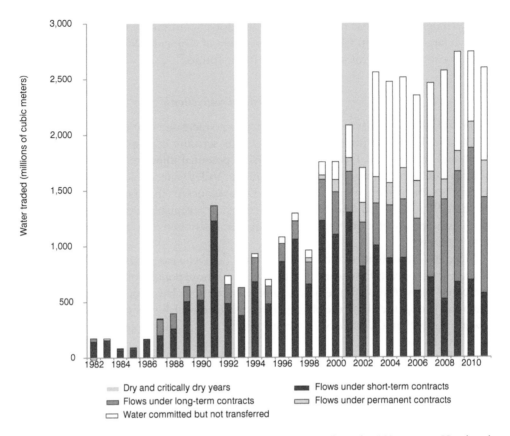

Figure 16.2 California's water market has grown substantially since the early 1980s (source: Hanak and Stryjewski, 2012).
Notes: The figure shows actual flows under short- and long-term lease contracts (black and dark grey bars), estimated flows under permanent sale contracts (light grey bars), and the additional volumes committed under long-term and permanent contracts that were not transferred in those years (white bars). The database includes transactions between water districts, federal and state agencies, and private parties that are not members of the same water district or wholesale agency. "Dry years" are those classified as critical or dry for the Sacramento Valley based on the California Cooperative Snow Survey.

It is useful to consider market development in three phases, characterized and shaped by different forces:

1 the early drought years (1988–1994);
2 an intermediate phase, when environmental concerns drove continued growth (1995–2002); and
3 the most recent period, marked by two distinct trends: a shift toward long-term and permanent trades and a slowing of overall growth in trades (2003–2011).

Following this review, the chapter examines the trends in the environmental water market in more detail, and reflects on the emerging experience of the critical drought of 2014.

Take-off spurred by drought purchases (1987–1994)

During the eight-year period from 1987 to 1994, California experienced only one "normal" precipitation year (1993); five of the remaining seven dry years were deemed "critically dry." These hydrologic conditions provided the opportunity for the state to help jump-start the market. DWR began making dry-year purchases to offset lower deliveries to SWP contractors and wildlife refuges in 1987. These early operations, which involved only a handful of Sacramento Valley water districts, quickly brought the total volumes traded in the state to over 615 million cubic meters. In 1991, when the dry-year market was opened up to any willing buyers and sellers, DWR purchased 1,012 million cubic meters for resale, bringing the overall market volume to over 1,350 million cubic meters.[23] Water banks and other dry-year purchases were also operated in 1992 and 1994. From 1987 to 1994, state and federal dry-year purchases for resale and environmental uses accounted for nearly half of a market that had increased more than five-fold from pre-drought levels.

Continued growth driven by environmental concerns (1995–2002)

Although the second half of the 1990s saw a succession of wet years, market activity remained strong, with volumes typically exceeding the drought-year levels, especially by the end of the decade. The only dips in a generally upward trend in purchases occurred in the exceptionally wet years of 1995 and 1998, when many areas of the state experienced flooding. Market growth in this period was largely driven by environmental concerns. The influence of environmental policy is most readily seen by comparing the patterns of water purchases during the drought years (1987–1994) to those in the subsequent eight years, when rainfall was generally above normal (Figure 16.3).

The most obvious element of the new role for the environment is the rise in direct purchases for in-stream uses and for wildlife refuges through federal and state programs, including USBR's new Water Acquisition Program (introduced under the CVPIA) and CALFED's new Environmental Water Account. As a beneficiary of DWR's drought purchases, the environment already accounted for 8 percent of purchases during 1987–1994. Between 1995 and 2002, this share rose to 21 percent. On an average yearly basis, environmental purchases increased more than six-fold, nearly three times faster than the market as a whole.

The less obvious component of demand related to environmental policy changes is the increase in water purchases by San Joaquin Valley farmers. Although this group's change in market share is less dramatic (growing from 31 to 41 percent between the first and second period), its increase in average volumes—by 400 million cubic meters per year—accounts for nearly half of total market growth. Much of this growth can be linked to the changes introduced under the CVPIA in 1992, which mandated that a portion of project water be returned to in-stream uses and wildlife refuges. Since then, CVP agricultural service contractors located south of the Delta have received full project deliveries in only three very wet years (1995, 1998, and 2006). One outcome has been an active water market, as some contractors sought to offset reductions in deliveries through market purchases.

These two components of the environmentally-related water market were not without tension. On the one hand, the environmental water purchase program could be

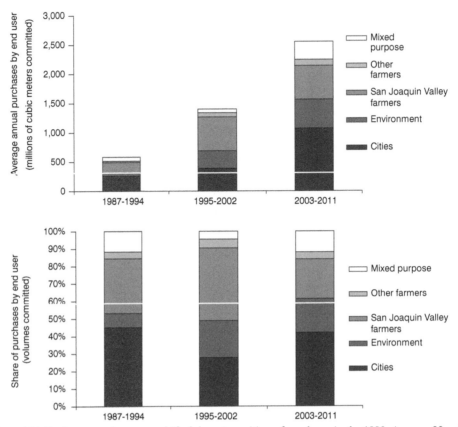

Figure 16.3 Environmental concerns shifted the composition of purchases in the 1990s (source: Hanak and Stryjewski, 2012).

Notes: The figure shows actual flows under all contracts and volumes committed but not transferred under long-term leases and permanent sales. "Mixed purpose" denote purchases by agencies with significant urban and agricultural uses.

viewed as a benefit to other water users, because it avoided additional uncompensated regulatory cutbacks to protect fish and other wildlife. On the other hand, the sheer size of these relatively well-funded programs meant that some farmers wishing to purchase make-up water viewed them as a source of tough competition.

The corollary of growth in environmentally-related demand was a decline in the relative importance of municipal and industrial users following the drought years. Whereas cities were the main recipients of traded water during the drought, accounting for 45 percent of all purchases from 1987–1994, their share in the following eight years fell to 28 percent. With the exception of 1991, when urban purchases reached nearly 620 million cubic meters, volumes remained relatively flat throughout the 1990s, averaging around 280 million cubic meters. This began to change by 2000, as some cities successfully negotiated long-term and permanent deals to purchase water.

Long-term transactions have risen (2003–2011)

The most recent phase of market development began in 2003, the first year of a package of long-term transfers among agricultural and urban Colorado River water-rights-holders, included as part of California's plan to reduce its overall use of this river. These contracts commit roughly 630 million cubic meters of annual transfers over a 75-year period,[24] and volumes flowing under these deals have increased steadily. These and numerous other long-term and permanent deals made between parties around the state have shifted the character of California's water market. Whereas short-term trades accounted for roughly three-quarters of all transfers in the 1980s and 1990s, they now account for less than half of all flows and only a quarter of total volumes committed.

These transfers have been supplied by a combination of system improvements (e.g., canal lining and operational efficiencies), agricultural land retirement, on-farm irrigation efficiency improvements (where improved efficiency generates net water savings), and releases of water from surface and groundwater storage. As with temporary transfers, agricultural water districts are the principal suppliers, originating roughly 80 percent of all long-term and permanent contracts and 95 percent of all committed flows. (They supply 85 percent of all short-term transfer flows.) Cities are the leading purchasers, with the largest overall volumes, average deal sizes, and average contract durations. However, this market is also serving other demands, including the environment and—increasingly—farmers, who are actively involved as buyers in mixed-use contracts as well as those destined purely for agricultural uses.

Nevertheless, farmers have declined in importance in overall market demand, accounting for only a quarter of all contractual commitments and 36 percent of actual flows from 2003–2011, with average purchases remaining virtually the same as in the prior period (Figure 16.3). Water purchases for the environment have remained important, increasing in absolute terms, and declining only slightly as a share of the overall market to 19 percent of commitments and 20 percent of flows. (As discussed below, environmental transfers declined considerably by the end of this most recent period, however, in response to a decline in funding.) The real demand growth has come from cities, with over three times more commitments and 2.4 times more actual flows acquired than in 1995–2002. As during the early drought years, urban agencies again account for over 40 percent of market demand, and this time most of this water is available to support longer-term growth, not just to compensate for shortages during droughts.

The growth of long-term and permanent transfers is a sign that the water market is maturing. These transfers generally involve more complex negotiations and more in-depth environmental documentation. They are particularly important for supporting economic transitions. By law, urban water agencies need to demonstrate long-term supplies to support new development, and transfers can provide this assurance.[25] Long-term commitments for environmental flows provide flexibility for environmental managers and can reduce the conflicts associated with regulatory alternatives to market-based transactions. Long-term commitments to make temporary supplies available during droughts are also an important way to enhance operational flexibility. A case in point is the recent 17-year transfer agreement between the Yuba County Water Agency and DWR, the so-called "Yuba Accord." In addition to making available some supplies for environmental uses, this transfer offers supplies to a pool of SWP and CVP contractors during dry and critically dry years. By working out the approval issues well in advance, such deals make it possible to act quickly during a drought or other supply emergency. This seems a more promising path in California than the use of short-term options contracts, which were initially considered promising, but which have been plagued by uncertainties related to approval procedures and other issues such as infrastructure access in more recent years.[26]

But overall market growth has slowed (2003–2011)

In spite of these positive market developments, there is also evidence that market momentum is weakening. There has been little growth in overall trading volumes since 2003 (Figure 16.2); and if the new package of Colorado River transfers is excluded, both committed and actual flows have actually declined since 2001. This weakening is particularly worrisome because drought conditions in the late 2000s might have been expected to boost trading.[27]

A variety of impediments—some longstanding and some new—appear to be at work. One new problem relates to conveyance infrastructure. Historically, California's sophisticated supply infrastructure has made it possible to transfer water either directly or through exchanges across most demand and supply centers. However, in the case of the Delta, a critical conveyance hub, new pumping restrictions since 2007 —introduced to mitigate conditions for endangered fish species—have impeded movements of both north-to-south and east-to-west transfers.[28]

In addition, a variety of impediments associated with the approval process—some longstanding and some more recent—are raising the transaction costs for parties wishing to engage in trading. For example, county groundwater ordinances, most of which have been adopted since 1996, tend to broadly restrict groundwater-related exports.[29] Although these ordinances were a useful stop-gap measure designed to prevent harm to local water users in the absence of state-level protections, they are not an ideal long-term solution for managing groundwater. Local basin management would be better served by more comprehensive plans and objectives that address locally generated overdraft as well as problems related to exports. Such programs could better manage local groundwater resources for all users without discriminating against potentially beneficial groundwater-related transfers.

The progressive tightening of DWR's approval process for north-to-south transfers represents another constraint. In addition to limiting "paper" transfers that might harm SWP contractors, the process—which involves frequent updates in its rules—creates uncertainties that are likely to create a chilling effect against legitimate transfers of wet water (Lund 1993).

A third constraint, noted earlier, is that transfers are now subject to environmental restrictions beyond the requirement of no injury to environmental flow conditions, such as fallowing limitations to protect habitat for the giant garter snake and Clean Air Act restrictions on the use of diesel pumps for groundwater-related transfers. In 2009, uncertainties over these new restrictions, combined with the inability to move water through the Delta, depressed drought water bank activity. Fewer than 100 million cubic meters were transferred, whereas DWR's goal was to acquire several times that volume.

Many observers have also pointed to high commodity prices—and particularly the price of rice, a crop that has often been fallowed to make water available for transfers—as a major reason for the recent slowing of the water market. While this may have been a factor for some farmers, it is not likely the primary explanation of market weakness. Real farm-gate prices of rice (like other field crops) have indeed been high in recent years relative to the early 2000s. But during most of the 1990s, real crop prices were not much lower than in recent years, and real prices for drought water purchases have also increased over time, with the potential to increase further to accommodate dry-year water demands.[30]

And finally, water market development has been hindered by the fragmentation of water management, with different types of water rights and contracts subject to different types of approval. These differences tend to limit market activity even when it would be economically and environmentally beneficial to engage in trading.[31] As a result, trades among agencies that have rights to use water within the same large projects (CVP, SWP, and Colorado River) continue to dominate the market, accounting for over 60 percent of all trades since the mid-1990s, and 80 percent of trades not involving direct state or federal government purchases. The "open market"—trades between agencies within different projects or not belonging to projects at all—still accounts for less than a fifth of all transactions.

The various new impediments to transfers have shifted the geographic patterns of trading. The Sacramento Valley—a primarily agricultural region in the northern half of the state with relatively abundant water supplies and lower value crops—now exports less than half as much water to other regions as it did during 1987–1994. Correspondingly, there has been an overall trend toward "regional self-sufficiency," with a higher proportion of total sales serving local demand. From 1987–1994, less than half of total sales were sold within the region of origin. In the most recent period, nearly two-thirds of all transfers (and 80 percent of non-environmental water trades) took place within the same region. Sales are also becoming more localized *within* regions. The share of non-environmental transfers to parties within the same county has progressively increased —climbing from 18 percent in 1987–1994 to 50 percent in 2003–2011. This shift reflects an increase in long-term transfers of water from nearby agricultural areas to urban areas within the San Joaquin Valley, as well as more localized farm-to-farm trading within both the San Joaquin and Sacramento Valleys.

Trades tend to be less contentious within regions, and especially within counties, than sales across longer distances, because the water stays in the local economy. When there is strong local pressure against exports, sellers may have fewer options to ship water to outside buyers, even if the outside buyer can offer a higher price.[32] A recent controversy along these lines occurred in Stanislaus County, where the Modesto Irrigation District sought to conclude a long-term transfer to the San Francisco Public Utilities Commission (which supplies the City of San Francisco and many Silicon Valley communities) that would help fund costly infrastructure upgrades. The City of Modesto and other local groups raised strong objections to the water leaving the county, even though San Francisco would have paid US$568 per thousand cubic meters for the water, roughly 70 times more than local farmers now pay (Holland 2012).

Environmental water acquisitions

Given the significant portion of the water market that involves acquisitions to support the environment, it is useful to consider these transactions in more detail. In all, over 5,550 million cubic meters of water was acquired for environmental purposes from 1982–2011, accounting for 14 percent of total commitments and 18 percent of total market flows. The environmental share of the market was highest during the early to mid-2000s, accounting for 30 percent or more of total flows in most years. Since 2008 total volumes have fallen, however, with the share of environmental acquisitions averaging around 15 percent.

Over time, the purpose and nature of these acquisitions has changed. Early purchases by the state and federal governments supported wildlife refuges and fish hatcheries; state efforts, overseen by the Department of Fish and Wildlife (DFW), were substantial during the late 1980s–early 1990s drought. In the mid-1990s, the CVPIA's Water Acquisition Program (WAP) created a systematic federal program for environmental water purchases, both for Central Valley wildlife refuges and for in-stream flows to support salmon runs in the San Joaquin River system. The WAP's in-stream flow program was a multiyear flow experiment and included a 12-year lease agreement, now expiring, with a consortium of San Joaquin Valley irrigation districts.

The next significant environmental acquisitions, beginning in 1999, involved the EWA of the CALFED program—a joint state–federal initiative focusing on the Bay-Delta region. The EWA was created to provide environmental managers with supplemental flows to support the Delta's threatened and endangered native fish species (delta smelt and Chinook salmon). The EWA accounted for half of all environmental water purchases between 2001 and 2007, with annual acquisitions from a variety of parties averaging roughly 220 million cubic meters. In 2008, this program was converted to a multiyear lease agreement (for eight years) with the Yuba County Water Agency as part of the multipurpose Yuba Accord, and it was scaled back to a third of this volume.

As EWA and WAP purchases have declined, two new types of environmental water purchases have emerged: acquisitions of water to mitigate the impacts of Colorado River water transfers from the Imperial Irrigation District to San Diego, and a variety of smaller transfers to in-stream flows under § 1707 of the Water Code. The Colorado River mitigation water aims to offset the reduction in irrigation drainage to the Salton Sea, which involves land fallowing. Irrigation drainage is the main source of water in this terminal saline lake, and fallowing reduces that drainage. The mitigation water was required under the terms of the transfer permit and is intended to prevent the acceleration of salt accumulation, which will eventually (even without the transfer) make the Salton Sea too saline to support food sources for migratory waterfowl. Under the current terms of the transfer, this mitigation water is to be provided through 2017.

Section 1707 of the Water Code, enacted in 1991, authorizes the dedication or transfer of water to in-stream flows by protecting it legally from inconsistent upstream uses by junior water rights-holders.[33] This statute has been used to dedicate flows to local watershed support measures in various parts of the state. Most such transfers involve relatively small volumes of water, are essentially permanent in duration, and are associated with non-profit organizations participating in watershed management.[34] In contrast to the other environmental water trades discussed here, the § 1707 acquisitions are generally made as donations rather than for monetary compensation (though some policy discussions are under way about making these donations eligible for tax deductions to create financial incentives to increase in-stream flows).

Because the bulk of the environmental market requires funding, an important question is: Where does the money come from? Roughly US$547 million (2011 dollars) were spent on the environmental water acquired between 1982 and 2011—costing an average of US$99 per thousand cubic meters. When these purchases have been made in the short-term market (as with the EWA prior to the Yuba Accord), the average prices paid have been higher than in the longer-term agreements, such as WAP purchases for in-stream flows. (As noted earlier, farmers wishing to buy water have at times been frustrated by the competition.) To date, the tab has principally been picked up by state and federal taxpayers, with the state paying the lion's share (52 percent state, 19 percent federal), mostly using general obligation bond funds secured with general tax dollars. Water users have paid for the remainder: 24 percent through an ecosystem restoration surcharge leveed on CVP contractors under the CVPIA (in 2012, roughly US$7.3 per thousand cubic meters for agricultural water districts and twice that for urban districts and power providers) and 5 percent by participants in the Colorado River water transfer.

Given general state and federal budget difficulties, the future of environmental water purchases is uncertain. Diminishing funds have already reduced volumes of environmental water acquisitions in recent years. Looking ahead, there is little money left from approved state bonds to fund these (or other) environmental programs, and federal budgets are equally constrained. The CVPIA restoration fund—supported by a surcharge on water users—is a more reliable funding source than taxes, but this fund has many potential uses, and now that the experimental in-stream flow program is winding down, the reduced budget for water acquisitions is concentrating on wildlife refuges. (The CVPIA set a quantitative goal for water deliveries to the refuges, through a combination of regulatory reallocations and purchases, but this goal has yet to be met.[35])

Apart from the general difficulties of raising taxes or introducing new surcharges on water users, it may be particularly hard to generate public support to expand environmental water purchases as a matter of policy in California.[36] Environmental water acquisitions are an alternative to uncompensated regulatory reallocations of flows.[37] Both occur when allocations to other sectors have left the ecosystem with too little water to function well. In some places, such as Australia, public policy has opted to principally use the market to buy back needed environmental flows.[38] California has operated with a hybrid policy, combining regulations and the market. Several state and federal statutes and the California constitution authorize the uncompensated cutback of water diversions when they cause environmental harm, and some water has been reallocated in this way in California watersheds.[39] The market was introduced as a way to generate environmental benefits while reducing the conflict associated with uncompensated cutbacks. Some have suggested that this market provides an added advantage of giving regulators the incentive to use environmental water more efficiently. But environmental water markets also have many detractors. Although an independent review team found the EWA—the largest taxpayer-funded program—to be moderately effective (meaning things could have been even worse without it), many observers view it as ineffective because it coincided with the collapse of native fish populations in the Delta.[40]

2014 drought update

As of this writing (April 2014), California has been making regular headlines in national and international news reports because of severe drought conditions facing the state. The calendar year 2013 was the driest on record (since 1895), and the 2013–2014 "water year" (measured from the beginning of October, when the rains typically start) looks to be driest since 1976–1977, the critically dry year that prompted initial gubernatorial and legislative policy consideration of launching a state-wide water market (Hanak et al. 2014b; Lund 2014). Among the actions called for in the governor's drought emergency declaration is the streamlining of short-term water transfer approvals.

While it is too soon to measure the success of these efforts, early indications suggest that California's water market may not be as capable of responding to extreme droughts as it is to more moderate shortages. The news stories have focused on the fact that many junior holders of surface water rights (notably those with CVP and SWP contracts) have been told they would receive a "zero" allocation this year. Perhaps more significant, 2014 marks the first year since trading began in earnest when senior holders of surface water rights were informed that they would not receive their full allocation. Faced with the prospect of significant (25 percent to 60 percent) surface water reductions, farmers in some senior-right irrigation districts that usually fallow to make drought water available have called the practice into question over concerns about local farm economy impacts, given the extent of fallowing that may already be required with available supplies (Hacking 2014). Meanwhile, tree crop farmers in the most severely affected areas of the San Joaquin Valley have been paying astronomical prices for water to keep their investments alive.[41] Interestingly, urban agencies have mainly been able to stay out of the market, thanks to significant investments in drought resiliency (including storage of conserved supplies in surface and underground reservoirs) since the prolonged drought of the late 1980s to early 1990s.

This experience highlights the contrast with Australia, where the prolonged, severe drought of the 2000s resulted in massive trading of irrigation water—in some years half of available supplies—within the Murray Darling basin in support of higher-return perennial crops. If critical conditions persist in California, we may yet see such a shift, but this will require a change in social norms within communities that still farm the lower-return commodities.

Lessons for responsible market development

Water marketing is an invaluable tool for helping California manage its scarce water resources more efficiently and sustainably over the long term. It augments the state's ability to cope with periodic droughts and facilitates the longer-term shift of some supplies to activities and regions with strong demand and insufficient water rights. Given the anticipated reduction in seasonal storage in the Sierra Nevada snowpack and the prospect of more frequent droughts due to climate warming, this tool is likely to become even more important in the future.

The state's water market is evolving in both expected and unexpected ways. Given agriculture's major share of total human water use in California (77 percent in 2005) and the relatively low value of some agricultural water uses, it is not surprising that agriculture is the primary source of market supplies.[42] Market demands have been quite diverse, with significant purchases by all three water-use sectors—farms, cities, and the environment. The market first grew in response to a major drought, spurred by direct state purchases. Subsequent growth was driven by changes in environmental policies rather than weather, with an increase in direct purchases of water for the environment and growing demand from farmers who lost supplies because of new in-stream flow requirements.

Most recently, urban demand has grown, as expected, with cities seeking to firm up supplies to support population growth and diminishing sources such as the Colorado River. This shifting demand has transitioned the market to one in which long-term and permanent trades predominate, rather than transactions negotiated annually. Perhaps surprisingly, the market has also become more localized, with half of all sales now occurring between parties within the same county—three times more often than when the market was launched. This shift likely reflects local pressures to sell locally, as well as growing infrastructure and institutional constraints on more distant sales. In particular, the ability to move water through a key conveyance hub in the Sacramento-San Joaquin Delta has been constrained in order to reduce harm to the Delta's compromised native fish populations, and hurdles to transfer approvals appear to be increasing. Finally, environmental water purchases, which were intended to reduce the conflicts associated with reallocations of water to the environment, are still a significant but now diminishing share of the market, confronted by declining revenues and a lack of public consensus that taxpayer dollars should be used to support such efforts. In recent years, infrastructural and institutional constraints have also diminished the market's ability to mitigate the costs of drought.

Both the progress to date and the constraints experienced suggest a number of policy implications for California and other regions facing freshwater scarcity. Further research can inform progress in all of these areas.[43]

Infrastructure matters

California's state-wide infrastructure network has been a boon to water marketing operations. However, problems in the Delta have limited the market's prospects for furnishing dry-year supplies. With a changing climate, flexible conveyance infrastructure is likely to become more critical for state-wide water management in the future. This is one of many arguments for improving the reliability of Delta pumping capabilities—now a major topic on the state's water policy agenda.[44] Optimization modelling can show the value of removing infrastructure constraints and inform cost–benefit comparisons of investment alternatives.[45]

The institutional review to prevent injury from transfers must be streamlined

Currently, some of the biggest problems seem to be those associated with temporary drought-related transfers, which need to happen quickly if they are to happen at all. Erring on the side of caution, California's regulators have progressively chipped away at the streamlined approach used in the early 1990s to manage drought-related transfers. The system could be simplified without requiring a complete overhaul of water rights—with "joint places of use" for water from different projects within the same basin, and more long-term contracts that do the necessary up-front analysis to enable dry-year transfers to occur expeditiously.[46] There also needs to be greater clarity and consistency in establishing conditions for trades (e.g., how much water is made available from fallowing) in some key export areas. Again, optimization models are a key tool for showing the potential gains from simplified trading rules.

Strong groundwater management is essential for a strong water market

In California, local management of groundwater basins is likely to be more effective than state-led management, but local officials need more incentives to get this right. One option would be for the state to use its authority under Article 2, § 10 of the constitution (requiring reasonable use of all water) to start requiring non-discriminatory protection of groundwater basins. This would be an improvement over the current, discriminatory county export ordinances, which are limiting transfers without instilling better local basin management. The state could also assert "no injury" protections for all groundwater-related transfers, thereby providing a more level playing field than the county ordinances currently provide for permit applications for export-related transfers. Outside pressure—with a credible threat that the state would step in if local officials do not—may be the best way to proceed. Ideally, this would be accompanied by positive financial incentives to improve basin management. Local officials can look to solutions that have been successful elsewhere in California, including special districts with pumping authority and fees and adjudications.[47]

One positive consequence of the latest drought—which has exposed the vulnerability of agricultural areas with overdrafted groundwater basins—is an effort to institute this type of groundwater reform, by putting in place enabling legislation that authorizes local entities to manage the resource, with a state backstop in the event locals fail to act (Hanak et al. 2014c). From an economic efficiency standpoint, the state should consider providing this backstop not only in basins already severely overdrafted, but also in those (such as most of the Sacramento Valley) where groundwater levels are still relatively high; these are the areas that can most readily serve the water market, but they are currently limited by a lack of basin management. This is an area where coupled hydrologic and economic models, taking into account the dynamic characteristics of basin behaviour, can yield especially valuable insights.[48]

Models are needed for mitigating the economic and environmental impacts of large-scale land fallowing deals

Over the long term, economic shifts make it likely that some cropland will be permanently retired, with its water supply becoming available for growing demand centres in other sectors and regions. In these cases, mitigation of community-related impacts of fallowing is likely to be important to ease economic transitions. Lessons may come from mitigation programs currently under way in the Palo Verde Irrigation District and the Imperial Irrigation District (for large water transfers in Southern California), as well as other cases such as the Northwest Forest Plan (which provided economic support to displaced workers and affected communities in areas where forestry lands were being converted to protected habitat).[49] In addition to retrospective and prospective economic analyses of local and regional impacts from transfers (e.g., Howitt 1994b; M. Cubed 2002), the development of successful models would benefit from political-economic analysis of the feasibility of different management alternatives.

In addition, consideration must be given to unintended environmental impacts of large-scale fallowing, both regarding air quality impacts associated with dust (a major concern in many agricultural areas) and the non-market ecological value of some lower-revenue crops as habitat. (In addition to the artificial wetlands function of rice fields noted earlier, crops like alfalfa provide valuable habitat for species such as migratory birds.) Ecosystem service payments may need to be part of the equation. Interdisciplinary research, taking into account agronomic, biological, and economic considerations, is needed to understand the least-cost options for dust prevention, the ecological benefits of various land use practices for wildlife, and the value to society of some non-market attributes of land management.

Environmental water purchases can reduce the conflicts associated with reallocating water to the environment

There is merit in pursuing water purchases alongside regulatory policies that reallocate some water, as California has done since the early 1990s. These transfers can also help improve the efficiency of environmental water management, by giving environmental water managers a budget that they can manage flexibly. Funding will be difficult, given the disappearance of state bond funds (responsible for half of all acquisitions to date) and on-going state and federal budget constraints. The CVPIA ecosystem restoration surcharge—funded by water users—is a potentially good model. The latest California drought has highlighted the potential value of also considering the environmental market in reverse: rather than continue the practice of uncompensated cutbacks in required environmental flows during severe droughts, the state could establish a payment system which establishes a fund to be used to support ecological hotspots during the drought and recovery actions afterwards (Lund et al. 2014).

In general, support from the public and water users will be much easier to muster if they can be presented with systematic evaluations demonstrating the effectiveness of environmental water allocations, whether acquired through purchases or regulatory reallocations. Constructing frameworks for evaluating the costs and benefits of environmental water allocations is a ripe area for research.

High-level leadership is needed to routinize water marketing transactions

If California's market has foundered in recent years, it has been partly due to the lack of priority attention to and understanding of the centrality of these tools in effective state-wide water management. Some innovations—such as streamlining the approval process—require risk-taking, and only high-level state and federal leaders have the position and authority to undertake such risks. State and federal leadership played a key role in developing drought water-banking in the early 1990s and in concluding the large package of transfers of Colorado River water in the early 2000s. A high-level coordinating committee from relevant agencies, with the ability to break through barriers, may be key to developing more durable solutions. Political-economic analyses of the conditions that contribute to effective policy reform in favour of market-based solutions—in California and elsewhere—can help shed light on feasible solutions for better water management.

Notes

1 Bjornlund (2003) provides a description of the electronic trading systems in Australia's market for irrigation water.
2 For a more in-depth analysis of many of these issues and detailed data descriptions, see Hanak and Stryjewski (2012).
3 California actually has a dual system of surface water rights, consisting of riparian and appropriative rights. Riparian rights can only be used on lands adjacent to the surface water source from which the water is diverted; this restriction limits transferability to other users. Appropriative rights—the predominant form of surface water rights in California and elsewhere in the American west—can be used on non-adjacent lands. Before 1914, these rights were acquired through demonstrated use ("appropriation"); since the adoption of the state's modern Water Code, they have been acquired through state permits. This chapter focuses on appropriative rights, which are transferable.
4 See Hanak et al. (2011), Chapter 2, for recent estimates of groundwater's share in California's overall water supply portfolio and in agricultural and urban uses by hydrologic region.
5 For a map of these areas, see Hanak et al. (2011), Figure 4.1. For a description of some trading activity within adjudicated basins, see Hanak and Stryjewski (2012), Appendix A.
6 Contract entitlements are often larger than the actual deliveries, depending on hydrologic and regulatory conditions. In the permanent sales of contracts, the buyer acquires the contract at face value and receives water deliveries in the same proportion as the original holder of the contract. Leases of contract allocations involve transfers of actual deliveries.
7 In 2005, the most recent year for which state-wide water-use estimates are available, agriculture used 77 percent of combined business and residential water use (Hanak et al. 2011, Chapter 2, using data from DWR).
8 Although we are not aware of any formal analyses of this proposition, one example that is often given is the relative ease of negotiating the large transfer of water to urban southern California from the property-owner-vote Palo Verde Irrigation District as compared with the popular-vote Imperial Irrigation District (Hanak 2003). Both transfers were part of California's response to the requirement to reduce its withdrawals from the Colorado River (discussed later in the text).
9 For instance, the State Water Resources Control Board reported that the theoretical amount of claims on water-use rights within the Sacramento-San Joaquin Delta watershed (which supplies water to most agricultural and urban users in the state) was eight times higher than the amount of flows available in an average year (Governor's Delta Vision Blue Ribbon Task Force 2008).
10 This is the case, for instance, for SWP and CVP contracts, most of which only receive full deliveries in very wet years.
11 For discussions of "gross" or "applied" and "net" or "consumptive" water use (the difference between which is usable return flow), see Hanak et al. (2009) and Hanak et al. (2011), Box 2.1.
12 In principle, only net savings constitute wet water that can be available for trading. For instance, because shifts in irrigation technology (e.g., from flood irrigation to drip irrigation) primarily reduce gross or applied water use, not the net amount consumed by crops, such shifts are generally not considered a valid method of making water available for transfer. What constitutes conservation for purposes of water transfers does sometimes diverge from net water savings, however, if those using the return flow do not have legal rights to use that water. For instance, canal lining reduces surface water losses from seepage into the groundwater basin, thereby reducing water that was available for neighbouring groundwater users. A large canal lining project was nevertheless authorized to support a long-term transfer from the agricultural users of the All American Canal (Colorado River water) to water users

in urban Southern California. The case was disputed in court, but the losing groundwater pumpers (on the Mexican side of the border) were deemed to not have the rights to the displaced water under the treaty that apportions Colorado River waters between the United States and Mexico (*Consejo de Desarrollo Economico de Mexicali v. United States*, 482 F.3d 1157, 9th Cir., 2007). Similarly, under the imported water doctrine, other users do not have the rights to use the return flow (including seepage from a canal) if the importer can show it intended to capture it all for reuse.

13 Legally, the Water Code requires permitting of groundwater found in "subterranean streams"—i.e., water that flows beneath the surface—which excludes most California groundwater (generally considered to "percolate" down from the surface). These distinctions are not technically correct by the standards of modern hydrology, which recognizes the interconnection between most bodies of groundwater and surface water.

14 These rules are described in the annual draft white papers on transfers, published by DWR in conjunction with the regional office of the USBR (California Department of Water Resources and U.S. Bureau of Reclamation 2012). As discussed below, these rules are becoming increasingly strict, and are not without controversy.

15 Transfers of post-1914 rights that are considered temporary (one year or less) are exempted from CEQA, on the grounds that they must go through SWRCB review.

16 The state Water Code does contain some restrictions on groundwater-related transfers. In particular, Water Code § 1220(a) states that no groundwater can be pumped for export from within the Sacramento and Delta-Central Sierra basins unless there is a groundwater management plan in place, and Water Code § 1745.10 limits groundwater substitution transfers to those that will not result in long-term overdraft. However, in 1994 an appellate court ruled that the state did not "pre-empt the field" in this area, and it granted counties the authority to exercise their police powers to protect public health and safety with groundwater ordinances (*Baldwin v. County of Tehama*, 31 Cal. App. 4th 166, 173-74 (1994, 3rd Dist.); review denied, Cal. Sup. Ct., March 17, 1995). See Hanak and Dyckman (2003) for a detailed discussion of the legal issues surrounding county ordinances.

17 Hanak (2003) conducted a detailed review and found few permit applications through 2002, suggesting that the ordinances were having a chilling effect. Anecdotal information from local agencies suggests that there has been little change in subsequent years.

18 See Hanak (2003) for a summary of these studies.

19 The transfer from the Imperial Irrigation District to the San Diego County Water Authority set aside US$40 million for socioeconomic mitigation, and the transfer from the Palo Verde Irrigation District to the Metropolitan Water District of Southern California set aside US$6 million (now over US$7 million with accumulated interest); see Hanak et al. (2011).

20 These conditions are presented in a draft white paper on water transfers prepared by DWR and USBR, which is updated annually (California Department of Water Resources and U.S. Bureau of Reclamation 2012).

21 Under current rules, farmers wishing to fallow land for transfers must limit each fallowed field to 320 acres (130 hectares) and surround these fields with cropped land (California Department of Water Resources and U.S. Bureau of Reclamation 2012). This practice, intended to provide corridors for safe passage of the snake, effectively limits the potential fallowed area to 20 percent at any given time, in highly specified patterns.

22 From 1998 to 2005, Californians used an average of 51,436 million cubic meters of water (40,706 million cubic meters in agriculture, 10,730 million cubic meters in urban uses). This total includes roughly 8 percent in conveyance losses (data from DWR; for details see Hanak et al. 2011, Chapter 2).

23 Wet conditions late in the 1991 rainy season (the "March Miracle") resulted in lower-than-anticipated purchases from the bank, so DWR was only able to sell about half of the water. The rest was melded into SWP supplies and paid for by all the project contractors.

24 The new transfer agreements under the Quantitative Settlement Agreement (QSA) include the movement of 374 million cubic meters per year of water from the Imperial Irrigation District (IID) to the San Diego County Water Authority and the Coachella Valley Water District, two canal lining projects that will move nearly 118 million cubic meters per year of

conserved water from IID and Coachella to San Diego and the San Luis Rey Indians, and the movement of up to 137 million cubic meters per year from the Palo Verde Irrigation District (PVID) to the Metropolitan Water District of Southern California. The QSA also recognizes an existing transfer of 136 million cubic meters per year from IID to Metropolitan, in place since 1987. In addition to these long-term agreements, some temporary transfers have taken place between PVID and Metropolitan during the recent drought.

25 Senate Bills 221 and 610, passed and signed into law in 2001, require large development projects (for larger agencies >500 new residential units or the equivalent in combined residential and non-residential demand, and for smaller agencies at least 10 percent growth in local water demand) to demonstrate at least 20 years of available supplies. For a discussion of how these laws are working, see Hanak (2005b, 2010).

26 In 1995 and again in 2003 and 2005, single-year "dry-year purchase" options trades were introduced for fallowing-based transfers. The idea of an options market is to make commitments between buyers and sellers early in the season (sometime in the fall), before the character of the water year is known, with sellers paid in instalments to maintain the commitment at successive call dates. The last call date was in the late spring—the latest point at which growers could plant if they did not part with the water. In 1995, DWR (which was buying the water for the Drought Water Bank) did not exercise the options because it ultimately proved to be a very wet year (growers thus received only the up-front instalment). In 2003, the Metropolitan Water District of Southern California (MWDSC) initiated a similar deal with Sacramento Valley rice growers, and it exercised most of those options, purchasing 153 million cubic meters, while not exercising 25 million cubic meters of options (Howitt and Hanak 2005; Hacking 2005). In 2005, under a similar deal, MWDSC did not exercise the options to purchase 160 million cubic meters, because there was ample rain, and growers again received just the instalment (US$8 per thousand cubic meters, versus a final purchase price for the water of US$101 per thousand cubic meters). Options have become less attractive in recent years, given the new operational restrictions on Delta pumping, which limits the attractiveness of fallowing-based transfers from growers using water stored in Lake Shasta. Springtime releases of cold water from Lake Shasta are required to protect salmon, and this same water can also be used by Sacramento Valley rice growers. With new Delta pumping restrictions, the water rice farmers would have used on their fields between April and June cannot be sent to users south of the Delta instead. Yet these farmers would still need to be compensated for this water in order to fallow their fields. As a result, the new pumping restrictions create an effective surcharge on water acquired through fallowing of about 40 percent. These constraints do not apply to some growers on the east side of the Sacramento Valley who use water stored in Lake Oroville, but infrastructure problems at this reservoir have made water deliveries more uncertain, and less amenable to options contracts.

27 Hanak (2005a) found that drought years between 1990 and 2001 were associated with significantly higher transfers after controlling for water allocations, crop prices, and other factors.

28 Some east-to-west transfers among south of Delta water users rely on sending water north (through the northward-flowing San Joaquin River and other eastside rivers) to the Delta pumps and then back south again to water users on the west side of the San Joaquin Valley. See note 26 for a description of some consequences of Delta pumping restrictions.

29 Hanak (2005a) showed that between 1996 and 2001, the ordinances reduced total county exports by roughly 20 percent, while increasing within-county trades by an even larger proportion (65 percent) but from a smaller base. On net, the ordinances reduced overall market sales by 11 percent. Interviews with water managers in the Sacramento Valley suggest that these ordinances remain important obstacles to groundwater-related transfers in some counties where there is significant potential, such as Butte.

30 For example, Sacramento Valley rice farmers were offered US$223 per thousand cubic meters during the 2009 drought water bank, 1.4 times higher in real terms than the price offered during the 1991 bank; the ratio of real farm-gate rice prices in those two years was 1.6. (South of the Delta, some farmers were reportedly paying much higher prices for water to sustain their perennial and vegetable crops.) Over the entire 1982–2011 period, Hanak

and Stryjewski (2012) found no evidence of a statistical correlation between rice prices and water trading. (Various regression analyses did not find a significant association between rice prices and either short or longer-term transfers.) Some local observers have suggested that increasing vertical integration within the rice industry—with more rice farmers now owning processing and marketing facilities—may have reduced some farmers' incentives to lease water, irrespective of trends in farm-gate prices.

31 For two years during the drought, an emergency measure to create a joint place of use between the CVP and SWP south of the Delta provided a temporary reprieve for some such transfers.

32 Hanak (2005a) found that within-county sales increased after counties adopted groundwater export restrictions. Local sales within the Sacramento Valley typically go for much lower prices than exports.

33 This protection is necessary because the environment does not have explicit water rights for in-stream flows under California law. Many of the environmental water transfer programs noted above operate with § 1707 permits, in addition to transfers discussed in this paragraph which are explicitly developed under this statute.

34 Hanak and Stryjewski (2012) provide a list of these transfers. Unlike other permanent transfers, the right to use the water under § 1707 transfers does not change hands. But to resume using it for its original purpose, the right-holder would have to go through a new review process and demonstrate that the change would not cause negative environmental impacts.

35 See the report by the CVPIA Independent Review Panel (2009). The panel found that the water the refuges have been acquiring is being managed well. It also noted Delta conveyance constraints as an impediment to adequate water acquisition.

36 Raising fees or taxes for water-related purposes faces numerous constitutional hurdles imposed by a succession of voter-approved fiscal reforms since the late 1970s. Locally-imposed surcharges on water bills to support environmental water trading would need to have a tight nexus with the service to the property owner or be subject to voter approval by a simple majority or possibly a two-thirds majority of voters (Hanak et al. 2014a). State-level surcharges would likely require a two-thirds vote by both houses of the state legislature.

37 These policies can be complementary. One purpose of § 1707 is to allow in-stream water-rights-holders to augment stream flows above the minimum regulatory requirements (Gray 1994).

38 See Garrick et al. (2009) and Australian Department of Sustainability, Environment, Water, Population and Communities (2010).

39 For a discussion of the legal issues, see Hanak et al. (2011), Chapters 5 and 7. Examples of large regulatory cutbacks include CVPIA reductions (which reduced CVP diversions to support salmon), recent Delta pumping restrictions (which more generally reduced Delta exporters' access to water), and the Mono Lake and Owens Valley decisions (which reduced Los Angeles's ability to draw water from its land holdings in the eastern Sierras).

40 An independent scientific assessment (Brown et al. 2009) considered the EWA's effectiveness to be modest at best: In the first five years, it likely increased the survival of winter-run Chinook salmon by 0 to 6 percent, adult delta smelt by 0 to 1 percent, and juvenile delta smelt by 2 to 4 percent. The gains could have been greater if the EWA water were allocated in a more focused way, so as to concentrate benefits on a single life stage of one species. For instance, concentrating on the spring season during dry years could have increased abundance of juvenile delta smelt by up to 7 percent and adult delta smelt by up to 4 percent. Chinook salmon runs could have been increased by 20 percent if water volumes were applied during the entire outmigration period.

41 An auction of stored groundwater in Kern County, which limited sales to in-county users, received bids for nearly five times the volume available, and cleared at roughly US$975/thousand cubic meters, twice the floor price (Henry 2014). Irrigation water sales in this region in 2009 were considered high if they reached that floor price.

42 See Hanak et al. (2012) for information on the marginal value of agricultural water uses and comparisons of gross state product per unit of water used in the agricultural and urban sectors.

43 For additional discussions of some of these issues, see Hanak et al. (2011), Chapters 6 and 7, and Hanak et al. (2012).

44 The state and water users reliant on Delta exports are proposing to build a new conveyance system for this region, with one or two tunnels that divert water upstream of the Delta and deliver the water to the conveyance network south of the Delta. This could add flexibility in environmental water management and reduce the susceptibility of the system to outages from levee failures. See http://baydeltaconservationplan.com/library/documentslandingpage/ BDCPDocuments.aspx (last consulted on November 24, 2013).

45 See, for instance, applications of the economic-engineering California Value Integrated Network (CALVIN) model of California's statewide water system, initially developed by Jay Lund and Richard Howitt from UC Davis in the mid-1990s. The model shows that that improved Delta conveyance is likely to be significantly more valuable with a dry form of climate change, because it facilitates the transfer of water from lower-value irrigation activities in the northern part of the state to areas of greater demand in the south (Hanak et al. 2011, Chapter 6; Ragatz 2013).

46 Deals such as the Yuba Accord, described above, are a model.

47 Many recent adjudications in Southern California have been relatively swift—accomplished within two to three years (sometimes less)—unlike the multi-decadal legal battles of the past that those in other parts of the state understandably wish to avoid. See California Department of Water Resources (2011) for a list of adjudicated basins through mid-2011 and the years in which the adjudication was filed with the court and finalized. (In some cases, there has been additional litigation to resolve issues with entities that are not party to the agreement following the finalization of the agreement.) Some recent adjudications have been voluntary (e.g., the Beaumont Basin adjudication, concluded in 2004). Blomquist (1992) describes some of the early southern California adjudications, some of which took 15 years or more to be resolved.

48 As an example, see Knapp et al. (2003). Hanak (2005a) expands on the economic and institutional implications of this analysis for groundwater regulation in a market environment.

49 See Hanak et al. (2011), Chapter 9.

References

Association of California Water Agencies (2011) Sustainability from the Ground Up: Groundwater Management in California, a Framework, Sacramento, CA, USA

Australian Department of Sustainability, Environment, Water, Population and Communities (2010) Water for the Future: Commonwealth Environmental Water Holder 2010-2011 Business Plan, Canberra, Australia

Bjornlund, H. (2003) 'Efficient Water Market Mechanisms to Cope with Water Scarcity', *Water Resources Development,* vol 19, no 4, pp 553–567

Blomquist, W. (1992) *Dividing the Waters: Governing Groundwater in Southern California*, ICS Press, San Francisco, CA, USA

Brandt, A. (2001) 'An Environmental Water Account: The California Experience', *Univ. of Denver Water Law Review*, vol 5 no 2, pp 426–456

Brown, L. R., W. Kimmerer, and R. Brown (2009) 'Managing Water to Protect Fish: A Review of California's Environmental Water Account, 2001-2005' *Environmental Management*, vol 14, pp 237–261

Burke, S., R. Adams, and W. Wallender (2004) 'Water Banks and Environmental Water Demands: Case of the Klamath Project', *Water Resources Research*, vol 40, no 9, DOI: 10.1029/2003WR002832

California Department of Water Resources (2011) Water Facts 3: Adjudicated Groundwater Basins, Sacramento, CA, USA

California Department of Water Resources and U.S. Bureau of Reclamation (2012) Draft Technical Information for Water Transfers in 2012: Information to Parties Interested in Making Water Available for 2012 Water Transfers, Sacramento, CA, USA

CVPIA Independent Review Panel (2009) 'Undelivered Water: Fulfilling the CVPIA Promise to Central Valley Refuges'. http://www.usbr.gov/mp/cvpia/docs_reports/indep_review/CVPIA_Final_Refuge_Report_2009-11-03.pdf

Garrick, D., M. A. Siebentritt, B. Aylward, C. J. Bauer, and A. Purkey (2009) 'Water Markets and Freshwater Ecosystem Services: Policy Reform and Implementation in the Columbia and Murray-Darling Basins' *Ecological Economics*, vol 69, no 2, pp. 266–379.

Governor's Commission to Review California's Water Rights Law (1978) *Final Report*, Sacramento, CA.

Governor's Delta Vision Blue Ribbon Task Force (2008) Our Vision for the California Delta, Sacramento, CA.

Gray, B. (1994) 'A Reconsideration of Instream Appropriative Rights in California, Revised Version with Postscript', in L.J. MacDonnell and T. A Rice (eds) *Instream Flow Protection in the Western States*, 2nd edition, University of Colorado Press, Boulder, CO, USA

Hacking, H. (2005) 'Agriculture Water Won't Flow Southward This Year', *Chico Enterprise Record*, April 1

Hacking, H. (2014) 'Water Transfer Proposal Ruffles Some Locals in Ag Community', *Chico Enterprise Record*, April 2

Hanak, E. (2003) Who Should Be Allowed to Sell Water in California? Third-Party Issues and the Water Market, Public Policy Institute of California, San Francisco, CA, USA

Hanak, E. and C. Dyckman (2003) 'Counties Wresting Control: Local Responses to California's Statewide Water Market', *University of Denver Water Law Review*, vol 6, no 2, pp 490–518

Hanak, E. (2005a) 'Stopping the Drain: Third-Party Responses to California's Water Market', *Contemporary Economic Policy*, vol 23, no 1, pp 59–77

Hanak, E. (2005b) Water for Growth: California's New Frontier, Public Policy Institute of California, San Francisco, CA, USA

Hanak, E., J. Lund, A. Dinar, B. Gray, R. Howitt, J. Mount, P. Moyle, and B. Thompson (2009) California Water Myths, Public Policy Institute of California, San Francisco, CA, USA

Hanak, E. (2010) 'Show Me the Water Plan: Urban Water Management Plans and California's Water Supply Adequacy Laws', *Golden Gate University Environmental Law Review*, vol 4, no 1, pp 69–89

Hanak, E., J. Lund, A. Dinar, B. Gray, R. Howitt, J. Mount, P. Moyle, and B. Thompson (2011) Managing California's Water: From Conflict to Reconciliation, Public Policy Institute of California, San Francisco, CA, USA

Hanak, E., and J. Lund (2012) 'Adapting California's Water Management to Climate Change', *Climatic Change*, vol 111, pp 17–44

Hanak, E., J. Lund, B. Thompson, W.B. Cutter, B. Gray, D. Houston, R. Howitt, K. Jessoe, G. Libecap, J. Medellin-Azuara, S. Olmstead, D. Sumner, D. Sunding, B. Thomas, and R. Wilkinson (2012) Water and the California Economy, Public Policy Institute of California, San Francisco, CA, USA

Hanak, E. and E. Stryjewski (2012) California's Water Market, By the Numbers: Update 2012, Public Policy Institute of California, San Francisco, CA, USA

Hanak, E., B. Gray, D. Mitchell, J. Lund, C. Chappelle, A. Fahlund, K. Jessoe, J. Medellín-Azuara, D. Misczynski, J. Nachbaur, and R. Suddeth (2014a) Paying for Water in California, Public Policy Institute of California, San Francisco, CA, USA

Hanak, E., J. Mount, and C. Chappelle (2014b) California's Latest Drought, Public Policy Institute of California, San Francisco, CA, USA

Hanak, E., E. Freeman, C. Chappelle, B. Gray, J. Lund, D. Misczynski, J. Medellín-Azuara, D. Mitchell, J. Mount, J. Nachbaur, R. Suddeth, and B. Thompson (2014c) 'Funding Sustainable Groundwater Management in California', UC Davis Center for Watershed Sciences, californiawaterblog.com, April 3

Hansen, K., R. Howitt, and J. Williams (2013) 'Water Trades in the Western United States: Risk, Speculation, and Property Rights', in J. Maestu (ed.) *Water Trading and Global Water Scarcity: International Experiences*, RFF Press, New York, USA

Henry, L. (2014) 'Thirsty Growers Bid Sky-High for Available Water', *Bakersfield Californian*, Feb. 5

Holland, J. (2012) 'Modesto Irrigation District Kills Proposed Water Sale', *Modesto Bee*, September 18

Hollinshead, S., and J. Lund (2006) 'Optimization of Environmental Water Purchases with Uncertainty', *Water Resources Research*, vol 42, no 8, DOI: 10.1029/2005WR004228

Howitt, R. (1994a) 'Empirical Analysis of Water Market Institutions: The 1991 California Water Market', *Resource and Energy Economics*, vol 16, no 4, pp 357–371

Howitt, R. (1994b) 'Effects of Water Marketing on the Farm Economy', in H. O. Carter, H. J. Vaux Jr and A. F. Scheuring (eds.) *Sharing Scarcity: Gainers and Losers in Water Marketing*, Agricultural Issues Center University of California, Davis, CA, USA

Howitt, R., and E. Hanak (2005) 'Incremental Water Market Development: The California Water Sector 1985-2004', *Canadian Water Resources Journal*, vol 30, no 1, pp 1–10

Israel, M., and J. R. Lund (1995) 'Recent California Water Transfers: Implications for Water Management', *Natural Resources*, vol 35, no 1, pp 1–32

Joyce, B.A., V. Mehta, D. Purkey, L. Dale, and M. Hanemann (2009) Climate Change Impacts on Water Supply and Agricultural Water Management in California's Western San Joaquin Valley, and Potential Adaptation Strategies, California Climate Change Center, Sacramento, CA, USA

Lund, J. R. (1993) 'Transaction Risk Versus Transaction Costs in Water Transfers', *Water Resources Research*, vol 29, no 9, pp 3103–3107

Lund, J. R. (2014) 'A Cheat Sheet on the California Drought', UC Davis Center for Watershed Sciences, californiawaterblog.com, April 24

Lund, J. R., E. Hanak, and B. Thompson (2014) 'Environmental Water Market Would Help the Losers in This Drought', *Sacramento Bee*, February 2

Knapp, K., M. Weinberg, R. Howitt, and J. Posnikoff (2003) 'Water Transfers, Agriculture, and Groundwater Management: A Dynamic Economic Analysis', *Journal of Environmental Management*, vol 67, no 4, pp 291–301

M. Cubed (2002) 'Socioeconomic Assessment of the Proposed Palo Verde ID Land Management, Crop Rotation and Water Supply Program', Prepared for the Palo Verde Irrigation District, Needles, CA, USA

Medellín-Azuara, J., R. E. Howitt, D. J. MacEwan, and J. R. Lund (2012) 'Economic Impacts of Climate-Related Changes to California Agriculture', *Climatic Change*, vol 109, no 1, pp 387–405

Nelson, R. (2011) Uncommon Innovation: Developments in Groundwater Management Planning in California. Water in the West Working Paper 1, Stanford University, Stanford, CA, USA

Phelps, C. E., N. Y. Moore, and M. H. Graubard (1978) Efficient Water Use in California: Water Rights, Water Districts, and Water Transfers, R-2386-CSA/RF, RAND Corporation, Santa Monica, CA, USA

Quiggin, J. (2006) 'Repurchase of Renewal Rights: A Policy Option for the National Water Initiative', *Australian Journal of Agricultural and Resource Economics*, vol 50, no 3, pp 425–435

Ragatz, R. (2013) 'California's Water Futures: How Water Conservation and Varying Delta Exports Affect Water Supply in the Face of Climate Change', MS thesis, Department of Civil and Environmental Engineering, University of California, Davis, USA

17

WATER TRADING OPPORTUNITIES AND CHALLENGES IN EUROPE

Gonzalo Delacámara, Carlos Mario Gómez, and Josefina Maestu[1]

Water trading essentially involves the voluntary exchange of rights or entitlements to use water (Hodgson, 2006; Grafton et al., 2010, 2011). To achieve the desired status of water bodies, quantitative constraints on abstraction must be set and converted into property rights over the use of water (Howe, 2000). There is therefore an overall cap on rights to use water. Water users can then trade these rights within the limits defined by the water authority. Water trading is different from water transfers despite implying the diversion of water, occasionally as part of major infrastructure projects.

As in Hanak (Chapter 16) and Crase et al. (Chapter 18), markets have mostly developed in the US western semi-arid states (Arizona, California, Idaho, Kansas, Montana, Nebraska, Nevada, New Mexico, North Dakota, Oklahoma, Oregon, South Dakota, Texas, Utah, Washington, and Wyoming), although in some cases trades are negligible; Australia (having the southern connected Murray-Darling Basin, the largest geographically defined market, with the highest numbers and volumes of entitlements); Chile (mostly in northern regions); and with a less mature status in countries such as Spain (central, south and southeastern basins) or Mexico and very incipiently in China, the UK, and South Africa.

In Europe, the recent EU Blueprint to Safeguard Europe's Waters [COM (2012) 673] emphasizes the importance of using water trading to tackle water scarcity and droughts, but it could also be seen as an option for integrating economic incentives in the programs of measures to attain good ecological status of water bodies as specified in Article 11 of the Water Framework Directive (2000/60/EC). Yet, experiences of formal trades are at an embryo stage in the EU and mostly restricted to some Mediterranean countries (Spain, and to a lesser extent France and Italy, the latter not yet being supported by national legislation) (Garrido et al., 2013; Gómez et al., 2013; Kahil et al., 2014) and also quite recently in England and Wales (OFWAT, 2010), with somewhat limited upstream markets.

What contribution can water trading make to cope with water policy challenges?

Water trading can be used to help address water scarcity and drought risk. It is especially relevant at a local level, when there is a suitable conveyance infrastructure for water (Grafton et al., 2014), costs are low and all use rights are defined over the same water source (Loch et al., 2013). Allocating water to its more valuable uses is a means to increase economic (or social) values and/or to meet environmental objectives in water bodies. It may be an opportunity to reduce water abstractions (i.e. producing more with less) (Libecap, 2010; Puckett, 2010), provided that other complementary instruments are put in place (Garrick et al., 2013).

Experience with water trading schemes shows their significant role in finding mutually beneficial agreements between buyers and sellers, thus increasing or insuring the production of goods and services and making water trades a convenient instrument to promote different economic activities (Young, 2010). These development objectives were actually the main driver in the original adoption of current water trading schemes and concerns about their environmental outcomes are still an emerging issue (Howe, 2000; Yoo et al., 2013).

The implementation of water trading schemes is also advisable, under certain conditions, as part of a policy mix aimed at regaining control over groundwater resources and harnessing the potential of water resources to provide higher levels of resilience and adaptive capacity for economic development (Gómez et al., 2013).

Voluntary trading can actually play a critical role in stabilizing the economy and in providing an effective drought management alternative, as long as all stakeholders are involved and provisions are made to compensate for third-party effects (as in the Tagus River Basin district, Central Spain).

Water quality trading could also be implemented, allowing emission rights to be traded among different activities, to allow increased economic activity, while insuring that water quality standards in water bodies are maintained or improved (Selman et al., 2009; Shortle, 2013).

Opportunities

Besides the evident interest of potential buyers and sellers in a water market, the actual opportunities of water trading lie in its potential to yield benefits for the purposes of water policy. In other words, success in water trading should not be judged by the number or the volume of transactions (the so-called market dynamism or market depth) or against prices paid, but according to their contribution to curb scarcity, reduce drought risk exposure, meet quality standards or increase resilience and/or adaptability. Designing and implementing a water trading scheme does consist of transforming a good business opportunity (on the grounds of private profitability) for those directly involved in the bargain, into a sound water governance opportunity for anyone.

These opportunities for water trading to improve water governance can be summarized by the following intermediate targets to which water trading is expected to make a significant contribution:

- Attain a Pareto improvement in the current water allocation among market activities in order to enhance the potential of the economy to increase the provision of goods and services within the limits of available water resources, given water quality requirements (such as in water markets in Australia, Chile, and Colorado, USA) (see for instance Howe, 2013).
- Improve adaptability by making water allocation to alternative uses contingent on available resources anytime in order to reduce welfare losses and to provide a better response to droughts (as in the Tagus River Basin district in Spain and Portugal; Garrido et al., 2013).
- Reallocate risks so that the vulnerability of water uses exposed to scarcity and droughts is diminished (Zuo et al., 2012).
- Create opportunities for water saving and water conservation, and to reduce pollution, thereby providing an alternative to traditional supply-side approaches to water management, deterring, for instance, costly investment in water infrastructures—including wastewater treatment infrastructures (NWC, 2011).
- Indirectly create incentives for research, technological development and innovation in water technologies and processes (Zilberman et al., 2011).
- Create a framework in which water users can make decisions based on local conditions and ad hoc needs (Garrick et al., 2013).
- Create a framework in which water users can independently adapt their practices to emerging issues (without relying on government action, but subject to its command) (Loch et al., 2013).
- Show water users the opportunity costs (i.e. those associated with foregone alternative choices) of some of their decisions on water use (Crase et al., 2013).

According to Maestu (2013), the reallocation of water use rights to voluntary agreements may actually be useful to improve the allocation of resources but may not necessarily lead to efficient water use. Allocative efficiency of water uses is far from being the only objective of water policy and needs to be factored in along with other objectives of public policy such as equity, spatial balance, preservation of rural activities (including rural income and employment), and the delivery of ecosystem services, inter alia.

The above-mentioned opportunities could be perceived as higher in the EU where potential for water trading is largely unexplored. As above, water trading generally increases the efficiency of water allocation but this admits countless nuances. Previous experience in non-EU countries represents a real learning experience that may help to improve design and implementation to make water trading a more effective, equitable and sustainable instrument of water policy as well as to avoid failures and to reduce the risks involved. The implementability of water markets is conditional on the existence of marketable water rights, freedom to agree on prices, and information such as an adequate price-revealing mechanism (as can be seen in experiences from the Murray-Darling Basin in Australia, Chile, California, and the Northern Colorado Water Conservancy District in the US, or even in the very incipient trades in the Tagus River Basin and other catchments in Spain). The lack of these structural requirements so far prevents water trading (as in China).

Assessing opportunities to trade water use rights and/or effluent emission permits is still an open issue (Garrick et al., 2013). Assuming zero transaction costs, opportunities to

Box 17.1 **Property rights: a pre-condition and the key to success**

Water use rights are pivotal for water trading. The effectiveness of private property rights hinges, however, upon the extent to which they can be unambiguously defined, well enforced, and readily transferred. Some resources and goods are more conducive to management via private property rights, while others are better suited to common property rights or state ownership. Institutional economics suggests that a private property right system has both pros and cons, and may not be appropriate in all circumstances. It is well recognized, for example, that private property rights can quell the race to use water resources before others do (i.e. the tragedy of the commons), but they can also be expensive to establish and enforce. If these costs outweigh the benefits of improved resource management, private property rights might not be justified. Similarly, private property rights can facilitate the reallocation of water to higher-valued uses. However, private water rights can also be costly to transfer when third-party impacts (externalities) must be controlled via administrative oversight processes. In this case, trade might not transfer water to its most valued uses, and the other benefits of private water rights (e.g. preventing overexploitation) might be underachieved. Thus, even completely defined and enforceable water rights are not foolproof paths to efficiency. No property rights system can remove all inefficiencies or resolve all disputes. It is therefore important to consider the relative strengths and weaknesses of each system. The system that maximizes society's net benefit from a particular water resource will largely depend on the characteristics and uses of that resource, some of which create the potential for market failures. These failures certainly reduce the effectiveness of water trading, but they do not imply that trading is outperformed by alternative policies. [See Griffin et al. (2013)].

reallocate water to its most productive uses will exist when the value of water is variable across water users (Bjornlund, 2007). When these differences exist there is then be an opportunity for individuals or stakeholders to engage in a bargaining process to reach mutually beneficial agreements at a price that must be set in between the maximum willingness to pay (WTP) of the potential buyers and the minimum compensation that potential sellers are willing to accept for water use rights (under temporary or permanent cession). In studies about the opportunities for water trading, this is typically the first step to identify the maximum potential for water markets.

However, these basic calculations do not answer the basic question: how much water can be negotiated and transferred between two stretches of a river basin or between river basins taking into account all the likely costs (conveyance and interim losses from infiltration and evaporation) and the environmental constraints. Even ignoring other transaction costs this is already an important outcome showing that opportunities for water trading decay with distance as transport costs increase.

Besides that, from a social perspective the key question to be solved is whether what is in the best interest of contracting parties is also in the best interest of all the potential people affected by the spatial reallocation of water and/or according to the different societal demands. Experience with water trading shows that finding a good answer to this question is very challenging as it requires due knowledge of all the likely effects and their welfare outcomes over the so-called third parties. Existing experiences with water

Box 17.2 **Market failures: a stumbling block to success**

Water trading achieves a socially optimal allocation of water only if the following conditions hold: the market is highly competitive (i.e. individuals cannot control price); buyers and sellers exclusively enjoy and incur all benefits and costs associated with their decisions (i.e. no externalities, public goods, or missing markets), and the market does not suffer greater transaction costs or information imperfections than other allocation mechanisms. These strong conditions clarify why water trading cannot be the panacea that its true believers want it to be. Nearly every water resource setting fails to meet at least one, if not several, of these conditions. The question is whether market failures are sufficiently severe to render water trading inferior to alternative approaches. Crucial elements of the answer are linked to the important matters of public goods, externalities, and transaction costs. [See Griffin et al. (2013)].

trading illustrate, to a significant extent, that these effects have not been always taken into account (Young, 2010; Hanak et al., 2011). Rather, they are often overlooked (along with the environmental impacts of water use right trades).

It is also well known that even when all water rights are legally defined, norms do not have a full picture of all the possible contingencies and all the ways decisions over water use and diversion might affect others' opportunities, not to mention all the environmental services potentially delivered by the water ecosystems along the entire river basin (Grafton et al., 2010; Adamson and Loch, 2013).

None of these problems (be it transport cost or potential third-party effects) are important when water is traded on a local basis among users of the same kind, as might happen when all farmers within the same irrigation district negotiate their granted individual rights so that the most productive ones can get a higher share of water instead of the quota allocated by the water authority (Kahil et al., 2014). This guarantees a better allocation of the overall water available without transactions being conditioned on significant transport costs or any relevant third-party effects.

Negative environmental effects can be avoided if potential sellers of water cede all their water use rights but all the physical returns to the environment (whatever their form) remain in the area and only depletion (instead of water use) can be transferred. The reduction in water use that would be needed to allow one unit of depletion to be transferred would depend on the technical efficiency with which water is used in the exporting basin and, in economic terms, this will result in an increase in the cost of water effectively transferred. The effect on water costs will be higher, the lower the technical efficiency (Adler, 2012). Of note, third-party effects on other activities dependent on former water users, such as local agro-industry, cannot be avoided in this way.

An environmentally neutral water market needs to anticipate these effects so that water trading does not have any impact on the environment. All these elements (transport costs, transient losses, and environmental constraints) result in substantial reductions in the amounts of water that might be transferred by an efficient market and an important increase in the (still hypothetical) equilibrium price.

This analysis has important implications regarding the use of water trading as an incentive to enhance water efficiency. In fact water saved by installing more efficient

---

Box 17.3 **Evidence of informal trading in the Segura River Basin district (southeastern Spain)**

Experience shows that even when water trading (rather than water use right trading) is not explicitly allowed, farmers are willing to spontaneously engage in such bargaining (Kuwayama and Brozovic, 2013). Evidence does exist that informal water markets may be trading substantial amounts of water in the Segura every year (Garrido et al., 2013). Since these transactions are uncontrolled, illegal trading might be supplying the market with water resources in excess of allowed quotas and this might be one of the emerging factors driving groundwater overexploitation.

If this is the case, water trading might be based on the capacity of some farmers to obtain additional (ground) water in excess of what they are actually allowed to use, instead of their willingness to use smaller amounts of water than those specified in their water entitlements.

Informal water transactions in the Segura River Basin may be trading significant amounts of water at the highest prices in Spain: evidence collected by Hernández-Mora and De Stefano (2013) shows that water prices during drought events hit 0.70 €/m^3 in the agricultural district of Campo de Cartagena (Segura River Basin).

This analytical approach to the opportunities of water trading consists of identifying the maximum amount of water that can be traded in a scenario where only the financial interest of the trading parties is considered (see Figure 17.1). Following a similar line of argument one may also build an extremely conservative scenario to consider the maximum water volume that might be traded in the presence of any possible provision to avoid detrimental environmental impacts or third-party effects (see Box 17.6).

---

devices (such as modern irrigation systems) might reduce water use but would not result in reduced consumption, which at best would remain constant. If the number of tradable water rights issued after a user (e.g. a farmer) proves to have upgraded the technology she uses is equal to the (actual or presumed) reduction in water consumption, this might be an effective way to put more water into the market. Yet, if the criterion to issue tradable water rights is the change in use then efficiency will never be a means to put more water into trade. This is but one example of how the criterion of expanding water trade as much as possible might be in contradiction with the criterion of guaranteeing that any water transaction should have at least a neutral effect over the environment (Garrick et al., 2013).

Design

Water trading may adopt different forms (see Hadjigeorgalis, 2009; Cui and Schreider, 2009; Grafton et al. 2010; O'Donnell and Colby, 2010):

- *Spot water markets*, both informal and formal (i.e. under legal arrangements), are common for the transfer of surface or groundwater resources for short-term trades in the context of a single basin. Spot, as opposed to long-term exchanges, stands for transactions in which water delivery is immediate or is meant to occur in the very near future.

Box 17.4 **Assessing the Potential for Inter-Basin Water Reallocation between the Tagus and the Segura River Basins**

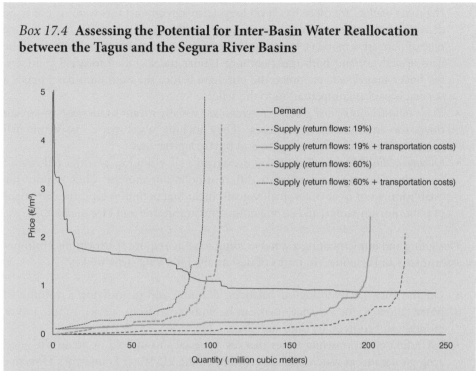

Figure 17.1 Focus on the environment: water trading in environmentally neutral markets

When transport costs and losses are factored in and only depletion instead of use is considered, the amount of water that might be efficiently transferred is only 85 instead of 240 million cubic metres, assuming no transaction costs and no environmental impact. In the absence of public intervention the market-clearing price may even be higher than the local financial cost of desalinated water. (see Gómez et al. 2013).

- *Water banks* are central institutions acting as a clearinghouse mechanism for users willing to purchase or sell water. A clearinghouse is an organization that collects and gives out information on supply and demand of water rights. Water is then sold at a price with a mark-up (i.e. an amount of money added onto the price) to cover the operating costs of the bank, which are often borne by the buyer.
- *Bulletin boards* are a type of water bank in which the price is not set by a central institution but rather the result of buyers and sellers posting bids and requests for water use rights on a central bulletin board (i.e. irrigation district authority) or through electronic platforms.
- *Auctions* are used to allocate rights between two or more users who compete for the same use right. Whereas in spot markets buyers and sellers occasionally interact, auctions allow as many trades as possible at a common price. In double-auction markets, buyers and sellers submit sealed bids for specific amounts of water rights. In all-in-auctions, bids are ordered during the auction session so that bidders see when their offer is accepted and have the opportunity to enter more bids.

- *Derivative markets* are those based on long-term agreements (i.e. water is not to be delivered either now or in the near future). In the so-called option markets, one type of derivative markets, buyer and seller agree on the quantity of water and the date of delivery and both must comply. Under the so-called forward contracts, the buyer may decide to forego the purchase before the expiration date; hence a deposit is paid as compensation to the seller.
- *Environmental leasing and purchase programs* are usually meant to increase in-stream flows for environmental purposes. They include water trusts, governmental leasing and purchase of use rights, and buyback programs.
- *Quality trading schemes* allow for the exchange of pollution permits (allowance or credits), among similar and/or different pollution sources. They require the establishment of clear water quality goals, including pollution caps, measurement of pollution emissions, and strong enforcement (Shortle and Horan, 2008).

Designing and implementing a water-trading scheme requires performing a number of assessments and activities. In terms of design, the following steps are key:

- On the basis of hydrological balances, defining and quantifying a volume of water (allowing for variance) that can be withdrawn from surface runoff and/or groundwater sources, by time and place.
- Excluding environmental flows that are necessary to uphold or attain the good ecological status of water bodies, according to the EU Water Framework Directive (200/60/EC); that is to say, the quantity of water that nature needs for the good

---

Box 17.5 Government responsibilities in water trading

Adequate human and regulatory means and measures to monitor transactions in order to avoid non-legal re-allocations and discharges are of utmost importance. Governments should guarantee that entitlements do not outweigh long-term renewable resources in the system – given environmental objectives, and making clear to all what will happen when unused / unlicensed water is activated. In order to control the volume of water effectively transferred, installation of meters and conversion from area-based licenses to a volumetric management system seems a binding requirement. Often other monitoring techniques like satellite information, real-time data and GIS could also be incorporated into the monitoring framework. Governments need to review trading requests based on the potential disturbance, externalities and other third-party effects. Annual assessment reports (as well as ideally term progress reports) about foreseeable effects of water trading and a post monitoring of final results and impacts may be carried out. Reports could contain analysis of legal, economic, environmental and social impacts of water trading. Improving transparency, accountability and access to markets is an important government responsibility. Equity and fairness objectives require careful attention about the way that trading opportunities, allocation decisions and policy changes are announced and set. The public regulator should conduct public studies about the value of water in the different uses and the real cost of water trading based on the opportunity cost, costs of conveyance, environmental and other third-parties effects, avoiding situations in which speculation and monopolist prices are set (see Maestu and Gómez-Ramos (2013).

ecological status to be achieved and the provision of ecosystem services to be maintained.

- Defining in a precise way water entitlements and rights. This includes how they relate to the physical resource and how to ensure a sustainable yield (temporally and spatially) that can be subject to trade.
- Setting up an institutional arrangement (i.e. legal reform, specific bodies within water authorities, official registry, arbitration and conflict adjudication procedures, etc.) to manage legal entitlements.
- Setting up an effective monitoring system, including metering and other devices to measure individual water use.
- Ensuring the enforcement of water use rights over all water sources.
- Setting up appropriate safeguard mechanisms (i.e. legal provisions, assessment procedures, etc.) to (i) guarantee environmental outcomes, (ii) to protect third-party potentially affected interests, (iii) to regulate the possibility of carrying over water between years, and (iv) to prevent hoarding of rights and speculative practices.

The structure and traits of water rights affect their performance. On one side, systems that limit marketable volumes of consumed water limit externalities and environmental threats (Brewer et al., 2008; Basta and Colby, 2010; Libecap, 2010). On the other, systems that allow the transfer of nominal entitlements without considering effective use face problems of over-allocation and, most importantly, externalities (Hadjigeorgalis, 2009; Loch et al., 2014). In addition, in some countries problems of water monopolization in non-consumptive use rights have emerged. And limiting transactions to agents already holding water rights (the so-called market incumbents) and to uses ranking higher than the seller, restricts the performance of markets.

To be effective, water trading requires making water use more flexible by allowing buying and selling instead of the strict use of water rights in the amounts, places and particular uses for which they are issued by the water authority. The definition of tradable water rights is a major change of the current institutions in place (as in Maziotis et al., 2013) where, contingent on the availability of water at each moment in time, individual users are granted with usufructuary rights that, unless an intricate authorization process is followed, cannot be used for another purpose or in another site than that authorized by the water authority.

Furthermore, evidence shows that trading schemes may have increased pressures on water resources (by putting into use water that might not have been used in the absence of markets). This may have been the case of the Murray-Darling Basin in Australia or the Copiapó aquifer in northern Chile, where available resources are claimed to be over-allocated (although there is no empirical evidence of this in the Andean country). Physical interactions between water bodies along a river basin and externalities that may arise make it difficult to find a set of property rights that can be efficiently traded. For instance, in Chile increased activity in consumptive water use markets has boosted conflicts with downstream users due to the effects of water use rights over return flows.

Pre-requisites (pre-conditions)

Water trading may only work if:

1 Opportunities do exist to reallocate water and to improve welfare at the same time (Bjornlund, 2007). This is the case when:
   - there is high variability among marginal returns from water among uses and places (i.e. profits obtained from water use), and when infrastructures can transfer water at a competitive cost;
   - water use efficiency and the contribution of water to social welfare can be substantially improved.
2 Property rights, in particular water use rights that become tradable, are properly defined and enforced (Crase et al., 2013). A critical issue in the implementation of markets is a clear but nonetheless full definition of water rights or entitlements and their associated risks. The structure and features of water entitlements affect the way in which those markets perform. Systems that limit marketable volumes to consumed water (effective use) limit externalities and environmental threats. Systems that allow the transfer of nominal entitlements without considering effective use face problems of over-allocation and, most importantly, externalities. In addition, evidence shows that trading schemes may have increased pressures on water resources (putting water to certain harmful uses that might not have otherwise occurred).
3 Environmental externalities and third-party effects are considered (Garrick et al., 2011; Gómez et al., 2013). It is also important to account for the interactions between surface and groundwater resources (no specific provisions can be found in many of the assessed systems). Main concerns, though, remain on third-party effects (for instance, linked to the definition of rights on water return flows) and other environmental externalities.
4 Trading water locally, e.g. among the members of a given irrigation district, does not raise important environmental concerns as long as all the parties directly involved in the agreement comply with the overall amount of water entitled to the group (such as in a "cap and trade" scheme).
5 This might not be the case when imperfectly controlled groundwater sources are involved. In this case, some farmers might be able to sell additional amounts of water without reducing their use accordingly. If that happens the demand for water for trading is not covered with the resources already available but rather through increasing short-term supply at the expense of higher water scarcity, and lower resilience to droughts, in the future.
6 Transaction costs are internalized (Garrick et al., 2013). Transaction costs should be minimized but not neglected, since they can be significant in certain situations.
7 Gaining consensus and making water trade politically acceptable requires transparent and effective encompassing measures to safeguard the environmental objectives of water policy (Wheeler et al., 2013). Otherwise the interest in water trading will not be able to overcome the risks associated with being an instrument to put more water into use, instead of reducing scarcity, and to extend scarcity throughout space and from scarce to less water scarce river basins. Worsening

---

Box 17.6 **Controlling outlawed overdrafts: a critical requirement for water trading**

An important question around the pervasive evidence of water trading in the Segura River Basin District is not whether it is the means by which local users avoid transaction costs imposed by prevailing regulations that prevent water from finding its more valuable use in the economy (which might be a legitimate function of markets), but a means to encourage outlawed water abstractions (which is a way to go deeper into the current unsustainable trends of water withdrawal).

---

overexploitation and scarcity trends are likely if water use rights do not match available water resources. This implies additional pre-conditions such as:

- Removal of excess (nominal) property rights when rights are allocated in excess of available water resources in order to avoid social disputes.
- Convincing and tested evidence on the dynamics of the water resource to show awareness and capacities to avoid undesirable effects on environmental flows, groundwater or other water related ecosystems (social perception of the potential effects of water trading on the environment depends on people's trust in institutional capacities to detect and control potential environmental threats).
- Trust in the ability to monitor and enforce water property rights. Risks are larger when monitoring and enforcement is poor and non-controlled or illegal rights are put into the market.

8 Additional safeguards are required to guarantee that the water trading scheme in place has properly addressed the following well known threats of trading with water:

- Leading to speculation with water rights when they are accumulated and not used. This can be addressed via charging permit fees for unused water, limiting applications for water use rights to the original needs and through water rights subject to forfeiture if not used (Hansen et al., 2013).
- Reinforcing social disparities and reducing spatial cohesion, as water is re-allocated to more valuable uses. This can be addressed through proper water planning decisions and specific assessments of major water diversions.

9 Water trading can only be part of the solution if the deficits that are covered in the importing basin—or area within a basin—are compatible with the closure of the exporting one; that is to say, if the capacity of the ceding basin to yield the surpluses that can be transferred at any time are compatible with the maintenance of environmental objectives.

10 Finding water-trading alternatives that fulfil all these conditions might be challenging but the real issue is that without transparent information showing that this is happening, social acceptance of water trading will remain difficult.

Outcomes

Experience with water trading shows how important this has been in helping to find mutually beneficial agreements between buyers and sellers, thus increasing or maintaining the production of goods and services and making water trading a convenient instrument to promote agriculture, manufacturing, hydropower, and other economic activities. These development objectives were the main driver in the original adoption of current water trading schemes, and concerns about their environmental outcomes (although not completely absent in origin) are still incipient in most cases.

Evidence also shows that trading schemes may have, in some instances, increased pressures on water resources (by putting into use water that might not have been used in the absence of markets). It is usual that in surface and groundwater systems where water entitlements and allocations are not tradable, a significant share of the entitlements issued are not used. Reasons for non-use in the case of groundwater include holding resources as a reserve to face drought events. This may have been the case of the Murray-Darling Basin in Australia (Crase, 2012), where the emergence of a complex and profitable water market has resulted in over-allocation that threatened the fulfilment of environmental goals. In order to solve this problem a series of measures have been implemented, including command-and-control policies (first via a decision to secure 500 GL of water for the environment under Living Murray Initiative and second by the transfer of basin-wide water planning responsibilities to an independent Murray-Darling Basin Authority) and financial instruments (the commitment of A\$3.1 billion for the purchase of water entitlements from irrigators and the commitment of A\$5.8 billion for investment in so-called water saving projects).

However, these policies may not be enough to face the challenge of increasing demand. In Chile, for instance, the entire river flows have been fully allocated since at least three decades ago, which has led to the deterioration of aquatic ecosystems in semiarid and arid regions of the country (with significant problems in the region of Copiapó, in northern Chile). This can be said to be gradually changing with a series of reforms implemented since then (such as forfeiture for non-use), but positive environmental outcomes are still uncertain, at best.

Physical interactions between water bodies along a river basin (including in-stream uses and the connection between surface and groundwater; the definition of property rights that can be efficiently traded in a market is still a challenge) and externalities (and third-party effects) that may arise, still make it difficult to find a set of property rights that can be efficiently traded in existing water markets. For instance, in Chile increased consumptive water use market activity has generated further conflicts with downstream users due to the effects of water entitlements over return flows. Almost all consumptive water use right holders generate significant return flows (leakage and seepage water) that are used by downstream customary right holders, but it is not known how these customary use rights are dependent on return flows (Ward and Pulido-Velázquez, 2008).

Voluntary trading can thus play a critical role in stabilizing the economy and in providing an effective drought management alternative, provided all stakeholders are involved and provisions are made to compensate for third-party effects. This was evidenced in the Tagus River Basin in Spain, where voluntary agreements allowed for the optimization of the use of existing water resources without building additional infrastructure, or engaging in massive groundwater abstractions or significant political costs.

Water trading is supposed to be a means to increase the overall allocation of water amongst places and economic activities. Provided transaction costs are not unreasonable (both the Chilean and Australian markets have a similar system of pro-rata share of water stocks, intended to reduce transaction costs and to eliminate opposition to transfers), the participation condition is more likely to be fulfilled when there are important differences in the marginal value of water among potential buyers and sellers and mutually beneficial agreements are feasible (so that the participation condition is met). Nevertheless, in many water right trading schemes, incentive compatibility is not guaranteed. Illustrative examples show that the option to trade water may put into use a substantial amount of resources that in the absence of trading opportunities would have remained in the environment. In this case, water markets can paradoxically contribute to increased environmental impacts to spread the water transfers in the Middle Tagus in Spain and also, at a much larger scale, in the Murray-Darling Basin in Australia.

Considering all of the examples discussed, the water-trading schemes being developed in Europe have many desirable properties (including provisions to account for third-party effects), but putting too much emphasis on those effects may also undermine the possibility of a dynamic market, which is not an end itself but rather a means to tackle water scarcity and drought risk, above all.

Note

1 *Disclaimer*: The views expressed in this document do not necessarily reflect the views of the United Nations Secretariat or the United Nations Office to Support the International Decade for Action (UNO-IDfA) 'Water for Life' 2005–2015. The designations employed and the presentation of the material in this document do not imply the expression of any opinion whatsoever on the part of the Secretariat of the United Nations concerning the legal status of any country, territory, city or area, or of its authorities, or concerning the delimitation of its frontiers or boundaries.

References

Adamson, D., and Loch, A., 2013. Possible negative sustainability impacts from "gold plating" irrigation infrastructure. *Agricultural Water Management*. http://www.uq.edu.au/rsmg/WP/Murray_Darling/WPM13_2.pdf

Adler, J., 2012. Water rights, markets, and changing ecological conditions. *Environmental Law*, 42, 93–113

Basta, E., and Colby, B.G., 2010. Water market trends: transactions, quantities, and prices. *The Appraisal Journal* 78(1), 50–66.

Bjornlund H., 2007. Do markets promote more efficient and higher value water use? Tracing evidence over time in an Australian water market. In G. Lorenzini, and C.A. Brebbia (eds) *Sustainable Irrigation Management, Technologies and Policies*. Southampton, WITPress, 289–302

Brewer, J., Glennon R., Ker, A., Libecap, G., 2008. Water Market in the West: Prices, Trading and Contractual forms. NBER Working Paper, No. 13002. Washington. DC, NBER.

Crase, L. 2012. Water markets, property rights and managing environmental water reserves, in. J. Quiggin, T. Mallawaarachchi, and S. Chambers (eds) *Water Policy Reform: Lessons in Sustainability from the Murray–Darling Basin* Edward Elgar, Cheltenham, pp. 37–48.

Crase, L., Pawsey, N., and O'Keefe, S., 2013. A note on contradictions in Australian water policy. *Economic Papers: A Journal of Applied Economics and Policy*, 32(3), 353–359.

Cui, J., and Schreider, S., 2009. Modelling of pricing and market impacts for water options. *Journal of Hydrology* 371 (1–4), 31–41.

Garrick, D., Lane-Miller, C., and McCoy, A., 2011. Institutional innovations to govern environmental water in the western United States: Lessons from Australia's Murray-Darling Basin. *Economic Papers: A Journal of Applied Economics and Policy*, 30(2), 167–184.

Garrick, D., Whitten, S.M., and Coggan, A., 2013. Understanding the evolution and performance of water markets and allocation policy: A transaction costs analysis framework. *Ecological Economics* 88, 195–205.

Garrido, A., Rey, D., and Calatrava, 2013. Water trading in Spain in water, in De Stefano L., and Llamas R. (eds), *Agriculture and the Environment in Spain: Can We Square the Circle?* The Netherlands, CRC Press, pp. 205–216.

Gómez, C.M., Delacámara, G., Pérez-Blanco, C.D., Ibáñez, E., and Rodríguez, M., 2013. Droughts and water scarcity: Tagus (Central Spain & Portugal) and Segura (SE Spain) interconnected river basins (Deliverable No. 4.3), Work Package 4 – Ex-ante Case Studies. 7th Framework Contract Project EPI-Water Project (GA 265213) European Commission. http://www.feem-project.net/epiwater/

Grafton, R.Q., Landry, C., Libecap, G., McGlennon, S., and O'Brien, B., 2010. An Integrated Assessment of Water Markets: Australia, Chile, China, South Africa and the USA. NBER Working Paper No. 16203. NBER (National Bureau of Economic Research), Cambridge (Massachusetts).

Grafton, Q., Libecap, G., McGlennon, S., Landry, C. and O'Brien, B. 2011. An integrated assessment of water markets: a cross-country comparison, *Review of Environmental Economics and Policy*, 5(2): 219–239.

Griffin, R.C, Peck, D.E., and Maestu, J., 2013. Introduction: myths, principles, and issues in water trading, in Maestu, J. (ed.), *Water Trading and Global Water Scarcity: International Experiences*. RFF Press, New York.

Hadjigeorgalis, E., 2009. A place for water markets: performance and challenges. *Review of Agricultural Economics* 31(1), 50–67.

Hanak, E., Lund, J., Dinar, A., Gray, B., Howitt, R., Mount, J., Moyle, P., and Thompson, B., 2011. Managing California's Water: From Conflict to Reconciliation (Report). Public Policy Institute of California, San Francisco (US).

Hansen, K., Howitt, J. and Williams, J. (2013) Water trades in the western United States: risk, speculation, and property rights', in J. Maestu (ed.) *Water Trading and Global Water Scarcity: International Experiences*, RFF Press, New York, USA

Hernández-Mora, N., and De Stefano, L. (2013) Los mercados informales de aguas en España: una primera aproximación. In: Embid Irujo, A. (ed) *Usos del Agua Concesiones, Autorizaciones y Mercados del Agua*. Thomson Reuters Aranzadi, Madrid.

Hodgson, S., 2006. *Modern Water Rights: Theory and practice*. FAO Legislative Study 92. FAO, Rome.

Howe, C.W., 2000. Protecting public values in a market setting: improving water markets to increase economic efficiency and equity. Univ. *Denver Water Law Rev.* 3, 357–372.

Howe, C.W., 2013. *Interbasin Transfers of Water: Economic Issues and Impacts*. Routledge, London.

Kahil, M. T., Dinar, A., and Albiac, J., 2014. Modeling Water Scarcity and Drought Severity for Policy Adaptation to Climate Change: Application to the Jucar Basin, Spain. Water Science and Policy Centre Working Paper 01-0114, University of California, Riverside, CA.

Kuwayama, Y., and Brozović, N., 2013. The regulation of a spatially heterogeneous externality: Tradable groundwater permits to protect streams. *Journal of Environmental Economics and Management*, 66(2), 364-382.

Libecap, G., 2010. Water Rights and Markets in the U.S. Semi Arid: Efficiency and Equity Issues. Prepared for the Conference on "The Evolution of Property Rights Related to Land and Natural Resources", September 20–21, Lincoln House, Cambridge, MA.

Loch, A., Bjornlund, H., and Kuehne, G., 2013. Water trade alternatives in the face of climate change. *Management of Environmental Quality: An International Journal*, 21(2), 226–236.

Loch, A., Wheeler, S., Boxall, P., Hatton-Macdonald, D., Adamowicz, W. L., and Bjornlund, H., 2014. Irrigator preferences for water recovery budget expenditure in the Murray-Darling Basin, Australia. *Land Use Policy*, 36, 396–404.

Maestu, J., 2013. *Water Trading and Global Water Scarcity: International Experiences*. RFF Press, New York.

Maestu, J., and Gómez-Ramos, A., 2013. Conclusions and recommendations for implementing water trading: how water trading can be part of the solution, in Maestu, J., (ed.) *Water Trading and Global Water Scarcity: International Experiences*, Taylor and Francis, London.

Maziotis, A., Calliari, E., and Mysiak, J., 2013. Robust Institutional Design Principles for Sustainable Water Markets. Water Scarcity and Droughts (The interconnected Tagus and Segura river basins, Spain) (Deliverable No. T4.2–Output 01), Work Package 4 – Ex-Ante Case Studies. 7th Framework Contract Project EPI-Water Project (GA 265213). European Commission. http://www.feem-project.net/epiwater/

NWC (National Water Commission), 2011. *Strengthening Australia's Water Markets*. NWC, Canberra.

O'Donnell, M., and Colby, B., 2010. Water Banks: A Tool for Enhancing Water Supply Reliability. University of Arizona, Department of Agricultural and Resource Economics. http:/ag.arizona.edu/arec/people/profiles/colby.html. (Guidebook for Stakeholders)

OFWAT, 2010. *A Study on Potential Benefits of Upstream Markets in the Water Sector in England and Wales*. London, OFWAT

Puckett, Paul W., 2010. Trading water: Using tradable permits to promote conservation and efficient allocation of an increasingly scarce resource. *Emory Law Journal* 59(4), 1001–1037.

Selman, M., Greenhalgh, S., Branosky, E., Jones, C., and Guiling, J., 2009. Water Quality Trading Programs: An International Overview. World Resources Institute Issue Brief. Washington, D.C.: WRI (World Resources Institute).

Shortle, J., 2013. Economics and environmental markets: lessons from water-quality trading. *Agricultural and Resource Economics Review* 42(1), 57–74.

Shortle, J. S., and Horan, R. D., 2008. The economics of water quality trading. *International Review of Environmental and Resource Economics*, 2(2), 101–133.

Ward, F.A., and Pulido-Velazquez, M., 2008. Water conservation in irrigation can increase water use. *PNAS* 105, 18215–18220.

Wheeler, S., Garrick, D., Loch, A., and Bjornlund, H., 2013. Evaluating water market products to acquire water for the environment in Australia. *Land Use Policy* 30(1), 427–436.

Yoo, J., Simonit, S., Connors, J.P., Maliszewski ,P.J., Kinzig, A.P., and Perrings C., 2013. The value of agricultural water rights in agricultural properties in the path of development. *Ecological Economics* 91, 57–68.

Young, M., 2010. Environmental effectiveness and economic efficiency of water use in agriculture, in *OECD Studies on Water*. Organisation for Economic Co-operation and Development, Paris (France), pp. 1–33.

Zilberman, D., Dinar, A., MacDougall, N., Khanna, M., Brown, C., and Castillo, F., 2011. Individual and institutional responses to the drought: the case of California agriculture. *Journal of Contemporary Water Research and Education*, 121(1), 3.

Zuo, A., Nauges, C., and Wheeler, S., 2012. Water-trading as a Risk management Tool for Farmers: New Empirical Evidence from the Australian Water Market. RSMG Working Paper Series (M12_2). School of Economics and Political Science. University of Queensland,

18

WATER TRADING IN AUSTRALIA

Understanding the role of policy and serendipity

Lin Crase, Sue O'Keefe, Sarah Wheeler, and Yukio Kinoshita

Introduction

The depth and sophistication of water trading in Australia, especially in the southern Murray-Darling Basin (MDB), is testament to the insight of policy makers and their hard work to create tradable property rights and systems that facilitate trade. The origin of water markets is also testament to the desire of Australian farmers to be able to source water during times of scarcity, and their demands for more flexible arrangements. By international standards, Australia's water markets are an exemplar of good water economics. Armed with empirical and theoretical evidence (e.g. Pigram and Musgrave 1989), and given the history of drought and water scarcity in Australia, policy makers had the foresight in the 1970s and 80s to realize that historical allocations of water and the institutions for managing them would be found wanting, especially in times of scarcity. A low-cost means of resource reallocation was required and markets for water entitlements and water allocations have ultimately proven up to the task.

Water markets have been widely acknowledged as central to the survival of agriculturalists during one of the longest and most severe droughts experienced in Australia at the start of this century (NWC 2012; Wheeler et al. 2014). In addition, the market has provided governments with a mechanism for securing so-called environmental water or e-water. In this context governments at the state and national level have successfully negotiated the transfer of water from extractive users to provide in-stream environmental gains at relatively low cost to taxpayers.

Regardless of these successes, questions remain about the extent to which the achievements of Australia's existing water markets can be transposed to other settings. Moreover, even within Australia there are doubts that the level of trade that typifies the water market in the southern MDB is replicable in other contexts. Is the success of water trading in the southern MDB a result of omniscient policy reform and hydrological factors or did other influences along with fate and circumstance play a part?

This chapter traces the development of water markets in Australia's southern MDB and extends the analysis to consider the feasibility of water markets elsewhere in the country. The chapter also critically examines some of the limits to the Australian approach

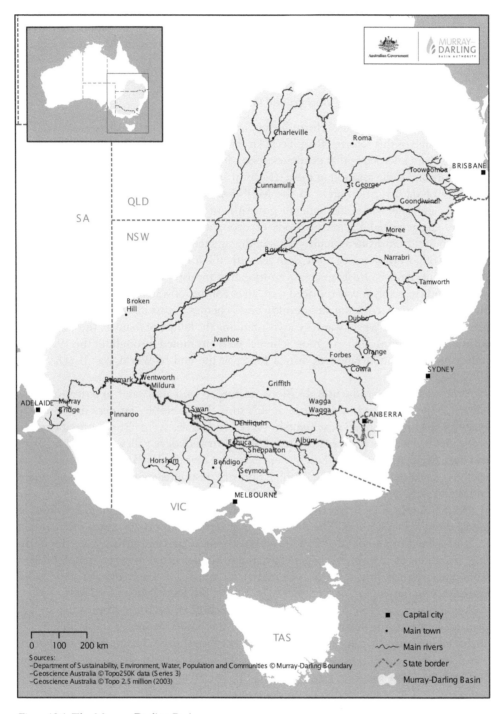

Figure 18.1 The Murray-Darling Basin

to water marketing. We touch on the historical backdrop to the development of markets in order to raise questions about the potential for duplicating market institutions beyond the MDB and also beyond Australia.

Background to water reforms

Australia has an active water reform history, particularly in the last three decades. The locus of much of this reform effort has been the MDB (Figure 18.1), an area of about 1 million square kilometres that generates about 40 per cent of the national income from agriculture and is the location of most irrigation activities in the country (MDBA 2012).

Many of the irrigation areas in the MDB were initially established as a result of colonial, and later state and federal, government policy. Australia's highly variable climate, characterized by long periods of drought, had promoted some interest in irrigation by adventurous agriculturalists in the nineteenth century. However, it was the ambition of governments to more densely settle the inland of the continent that drove the most substantive efforts to create an irrigation industry. State irrigation trusts were created and riparianism was replaced with state control of water. This was followed by attempts to expand irrigation and develop new schemes as part of a "soldier settlement" process at the end of World War I. Notwithstanding the failure of many of these schemes, additional state-controlled irrigation schemes were promoted throughout the 1930s and following the demobilization of armed forces at the end of World War II (Davidson 1969; NWC 2011a).

Almost all of the production decisions relating to land and water in irrigation areas were heavily influenced by the state at this time. For example, farmers in the newly established Murray Irrigation scheme in New South Wales were not permitted to grow crops that had already been established in the Murrumbidgee valley. The size of farms had been decided centrally and there were extensive caveats placed on farmers who took up those lands (described in detail later). Similar constraints applied to the assignment and use of water within irrigation districts.

On the output side, markets played a relatively muted role in shaping decisions, at least for individual farmers. The distribution of agricultural products was often vested in state-owned monopolies or marketing boards and mandated prices were the norm for generations of farmers. More generally, the Australian economy was largely shielded from international prices, with high tariffs on imported manufactured goods and embargoes on the importation of many agricultural products (Musgrave 2008).

A number of key developments led to a redirection of the focus from water supply management towards demand management in the MDB. Some of these include: a) anecdotal evidence that during Australia's World War II drought unofficial temporary water trades occurred; b) some states placing moratoriums on the issue of new entitlements from the late 1960s onwards; c) informal MDB markets for seasonal water in the 1960s and 1970s; d) an expansion of the mandate of the River Murray Commission to include water quality issues (and hence water markets); and e) formal transfer of water entitlement and allocations in the 1980s (Wheeler et al. 2014).

Water market development proceeded gradually and the 1990s saw a broadening of water reform and trade. Following federal–state negotiation, a new MDB Agreement was crafted in 1992; it was supported and maintained by the creation of the Council of Australian Governments (COAG) in 1994. In the context of water, the *Water Reform Framework* of 1994 represents a significant milestone and shares much with the underlying policy discourse of this time. The Framework was adopted by all state governments in return for a share of the proceeds from the sale of the nationally-owned telecommunications provider. States agreed that water prices that recovered costs and removed subsidies should be introduced. Important functions like regulation, service delivery and resource management were separated and all future investments in infrastructure were to be subject to economic and environmental scrutiny. Volumetric and tradable water rights that were separated from land also formed an important part of this commitment (NWC 2011a).

The *Water Reform Framework* was subsequently updated and strengthened in the form of the *National Water Initiative* (NWI) in 2004. The NWI expanded the commitment to water trade and further strengthened individual rights to water. More specifically, water was to be defined as a perpetual share of a variable consumptive pool and rights were to be harmonized across states to favour an expansion of inter-state water markets. In addition, price signals for the use of water services were to be enhanced with arrangements put in place to improve the pricing of storage and delivery (Crase et al. 2013; NWC 2011a).

Whilst the basic tenets in the NWI remain in place, in the past ten years much attention has focused on the notion of over-allocation. This stems from concerns that the ecological status of many of the MDB's rivers had been undermined by excessive extraction, especially during dry years. Whilst some progress was made to address this issue by using planning and state-controlled infrastructure projects, the largest portion of water redistributed to meet environmental need was sourced from the market. This was driven by the Commonwealth's 'Water for the Future' programme, introduced in 2007–08. For further discussion on Commonwealth government policy, see Wheeler et al. (2014) and Crase et al. (2012). Water buy-backs have involved sourcing water rights from willing sellers. The most common method entails potential sellers indicating their reservation price for water entitlements. Such entitlements are held in different locations within the MDB and can vary in reliability. The Commonwealth has chosen to accept bids on the basis that they represent value for money. Most of these sellers have been individual farmers making choices about the future structure of their farming businesses. These farmers chose to sell because they wanted to exit irrigation, to reduce debt, had surplus water or needed money to restructure their farm business (Wheeler and Cheesman 2013). Clearly, this stands in stark contrast to the centralized decision-making apparatus that typified irrigation in the mid-twentieth century including the preoccupation with controlling land, which we discuss in more detail later.

The status of water trade

Institutional fundamentals

In order to comprehend the extent of water trading it is first necessary to appreciate the types of water products that now exist in Australia, especially in the MDB. Although the Commonwealth has sought to provide a common ground for policy formulation (e.g. the NWI), states have been able to adapt this policy framework to jurisdictional circumstances and histories. The upshot is that water rights can vary significantly by location, although trading of rights between jurisdictions is possible, provided there is a recognized hydrological link between buyer and seller. For illustration we briefly focus on differences between the largest irrigation states of New South Wales, Victoria and South Australia in the southern MDB.

In line with the NWI, all of these states have separated water rights from land to now represent a variable but perpetual share of the available water resources in a given period. The long-term right is termed a water entitlement (euphemistically called "permanent water") and the water that accrues against that entitlement in a given period (usually a season) is termed the water allocation (euphemistically called "temporary water").

An important reform in the last decade was the unbundling of water entitlements to distinguish water access from water use. Thus, water access entitlements represent the long-term rights to access available water but the resource cannot be deployed without use rights. In many cases the use rights relate to employing infrastructure held in common. One of the complications in New South Wales and South Australia particularly has been the separation of access and use rights in communal irrigation districts. Irrigation infrastructure operators (IIOs) have historically held water access rights on behalf of individual irrigators such that individuals have rights to access water only from the IIO. These arrangements do not prohibit trade within irrigation districts, but for an individual to trade their water to an area outside that controlled by the IIO requires that irrigation rights (i.e. the arrangements struck between the IIO and the irrigator) be "transformed" to standard water access entitlements. In some instances IIOs have sought to limit this trade, based on concerns that the de-watering of the district would render it unprofitable for others. In response, a set of Water Market Rules were introduced by the Australian government under the Commonwealth Water Act 2007 to ensure that IIOs were not unreasonably delaying or prohibiting transformation of irrigation rights to water access rights. In most cases this has meant that irrigators now have transformed rights into water access entitlements and delivery rights – the latter relating to access to irrigation infrastructure held by an IIO (NWC 2013, p. 53).

Most of the streams in the southern MDB have large water storages that regulate flows to downstream users. Overall, the storage capacity for the Basin is about 23,000 GL[1] and around 16,000 GL of this is located in the southern portion of the Basin. Average annual inflows to the Basin as a whole are around 32,800 GL although this is highly variable by location and temporally (MDBA 2012). Notwithstanding this variability, many water rights held in the southern part of the Basin relate to highly regulated river systems such that the probability distribution that attends the delivery of water is reasonably well known.

In New South Wales most water access rights were previously of a form termed "general security". This means that most water access rights in this state are less reliable than what was called "high security" rights. In simple terms most New South Wales water access rights are likely to be characterized by dam management algorithms that allocate more water in the current season such that less is held in reserve against rights for the following season. In contrast, Victorian and South Australian water access rights are allocated with a more conservative algorithm so that the following year's allocation is secured before the current season is allocated. Because water can be traded between states these algorithms attend any trade. For example, a Victorian farmer cannot buy New South Wales water access rights and expect a high reliability product – the rights are effectively "tagged" with their initial characteristics plus some allowance if water is moved over large distances and "losses" accrue.

In New South Wales, Victoria and South Australia, holders of water access rights can also carry over unused water allocations to the following season.[2] However, carryover provisions are not the same within and between jurisdictions. For example, in 2011–12 and 2012–13 the New South Wales Murray carryover was limited to 50 per cent for general security access rights whilst Victorian access rights in the same valley have 100 per cent carryover, albeit subject to spill rules for the main upstream storage (Hume Dam) and a new cap on carryover that applied to the Murray from June 2014 (NWC 2014). Differences in carryover arrangements have led to the development of markets around temporary carryover, although states have been keen to close off some elements of this trade since it can result in perverse third party effects[3] (see, NWC 2012, p. 59).

Given that water rights are relatively complex and can be traded across jurisdictions, an administrative architecture is required to manage and monitor trade. This takes the form of water registers controlled by each state. To assist the flow of information about water trading, details of water prices and trades are published on state registers. The National Water Market System was initiated in 2009, with the purpose of improving the efficiency (and transparency) of state and territory water registers, transactions and market information functions. Water brokers are also active in this environment providing both public and proprietary data on water trade (see, for example, Waterfind 2013).

Trends in water trade

This section draws heavily from the National Water Commission's (NWC 2013, 2014) reports on the trends of Australian water markets. The Commission was initially established under the NWI and it provided oversight and reports progress against the objectives of the NWI and related matters. Due to budgetary reasons, the Commission is due to be terminated at the end of 2014.

The basic tenet that underpins water trading is that water resources should be able to be reallocated to the highest value uses. As we have already noted most of this trade has occurred in the southern MDB and initially much of this was in the form of allocation (or temporary) water trade. Whilst there are numerous explanations for this phenomenon (see, for example, Crase et al. 2000), entitlement trade has now grown to be a more prominent feature of the water trading landscape, especially in the MDB.

In 2012–13 a total of 1343 GL of water entitlement was traded, most within the MDB (Table 18.1) and much involving "regulated" entitlement.

Table 18.1 Volume of water trading in major market segments, 2012–13 (GL)

	Allocation trading	*Entitlement trading*
Southern MDB	5478	678
Northern MDB	580	365
Outside the MDB	126	300
Australia total	6184	1343

Source: Adapted from data in NWC (2014).

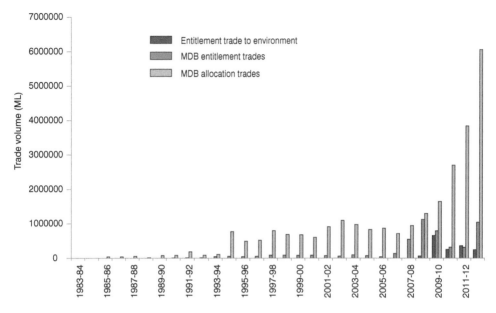

Figure 18.2 Water trade in the southern MDB from 1983–84 to 2012–13 (source: adapted from data in NWC 2011b, 2014).

The prices paid for water entitlements have been less volatile over time than the prices for water allocations. This might not be that unexpected given that water entitlements constitute a long-term investment to secure water access rather than a response to dealing with short-term variations in input availability. Data from the NWC suggest that high reliability water entitlements have generally fetched prices ranging between $AUD1500/ML and $AUD2000/ML between 2007–08 and 2011–12 while most general security entitlements sold for between $AUD750/ML and $AUD1000/ML (NWC 2013, p. 18).

The trend in entitlement trade in the southern MDB shows a spike in 2007–08 as the severe drought in the southern states continued and the Commonwealth buyback program was initiated. In subsequent years this level of entitlement trade has coincided with increased activity on the part of the Commonwealth to secure water entitlements to bolster long-term environmental water reserves playing a part (discussed in detail later). Notwithstanding that the activity of the Commonwealth in the entitlement market has varied since 2007–08, water entitlement trade is now clearly more active in this zone than was the case in previous decades (Figure 18.2).

Water entitlement trade also occurs in the northern portion of the MDB and this is where most unregulated entitlement (i.e. no upstream controlling water storage) is traded (albeit with modest volumes – circa 365 GL in 2012–13). Because the trading zones tend to be more confined in the northern part of the Basin, prices tend to be more volatile and subject to fluctuations with local conditions. Nonetheless, trade grew significantly from 2011–12 to 2012–13. The entry of the Commonwealth to purchase water entitlements in northern catchments stands to represent a proportionately larger intervention compared with the southern connected catchments.

Outside the MDB, water entitlement trade has historically been most active in Queensland, although trade in Tasmania's irrigation areas has expanded, especially since 2007–08. In volumetric terms the trade of water entitlements outside the MDB has stood at about 200 GL per year for the period 2007–08 to 2011–12 (e.g. 218 GL in 2011–12).

The trade in Tasmania is in part a reflection of property sales, as much as an expansion in interest in water trading per se. Trading of water entitlements is also reported in districts like the Hunter Valley in New South Wales and the Macallister, Bacchas Marsh and Werribee districts in Victoria. Water licence transfers have occurred in the Harvey Irrigation district in Western Australia, but the volumes involved are usually 1–3 GL per year. Again, prices in these regions reflect the local conditions within the relatively small trading areas. The geographic proximity of water users also means that buyers and sellers are more likely to be involved in similar farming enterprises, thereby reducing potential heterogeneity to differences in management expertise and the like.

A similar scenario applies to the trading of groundwater entitlements. The limited connectivity between groundwater resources limits the scope for trade between heterogeneous users. The NWC (2013, p. 25) also notes that groundwater trade has been limited because in some cases, jurisdictions have been slow to unbundle water rights, the trading rules are relatively new in places while groundwater access is still available and water shortages rare in other cases. Collectively, this accounts for the modest trade of groundwater entitlements (around 196 GL nationally in 2012–13) even though around half of all entitlements by number and about a fifth by volume apply to national groundwater resources (NWC 2013, 2014).

We noted earlier that entitlement trade historically lagged behind the trade in water allocations. Notwithstanding the growth in water entitlement trade described above, it remains the case that allocation trade continues to significantly outstrip entitlement trade (Table 18.1). Across Australia 6184 GL of water allocation was traded in 2012–13 with 6058 GL of this in the form of allocation in the MDB. Figure 18.2 illustrates the continual growth in allocation trade in the southern MDB, with 71 per cent of water allocations traded in 2012–13 (compared to 38 per cent in 2011–12 and 8 per cent in 2001–02). There are several important elements to allocation trade that are reflected in both the volumes of water on offer and the prices paid. During wetter periods allocations received against access entitlements tend to be higher. This means that there is generally more allocation to be sold but demand is also much lower and prices usually lower. The resulting inverse relationship between allocation trade and the volume of allocations/proportion of allocations to entitlement was clearly evident during the extended dry period between 2001–02 and 2007–08. Over this period the percentage of allocations traded rose markedly from 8 per cent to 30 per cent (NWC 2013, p. 29). An important driver of this trade was the differing water requirements for perennial uses versus annual activities. For example, producers of annual crops in New South Wales, when faced with very low allocations at the commencement of the irrigation season, opted to suspend production altogether and sell their limited water allocations to those undertaking perennial agriculture, like horticulture. During this period, dairying, once regarded as a perennial activity, was also temporarily transformed with many producers opting to sell water allocations and purchase fodder in preference to irrigation (NWC 2012). These types of innovations arguably resulted in more agricultural enterprises surviving the prolonged drought than would have otherwise been the case.

Notwithstanding the steady improvement in water availability since 2008–09, especially between 2010 and 2013, the previous inverse relationship between allocation trade and the proportion of allocation against entitlement has not persisted. More specifically, there has been a marked increase in late-season trade in the allocation market.

A key influence here has been the refinement of rules around carryover arrangements and differences in those rules between states.[4] With higher allocations available, farmers (and environmental water users) are increasingly seeking to use the allocation market in tandem with carryover rules to optimize their water use between years.

The large increase in the percentage of allocations traded in 2012–13 is due in part to activity by the Commonwealth Environmental Water Office (see next section).

Governments and trade

It should come as no surprise to economists that trade of water rights requires strong support from government. Entitlements and allocations cannot be traded without government first defining property rights and establishing the mechanisms to legitimize exchange. In the case of Australian water markets, governments have also exerted other influences and we deal specifically here with the constraints imposed around inter-state trade and the role of the Commonwealth in establishing significant environmental water holdings in the MDB.

The Commonwealth Environmental Water Office

We noted earlier that the NWI placed additional focus on the notion of over-allocation. Whilst over-allocation was not defined with precision at the time (see, Crase 2012), the idea of restoring a balance between consumptive and environmental demands on water was heavily promoted during the millennium drought and ultimately manifested in the Commonwealth's Water Act 2007. The Act required the formulation of a Basin Plan that would then indicate the sustainable level of take from rivers and groundwater sources and ensure a reserve was available to met environmental demands. A full description of the water planning that attended this exercise is beyond the scope of this chapter but it is worth noting that planning on this scale was necessarily complicated. Key complications related to the sheer size of the Basin, the limits of scientific knowledge, the large hydrological and climatic variability across the Basin and the divided political ambitions of state jurisdictions and the Commonwealth. The exercise also involved a substantial shift in water resource planning expertise since this had previously been undertaken by states with oversight from a Basin-wide Commission and was now vested in a single Commonwealth Authority (the Murray-Darling Basin Authority).

In the context of the Basin Plan a key (and contentious) feature was the sustainable diversion limits that were set for each valley in the Basin. Since the Water Act largely superseded state-based water plans (which largely complied with the NWI at the time) the processes for meeting the new limits required attention. Under the NWI provisions existed in water plans to adjust some extractive use without compensation. In contrast, the Commonwealth reached a decision that the reduced water access for consumptive users that occurred as a result of the Basin Plan would be fully compensated. The process of managing the environmental water holdings rests with the Commonwealth Environmental Water Holder.

The water entitlements that make up the environmental water reserve have been sourced in two main ways. First, from publicly-funded infrastructure projects that were undertaken from the Water for the Future programme and any water purportedly saved was then shared with the Commonwealth Environmental Water Holder. This approach has attracted criticism on several fronts (see, Crase et al. 2012) not least because of the cost to taxpayers and the modest amounts of water accrued by this approach. Second, entitlements have been purchased from willing sellers.

Between 2007–08 and 2012–13 the Commonwealth Environmental Water Holder had registered 1599 GL of water entitlement using water buybacks (NWC 2013, p. 62). This makes the Commonwealth Environmental Water Holder the largest current single holder of water entitlements in the MDB.

In order to acquire this amount of water entitlement the Commonwealth has clearly been active in the entitlement market. On the one hand, some irrigators and states have expressed concern that this volume of activity stands to crowd out private interests but on the other, some irrigators have benefited from the increase in water market prices that have assisted agricultural adjustment. In this context it is worth noting that in 2011–12, 80 per cent of irrigators who transformed an irrigation right into a tradable water access entitlement did not terminate the accompanying delivery rights immediately after transformation (NWC 2013, p. 53). Wheeler and Cheesman (2013) found that of the farmers who stayed farming after selling water entitlements, 94 per cent kept their delivery rights. This supports the view that irrigators are maintaining agricultural production through the allocation market and/or keeping their options open for future irrigation.[5] In addition, surveys of farmers selling water access entitlements to government reveal that the majority have remained active in agriculture, using funds to pay down debt and/or to establish other enterprises (e.g. NWC 2012; Wheeler et al. 2014). Notwithstanding this evidence, states have remained cautious about the sale of "their water" to the Commonwealth.

Having assembled such a large water entitlement holding the Commonwealth Environmental Water Holder is well positioned to play a major role in allocation trade. In 2012–13 there were no sales from the Commonwealth to irrigators, unlike trade activity that occurred in early 2014 (NWC 2014). As noted earlier, understanding and delivering environmental gains in such a complex river Basin is no simple task. This is further complicated by the fact that some environmental objectives are achieved by using water held by state environmental agencies (e.g. the Victorian Environmental Water Holder) or the activation of what is known as "rules-based" water. Coordination is thus critical and by using allocation trade the Commonwealth Environmental Water Holder is able to better deploy allocations to specific sites. Water allocation is technically "traded" when it moves from one location to another. Thus, a within-environment trade appears as increased market activity, even though payment is not usually involved. In 2010–11 and 2011–12 Victoria's environmental allocation trade equated to more than 1200 GL, representing about 37 per cent of that state's allocation trade in those years (NWC 2013, p. 65).

Whilst environmental trade of allocations has hitherto involved limited interaction with non-environmental interests, there is scope for expansion on this front. Recent economic modelling has also shown that there are possible environmental and financial benefits from the Commonwealth Environmental Water Holder using allocation trade that involves non-environmental interests (CEWHA 2011; Wheeler et al. 2013).

Limits to trade

Notwithstanding the commitment by all jurisdictions to facilitate trade in both water entitlements and allocations, there is ample evidence that states are reluctant to allow individuals from other jurisdictions unfettered access to what is perceived as "their water". This is especially the case for entitlement trade and includes constraints on trade associated with the Commonwealth's environmental water reserve.

At the outset, Victoria imposed a 4 per cent cap on the trade of water access entitlements out of designated communal irrigation districts (the cap is defined as a percentage of the volume of water access entitlement associated with land in an irrigation area). Initially, this constraint was not binding, insomuch as entitlement trade fell short of the limit. However, with the growth in entitlement trade, the cap now often binds, with seven districts forced to operate within the cap in 2011–12. The operational arrangements for dealing with circumstances when sell offers exceed the 4 per cent limit involve a ballot of sellers (where usually the first received requests to sell water entitlements are the ones granted). Once the limit of water sales is reached, no more water entitlement sales are allowed for that year in that district. Clearly, this approach is in breach of the notion of markets generally and the ambitions of the NWI to have water resources move to the highest value use. The NWC (2012) observed that these arrangements also arbitrarily disadvantaged both buyers and sellers and Crase et al. (2013) have queried the true motives of those who continue to advocate for such interventions.

In New South Wales a 3 per cent per valley cap was applied to water access entitlement trade in January 2013. This applies specifically to buybacks by the Commonwealth for environmental purposes. The purported aim of this cap is to better facilitate adjustment, although the NWC notes that "the cap could penalise individual water holders who are looking to manage their risks through the marketplace" (NWC 2013, p. 60).

Other interventions by state governments include the suspension of water sharing plans and the qualification of rights by ministers, especially in severe drought. As previously noted, states have also intervened to halt trade when it could result in unintended third party effects, insomuch as it alters the capacity of water storage managers to then deliver water allocations within perceived acceptable parameters. Notwithstanding these interventions water markets have been shown to be particularly robust and adaptable. More specifically, participants in the markets have been able to adapt within the varying constraints imposed by government, even as those constraints change. The upshot is that water markets are delivering substantive gains to economic welfare in a range of settings (see, for example, ABARES (2011); Grafton and Jiang (2011); Mallawaarachchi et al. (2010); NWC (2010, 2012); Peterson et al. (2004); and Wittwer (2011)).

That the water market should be operating so well, particularly in the MDB, is not particularly surprising to economists who were familiar with the constraints faced by farmers when water rights were tied to land in the 1980s (see, Hall et al. 1994). However, there is also a risk that the impact now being witnessed from the creation and activation of water markets is overstated. Accordingly, to advocate solely for the creation of water markets akin to those in Australia as a panacea for improving all manner of water resource allocation problems runs the risk of disguising important nuances.

Some reflections on the scope for replicating water markets

From a historical perspective, many other important reforms have coincided with the creation of water markets in Australia and, similarly, the successes of water markets in Australia might be attributed to a number of coincidences tied to hydrology and/or agriculture and/or economic factors. In the interests of space, in the following section we draw attention to changes in the macroeconomic setting and also address some of the developments in legislation relating to the use of farm land.

The macroeconomic setting

A major transformation of Australian water policy occurred in the 1980s and 1990s[6] at the time water markets were emerging as a serious policy alternative for facilitating allocation. Importantly, this coincided with a substantive shift in the accepted role of markets within the economy more generally. The national government of the day had taken significant steps to make the economy more open, concerned at the time with the potential malaise derived from falling terms of trade and mounting current account deficits. The exchange rate was floated in 1983, money markets were liberalized and labour markets were progressively deregulated over a decade. Similarly, the National Competition Policy and related policy instruments saw the abandonment of many government-controlled activities, including those involving agricultural marketing. Whereas Australians once paid little attention to international markets at the farm level, movements in the exchange rate and interest rates became regular features of daily news broadcasts watched with interest (and some understanding) by the general population.

Our contention is that these events provided an important, but potentially under-emphasized setting for the establishment of Australian water markets. In addition, Australia has a federal system of government and, as noted earlier, the initial reforms that dealt with water marketing provided a unique set of political and financial rewards for encouraging reform. States were offered and received financial payments from the Commonwealth in return for meeting specific targets. One of these was the requirement to separate water rights from land and to then establish mechanisms for trade. Whilst the financial benefits to states are clear, it was also the case that such arrangements offered scope for defraying some of the political costs associated with market reform. More specifically, states could assign some political costs to the Commonwealth whilst simultaneously accepting financial benefit. Once the market began to operate, participants were able to appreciate the benefits to the extent that political costs were mitigated. It is not clear that similar mechanisms would be in place elsewhere to underpin the establishment phases required for water markets.

It is also important to appreciate that agriculture, whilst important to the Australian economy, is a relatively modest employer nationally. More specifically, agriculture employs around 3 to 4 per cent of the national workforce (ABS 2008) and often the most profitable agricultural enterprises are not directly related to irrigation. This arises because broad scale, dry land agriculture exploits Australia's most abundant resources: i.e. arable land, not water (Davidson 1969). Like most countries, agriculture uses most of Australia's water resources, standing at around 54 per cent nationally (ABS 2012) but this is proportionately much higher in places like the MDB. The point is that market-oriented policies for the reallocation of water resources may not attract the levels of criticism that might be expected in nations where employment in irrigated agriculture is more prominent. This is not to say that rent seeking by those involved in irrigation is not common – rather the scale of irrigated agriculture against the size of the Australian economy generally means that rents can be accommodated if required to achieve policy reform.

Land reforms

The intellectual and technical challenges of creating property rights and markets for water are arguably more novel than with land, especially given the fugitive nature of water resources. Nonetheless, the establishment of water markets in the MDB had been preceded and accompanied by the progressive relaxation of the controls on land tenure, use and transfer in irrigation areas. The history of irrigation in most states in Australia and especially in the MDB is in reality a social history. We use the case of New South Wales to illustrate the general trends in land reform.

Land rights in New South Wales irrigation farms were crafted in a bid for closer settlement to invigorate the "vast interior". Agriculture was seen as the vehicle to achieve these social goals, as in the time before the existence of welfare payments, the ownership of land had been seen as a form of insurance against poverty. In the spirit of egalitarianism, measures were also introduced to prevent the undesirable concentration of property in the hands of a few. A 1985 discussion paper produced by the Water Resources Commission characterized this era as one in which the emphasis was on how many people populated each acre, rather than how much production was generated (p. 39). These social goals were pursued with almost religious fervour, and it appears with the benefit of hindsight that attention to economic realities was secondary.

The result was that irrigation regions were conceived of and managed under the paternalistic and protectionist auspices of the state which exercised considerable power over property rights in these areas. The 1912 Irrigation Act ensured that irrigation settlements were severely constrained and "…under a control almost as absolute as that of an eighteenth century 'enlightened despot'…" (Langfield-Smith and Rutherford 1966, p. 28). The legacy of this approach to farming was felt in these regions for decades (WRC 1985), and has played out in some additional reluctance to change.

Irrigation schemes in New South Wales were divided into "Areas" and "Districts". Irrigation Areas were those based on intensive irrigation on small blocks and these were particularly constrained in relation to their property rights. Irrigation Districts were comprised of more extensive land holdings with supplementary irrigation often used to bolster grain production and pastures. Notwithstanding the additional autonomy in Irrigation Districts, in reality, the property rights to land under irrigation were severely attenuated on a number of dimensions, including duration, transferability, divisibility and flexibility.[7]

A key element of the drive for closer settlement was the application of the concept of the Home Maintenance Area (HMA), which in effect limited the divisibility of the property right. An early reference to this general concept appears in the Crown Lands (amendment) Act or Conversion Act of 1908. According to Lewis (1963, p. 98), a HMA is "an area which when used for the purpose for which it is reasonably fitted would be sufficient for the maintenance in average seasons and circumstances of an average family".

The intention was that the holdings should be of sufficient size to allow for a reasonable return to labour, but at the same time should act as a restriction on the amount of land that an individual could hold. The HMA was thus used both as a maximum in restricting transfers of land to prevent the undesirable aggregation of holdings and as a minimum in determining the size of farms to be allotted in closer settlement programmes (Lewis 1963). However, the upshot of this approach was that land holders were constrained in their ability to adjust to changing markets and technologies (WRC 1985, p. 9), and large numbers of non-viable economic units were created and maintained, many of which were over-stocked or "recklessly grazed" (Benecke 1998, p. 64).

The HMA concept applied to all land in Irrigation Areas, irrespective of whether it was leasehold or freehold. In Irrigation Districts, where most of the land was freehold, the only land to which the HMA concept applied was holdings under the Closer Settlement Act administered by the Crown Lands Office (WRC 1985, p. 8). Nonetheless, District landholders who wished to acquire land in an Irrigation Area were subject to HMA restrictions (WRC 1985, p.20). The Commission had powers over the aggregation of any land supplied with irrigation supplied by the state. Although subject to some minor adjustments in acreage, these limitations persisted until the late 1990s, after the Water Reform Framework had stipulated the requirement for a water market to be operational in each state.

In Irrigation Areas, all farming land was held under restricted title and subject to a series of controls on both transactions and ownership that had proven particularly enduring. Under Crown Lands legislation, the Crown Lands (Continued Tenures) Act 1989, and its predecessor, the Crown Lands Consolidation Act 1913, transfer restrictions had persisted for nearly 90 years until enactment of the Crown Lands and Irrigation Legislation Amendment (Removal of transfer restrictions) Act, 1997 (Benecke 1998, p. 62). The transfer of a farm was subject to the consent of the Minister for Land and Water Conservation; companies and corporations were prohibited from acquiring land; the HMA applied; the maximum number of persons allowed to hold a farm unit was two (where husband and wife are counted as one) and no individual under 18 could hold any form of title (Benecke 1998, pp. 62–3). Land use was also circumscribed by regulations and limitations driven by the state's assessment of suitability or to prevent the undesirable spread of rice production or dried vine fruits (Langford-Smith and Rutherford 1965, p. 4).

The investments of public funds and attenuated rights that have historically characterized irrigation settlements in New South Wales are indicative of the notion that agricultural pursuits have a perceived value that supersedes mere economic interests. The history of irrigation in New South Wales and indeed throughout Australia shows how agricultural policy has often been designed with social goals in mind (Davidson 1969; Musgrave 2008). By the time water markets were introduced in Australia, the shackles associated with land tenure legislation were being loosened, if not yet completely removed. Farmers in Irrigation Areas and Districts (and elsewhere) were increasingly being asked to pursue their own adjustment activities, including those that were being driven by changes in related markets, including financial. Clearly, these observations are only possible with the benefit of hindsight and should not be taken as criticism of different generations of reformists. But the point is that the devolution of individual rights to water was being accompanied by a sequence of similar devolutionary reforms, in land in particular, and it is our contention that these influences have played an important part in shaping and deepening water markets in the MDB. It has also raised interest in water marketing in other states, but the aforementioned conditions may ultimately prove unconducive to active and sophisticated water markets to rival the MDB.

River regulation and heterogeneity of water users

To reiterate, Australia's most active water markets are in the southern MDB. The hydrology in this region is highly variable but reasonably well understood. Significant infrastructure is also in place to harvest water and to facilitate redistribution to meet varying demands in different locations, especially those demands that arise in summer. Whilst this has necessarily caused significant disruption to the natural ecosystem (see, Hillman 2009) it has also supported the development of industries with markedly different demands for water.

In New South Wales the structure of the initial water entitlements as "general security" rights favoured annual agricultural production. In South Australia and Victoria, high reliability rights favoured the development of perennial agriculture. Subsequent urban expansion in Canberra, Adelaide and large regional cities has meant that there is now another competing and high-value demand in the form of households and industrial water users. Added to this is the demand for environmental water, expressed through passage of the Commonwealth Water Act with the support of all major political parties in 2007.

The extent to which similarly diverse demands and accompanying supply infrastructure attends other water resources is questionable. As we have noted, many groundwater systems operate as the supply for confined economic activities. The main exception is Western Australia, where groundwater is the main source of supply for growing urban populations around Perth and horticulture and other agriculture continues to draw from this source. Despite initial reticence on the part of the Western Australian government to invoke trading (see, Crase and O'Keefe 2008), substantial volumes of entitlement are now being exchanged with 100 GL exchanged in 2012–13, up from 14 GL in 2011–12 (NWC 2014).

Experience in the southern MDB suggests that the uptake of water markets was in no small measure enabled by a political shift in the acceptance of the role of markets in the economy more generally. In examining the extent to which the Australian experience can be transferred to other contexts, this may be seen as suggesting that countries that are more market-orientated will be more amenable to water marketing; however, the pre-conditions are more complex than this. In the southern MDB, both existing conditions relating to hydrological realities, the heterogeneity of users, and institutional development and water scarcity along with an appetite for change favoured reform. In effect, the momentum produced by moves towards market-based approaches in other policy contexts smoothed the path towards the establishment of water markets; yet it is unlikely that this in itself would have been sufficient in the absence of a host of other conditions.

Concluding remarks

In this chapter we have endeavoured to summarize important elements of water marketing in Australia. We have described the institutional changes that ushered in water markets and provided insights into the depth and sophistication of those markets, especially in the southern portion of the MDB. In comparison to the operation of water markets elsewhere, these markets continue to facilitate the exchange of large volumes of water access entitlement and large allocations that attend those entitlements.

The growth and depth of water markets in the MDB illustrates the magnitude of welfare gains that can attend institutions of this form. However, whilst encouraged by their continued evolution, we have also endeavoured to highlight important contextual nuances that might be overlooked by potential reformists.

The macroeconomic setting in Australia at the time of reform and subsequently, offered an arguably unique opportunity for policy makers. The system of federal governance also afforded particular political advantages that may not attend other instances where reform is being sought. The underlying differences in demand and the readily available supply infrastructure that support interconnection of users may also be relatively unique. Importantly, we have also noted that land reforms in agriculture occurred simultaneously, notwithstanding that these changes attracted less interest from scholars and analysts. Again, this raises some questions about the extent to which the success of water markets can be duplicated elsewhere, at least in the form promoted in the MDB.

These comments should not be taken as advocating against reform. Rather we hope by exposing these nuances that policy makers can distil lessons that make water markets operational but crafted to suit specific settings.

Notes

1 Australia uses megalitres (ML) and gigalitres (GL) to account for water in irrigation and environmental settings. Kilolitres are most common in urban use. A kilolitre is equivalent to a cubic metre (1000 litres), a megalitre is equal to 1000 kilolitres (1 million litres) and a gigalitre is 1000 megalitres (1000 million litres). A megalitre is about 0.812 acre-feet.

2 In many cases carryover arrangements were an innovation driven by the drought at the start of the 2000s.

3 It might come as no surprise to many economists that the rules that constrain the market are often tested by participants. There are numerous other examples of the requirement for rule-makers to play 'catch-up' in order to close out opportunistic behaviour. The difficulties associated with so-called sleeper and dozer rights are discussed elsewhere (see Young 2013).

4 We noted earlier that there is a continuing theme around changes to market rules and the necessity for rule-makers to close off resulting loopholes. In mid-2010 the Victorian government removed limits to carryover which resulted in many Victorian farmers purchasing New South Wales allocation. The suspension of this trade by the minister promoted farmers to trade water allocations through South Australia and back into Victoria to work around the limits (NWC 2013, p. 27).

5 It also needs to be acknowledged that the costs of exit fees (i.e. a sum paid to IIOs to quit water delivery rights) also play a part.

6 This is not to imply that other episodes of reform were insignificant, and in particular we draw attention to the 1915 River Murray Waters Agreement. Rather, in the interests of space we focus primarily on this more recent period of change.

7 These issues are discussed in more detail in Crase et al. (2012).

References

ABARES (Australian Bureau of Agricultural and Resource Economics and Sciences) (2011) "Modelling the economic effects of the Murray-Darling Basin Plan", Report prepared for the MDBA, ABARES project: 4311 (November)

ABS (2008) "Australia at a glance", Australian Bureau of Statistics, Canberra

ABS (2012) "Water Account, Australia", Australian Bureau of Statistics, Canberra

Benecke, I. (1998) "Crown Land Titles- More Legislative Refinements", *Law Society Journal*, vol 36, no 3, pp. 62–67

CEWH (Commonwealth Environmental Water Holder)(2011) "Commonwealth environmental water – trading arrangements discussion paper", CEWH, Canberra

Crase, L. (2012) "The Murray-Darling Basin Plan: An Adaptive Response to Ongoing Challenges?", *Economic Papers*, vol 31, no 3, pp. 318–326

Crase, L., O'Reilly, L. and Dollery, B. (2000) "Water Markets as a Vehicle for Water Reform: The Case of New South Wales", *Australian Journal of Agricultural and Resource Economics*, vol. 44, no 2, pp. 299–321

Crase, L. and O'Keefe, S. (2008) "Acknowledging Scarcity and Achieving Reform", in L. Crase (ed.) *Water Policy in Australia: The Impact of Change and Uncertainty*, RFF Press, Washington, DC, pp. 166–183.

Crase, L., O'Keefe, S. and Kinoshita, Y. (2012) "Enhancing Agri-Environmental Outcomes: Market-Based Approaches to Water in Australia's Murray-Darling Basin", *Water Resources Research*, vol 48, no 9, pp. 1–8

Crase, L., Pawsey, N. and O'Keefe, S. (2013) "A Note on Contradictions in Australian Water Policy", *Economic Papers*, vol 32, no 3, pp. 353–359

Davidson, B. (1969) *Australia Wet or Dry? The Physical and Economic Limits to the Expansion of Agriculture*, Melbourne University Press, Melbourne

Grafton, R. and Jiang, Q. (2011) "Economic Effects of Water Recovery on Irrigated Agriculture in the Murray-Darling Basin", *Australian Journal of Agriculture & Resource Economics*, vol 55, no 4, pp. 487–499

Hall, N., Poulter, D. and Curtotti, R. (1994) "ABARE Model of irrigation farming in the southern Murray-Darling Basin", ABARE research report, vol 94, no 4

Hillman, T. (2009) "The Policy Challenge of Matching Environmental Water to Ecological Need. Policy and Strategic Behaviour", in A. Dinar and J. Albiac (eds) *Water Resource Management*, Earthscan, London

Langfield-Smith, T. and Rutherford, J. (1965) *Water and Land: Two Cases in Irrigation*, Australian National University Press, Canberra

Lewis, J.N. (1963) "Is the Concept of the Home Maintenance Area Outmoded?", *The Australian Journal of Agricultural Economics*, vol 7, no 2, pp. 97–106

Mallawaarachchi, T., Adamson, D., Chambers, S. and Schrobback, P. (2010) "Economic analysis of diversion options for the Murray–Darling Basin Plan: Returns to irrigation under reduced water availability", report for the Murray–Darling Basin Authority, Risk and Sustainable Management Group, School of Economics, the University of Queensland

MDBA (Murray-Darling Basin Authority) (2012) "About the Basin", http://www.mdba.gov.au/explore-the-basin/about-the-basin, accessed 2 January 2012

Musgrave, W. (2008) "Historical 'Development' of Water Resources in Australia: Irrigation in the Murray-Darling Basin", in L. Crase (ed.) *Water Policy in Australia: The Impacts of Change and Uncertainty*, RFF Press, Washington

NWC (2010) "The impacts of water trading in the southern Murray-Darling Basin: an economic, social and environmental assessment", National Water Commission, Canberra

NWC (2011a) "Water markets in Australia: a short history", National Water Commission, Canberra

NWC (2011b) "Australian water markets: trends and drivers, 2007-08 to 2009-10", National Water Commission, Canberra

NWC (2012) "Impacts of water trading in the southern MDB between 2006/7 and 2010/11", National Water Commission, Canberra

NWC (2013) "Australian water markets: Trends and drivers 2007-08 to 2011-12", National Water Commission WC, Canberra

NWC (2014) "Australian water markets: trends and drivers, 2007-08 to 2012-13", forthcoming, National Water Commission, Canberra

Peterson, D., Dwyer, G., Appels, D. and Fry, J. (2004) "Modelling water trade in the southern Murray-Darling Basin", Productivity Commission Staff Working Paper, Productivity Commission, Melbourne

Pigram, J. and Musgrave, W. (1989) "Transferability of water entitlements in Australia", Annual meeting of the university council on water resources, Minneapolis, MN

WRC (Water Resources Commission New South Wales) (1985) "Land Tenure and Transferability of Water Entitlements in Irrigation Areas and Districts", Discussion Paper, Water Resources Commission

Waterfind (2013) "Water market specialist services", http://www.waterfind.com.au/index.html, accessed 1 October, 2013

Wheeler, S. and Cheesman, J. (2013) "Key findings from a Survey of Sellers to the *Restoring the Balance* programme", *Economic Papers: A Journal of Applied Economics and Policy*, vol 32, no 3, pp. 340–352

Wheeler, S., Garrick, D., Loch, A. and Bjornlund, H. (2013) "Evaluating Water Market Products to Acquire Water for the Environment in Australia", *Land Use Policy*, vol 30, no 1, pp. 427–436

Wheeler, S., Loch, A., Zuo, A. and Bjornlund, H. (2014) "Reviewing the Adoption and Impact of Water Markets in the Murray-Darling Basin, Australia", *Journal of Hydrology* vol 518, pp. 28–41

Wittwer, G. (2011) "Confusing Policy and Catastrophe: Buybacks and Drought in the Murray–Darling Basin", *Economic Papers: A Journal of Applied Economics and Policy*, vol 30, no 3, pp. 289–295

Young, M. (2013) "Trading into and out of trouble: Australia"s water allocation and trading experience", in J. Maestu (ed.) *Water Trading and Global Water Scarcity: International Perspectives*, RFF Press/Taylor and Francis/Routledge

19

TRADEOFFS

Fish, farmers, and energy on the Columbia

Ray G. Huffaker

This chapter investigates how effectively water has been allocated among competing uses in the Columbia River Basin. The major challenge is that river system management compatible with one use imposes large economic losses on others. The chapter sets the stage by describing competing water uses, and the federal, interstate, and state institutions managing water in the Basin. It shows how their efforts to coordinate water policy have been frustrated by (1) conflict between federal and state laws regulating water use; (2) the inflexibility of the prior appropriation water rights system in allocating water to modern-day non-appropriative uses such as hydroelectric power generation and salmon habitat; (3) rent-seeking behavior by traditional appropriators; and (4) judicial decisions restricting regulatory authority that state water agencies require to police water use. The chapter recommends that Columbia Basin states improve water allocation by recognizing non-appropriative water activities as beneficial uses so that the holders of these rights can compete on par with traditional appropriative rights holders in water markets. State water agencies would need to protect the integrity of non-appropriative rights against rent-seeking activities of traditional appropriators.

The Columbia Basin

The Columbia Basin is North America's fourth largest and drains about 250,000 square miles in the Pacific Northwest. It is framed to the east and north by the Rocky Mountains, to the west by the Cascade Range, and to the south by the Great Basin. The Columbia River begins its 1214 mile journey through the Basin from Columbia Lake in Canada's Rocky Mountain Range (Figure 19.1).

It eventually turns southward across central Washington where it is joined by its major tributary, the Snake River. The Columbia then turns westward to form Washington's southern border with Oregon until it enters the Pacific Ocean. The Snake River originates from Jackson Lake in Wyoming, flows across Southern Idaho, and then turns north to form the border between Idaho and Oregon.

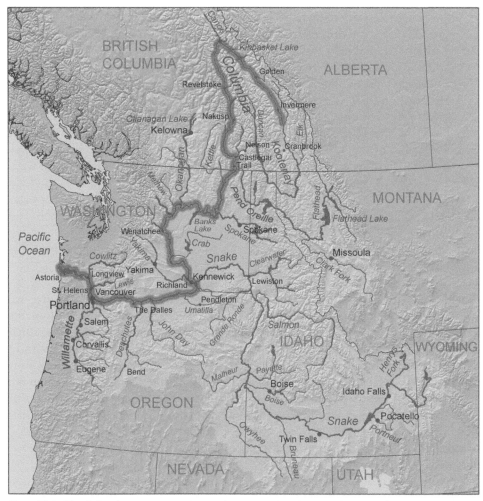

Figure 19.1 The Columbia River Basin (source: http://en.wikipedia.org/wiki/File:Columbiarivermap.png)

Columbia Basin rivers have been the cornerstone of the region's economy for millennia. They provided transportation to enable trade and social exchanges among many diverse cultural groups. Indigenous populations relied on abundant salmon runs for core subsistence. Since the late 19th century, the construction of 18 hydroelectric dams transformed the Columbia and Snake Rivers from free-flowing rivers to a series of backed-up slack water pools. The federal government owns ten of these dams, of which the U.S. Army Corps of Engineers operates nine, and the U.S. Bureau of Reclamation operates one. The federal projects produce several regional economic benefits. These include:

- *Hydroelectric power.* Federal dams generate about 60 percent of regional hydroelectric capacity for municipal, rural, and industrial use (FCRPS 2013). The large supply of inexpensive electricity has attracted several energy intensive industries to the region including aluminum, food processing and mining.

315

- *Navigation*. A series of eight locks along the Columbia-Snake Inland Waterway provide inland barging averaging 10 million tons of commercial cargo valued at US$3 billion annually. Inland barging accounts for half of all Columbia River wheat exports (Pacific Northwest Waterways Association 2010).
- *Irrigation*. The Columbia Basin Project (CBP)—operated by the Bureau of Reclamation—provides irrigation water to about 671,000 acres in east-central Washington for the production of grains, vegetables, fruits, beef, and dairy products. The CBP was developed mainly in the mid 1900s, and covers about 65 percent of the 1,029,000 acres that Congress originally authorized. Primary irrigation facilities include Grand Coulee Dam, Banks Lake, and the Main, West, East High, and East Low Canals (Bureau of Reclamation).
- *Flood control*. Dams and a series of reservoirs provide 55.3 million acre-feet of water storage that reduces the risk of flooding during winter and spring snowmelts and rain (FCRPS 2013).
- *Recreation*. Slack water pools provide opportunities for swimming, waterskiing, boating, fishing, and so on.

The commercialized Columbia Basin continues to provide some habitat for five anadromous salmon species (chinook, coho, chum, sockeye, and pink). Anadromous salmon hatch and spend their first few years in fresh water tributaries. They then migrate to the ocean where they spend from two to five years. Finally, they return to their places of birth to spawn. These stocks have declined precipitously. Annual runs that historically totaled approximately 10 to 16 million fish fell to near 1 million in the 1990s. Runs for specific species declined to a handful. For example, annual counts of Idaho-bound sockeye dropped from 1,276 in 1964 to 4 in 1989 (Natural Resources Law Institute 1990). The decline of anadromous salmon populations prompted the U.S. National Marine Fisheries Service to list the Snake River sockeye salmon as endangered under the Endangered Species Act (ESA) (16 U.S.C. §§1531–1543) in November 1991 (50 C.F.R. §222 (1991)), and the Snake River spring, summer, and fall Chinook salmon as threatened in April 1992 (50 C.F.R. §227 (1992)). In 2011, NOAA Fisheries extended listing for all 13 salmon and steelhead species listed in the Columbia River Basin (NOAA Fisheries, Salmon and Steelhead Listings).

All of these activities compete for scarce Columbia Basin water. Hydroelectric power generation uses the potential energy stored by the hydraulic head at the dam—the difference in elevation between the upstream reservoir surface and the river surface on the downstream side of the dam—to drive turbines and generators. Alternatively, reservoir level and hydraulic head decline as impounded water is withdrawn for irrigation or evacuated to create storage space for downstream flood control.

Hydropower generation also places demands on the river system that are out-of-phase with the migratory needs of salmon. In general, the system of dams along the Columbia and Snake rivers stores water in the spring and summer to save hydroelectric power generation for the winter when Northwest residents consume most energy. Unfortunately, this impoundment period occurs when smolts require water to wash them downstream. Consequently, municipal and industrial consumers of hydroelectric power may face higher rates if the operation of the river system is "reshaped" to meet the migratory needs of salmon. Along the same lines, increasing river flow velocities for migrating salmon may require that instream flow be increased by limiting water diversions (e.g., irrigated agriculture); or that river channels be narrowed by drawing down impounded reservoir water at the expense of hydroelectric power generation, barging, and irrigation pumping from shoreline pumps. Dumping water over dam spillways so that migrating fish can avoid turbines, or breaching dams altogether, create the same tradeoffs.

Federal and interstate agencies

The key federal and state agencies engaged in regulating water use in the Columbia Basin are:

- *NOAA Fisheries*. The Secretary of Commerce—acting through NOAA Fisheries— is the federal agency authorized by the ESA to list marine species as threatened or endangered (§1533). Listing requires NOAA Fisheries to designate the species' critical habitat and prepare recovery plans. Species recovery is elevated to top priority and other federal agencies must consult with NOAA Fisheries to ensure that proposed actions do not jeopardize recovery (§1536). NOAA Fisheries can require that other federal agencies adopt mitigation measures to minimize negative impacts. The ESA also bans the "taking" (e.g., harassment, killing, or capturing) of listed species by private parties (§1538).
- *Federal Water Management Agencies*. The Bureau of Reclamation (U.S. Department of the Interior) operates the Central Basin Project. It is the largest water wholesaler in the U.S., and second largest producer of hydroelectric power in the western U.S. (Bureau of Reclamation). The U.S. Army Corps of Engineers (Corps) operates nine dams in the Columbia Basin, wholesales water, produces hydroelectric power, and dredges waterways (U.S. Army Corps of Engineers).
- *Bonneville Power Administration (BPA)*. Hydroelectric power generated by federal dams is wholesaled by BPA, a federal nonprofit agency in the U.S. Department of Energy (Bonneville Power Administration).
- *Federal Columbia River Power System (FCRPS)*. The Bureau of Reclamation, the Corps, and BPA collaborate in the FCRPS to "maximize the use of the Columbia River by generating power, protecting fish and wildlife, controlling floods, providing irrigation and navigation, and sustaining cultural resources" (FCRPS, 2003, p. 1). In particular, the FCRPS claims that collaboration has resulted in more reliable power and efficient operations.
- *Northwest Power Planning Council (Power Council)*. The Power Council, established pursuant to the Northwest Power Planning and Conservation Act of 1980 (16 U.S.C. §839b), is composed of eight members, two each appointed by the governors of Idaho, Oregon, Montana, and Washington. The Power Act instructs the Power Council to formulate a Regional Energy and Conservation Plan that includes a Columbia River Basin Fish and Wildlife Program (§839b(h)). The BPA must conform to the plan "to the fullest extent possible" in marketing federal power (§839b(h)1(A)). The Fish and Wildlife Program operates parallel, but subservient, to NOAA Fisheries policy.

This regulatory apparatus has struggled to devise a long-term and coordinated water management plan to resolve the tradeoffs among hydropower generation, irrigation, barging, and salmon preservation.

A leading reason is unresolved conflict between federal environmental laws and state water laws. To what extent can federal regulators exert ESA authority to disrupt state appropriative water rights to shape flows for improved salmon habitat? Some argue that federal environmental programs establish "federal regulatory rights" empowering the federal government to "cancel the historic de facto assignment of property rights in commons to exploiters and reassign them to the government as agent for the public generally" (Tarlock 1985). Alternatively, a federal court holding requires the federal government to defer to state water rights without explicit congressional intent to pre-empt them (*California* v. *U.S.* 1978).

The federal government has not used the ESA as leverage to establish federal water rights, but it has invoked the ESA to curtail state water rights on isolated occasions. After Section 1536 consultation with the U.S. National Marine Fisheries Service (NMFS), the U.S. Forest Service closed irrigation ditches in the Methow Valley in Washington State for much of the 1999 irrigation season (Hicks 1999), and the Bureau of Reclamation cut off water to 90 percent of the 220,000 acres in the Klamath Project in Oregon for much of the 2001 irrigation season (Bernard 2001). On another occasion, NMFS invoked Section 1538 authority to threaten to curtail the rights of irrigators using non-federally developed or delivered water in the Methow Valley unless the water district adopted distribution technologies less harmful to fish (Hanson 2001).

Federal agencies pay a potentially large political price for curtailing state water rights. Curtailments generate substantial ill-will from irrigators and rural communities suffering economic losses. For example, 100 irrigators risked arrest when they diverted water around a canal head gate that the federal government had closed in the 2001 Klamath Project curtailment (Associated Press 2001). Enraged rural communities presumably contact their local elected federal representatives, who can express their displeasure with the federal agencies during budget hearings (Gwartney and Wagner 1988).

For the same reasons, federal agencies have been reticent to impose a coordinated water management plan on the region. Unfortunately, competitive user groups have not succeeded in reaching the consensus needed to formulate a plan on their own. Groups relying on salmon for cultural, commercial, recreational, and environmental purposes tend to support measures replicating natural stream flow conditions preferred by wild salmon including flow augmentation, draw-downs, dam breaching, and spilling smolts over dam spillways (see e.g., Save Our Wild Salmon Coalition; PCFFA). These measures require that dams be operated to some extent are inconsistent with traditional industrial/agricultural uses and thus are opposed by these user groups (see e,g., CSRIA, PNWA). Moreover, since the salmon life cycle reaches from tributary to ocean and back, competitive user groups blame salmon decline on a part of the system that they don't use. For example, industrial/agricultural groups blame decline on ocean conditions and over-harvesting, and thus favor harvest controls (CSRIA). Alternatively, fishing interests blame hydroelectric dams (PCFFA). Finally, competitive user groups can mask self-interest behind an avalanche of conflicting scientific evidence that allows them to pick and choose the science that best fits their positions (Brinckman 1999).

State water rights

The limited success of the federal and interstate regulatory apparatus in formulating an enduring coordinated water management plan, and the failure of the public to agree on a plan themselves, shifts substantial burden on states to resolve tradeoffs among competitive uses.

Prior appropriative doctrine

The prior appropriation doctrine provides the foundation for water rights in the western U.S. We now consider its central tenets, and their effectiveness in allocating water among competing uses in the Columbia Basin. The central tenets are:

> A person attains the right to use the quantity of publicly owned water ("water duty") first diverted to a beneficial use on a fixed tract of appurtenant land. A water duty earmarked for irrigation typically includes sufficient water to irrigate an average mix of crops on the appurtenant land with the irrigation technology prevailing when the water right was granted. The priority of the right reaches back to the date of first diversion. The most recent ("junior") appropriators are first cut off when water supplies are short. More senior appropriators receive their full entitlements until no water is left. Irrigators must execute a new water right having the most junior priority to expand their water duty ("no expansion of use"). Water that is not beneficially used is forfeited and available for re-appropriation by another user ("use it or lose it").

The prior appropriation doctrine did not accommodate non-diversionary uses of water (e.g., hydroelectric power generation and salmon habitat) since they were not recognized as beneficial uses when traditional appropriative rights were cemented into irrigated agriculture in the late 19th and 20th centuries. Irrigation retains priority to appropriate the vast bulk of dependable river flows in the West (80–90 percent) regardless of how little water remains for non-diversionary uses when streamflow is low (Wilkinson 1992). Modern western appropriation doctrines continue to prevent a person from applying for an appropriative right for an instream use, and thus from engaging directly in market transactions.

Public trust doctrine

The prior appropriation doctrine opens the door to state intervention to complete the job of allocating water to instream flows because it recognizes water as publicly owned. States are obligated to manage water in trust for the public benefit (Stevens 1980). For example, states could invoke public trust authority to establish instream flow targets with seniority over appropriative rights. They could reject applications for new appropriations that would jeopardize instream flow targets, and/or condition new and existing appropriations to ensure compatibility.

In practice, western states have recognized instream flows as a beneficial use, but not on par with irrigation. For example, only the Idaho Water Resources Board can apply for instream flow rights which are junior to pre-existing rights. In Washington

State, legislatively mandated state water trusts broker the purchase of instream flow rights that also are subordinate to existing senior rights (Washington State Department of Ecology, "Instream flows"). Consequently, the instream flow rights established by Washington's 1980 Columbia River instream flow rule (WAC 173-563) enjoy seniority only over appropriations with priority dates after 1980. These appropriators hold "interruptible rights", which means that the Washington State Department of Ecology (WDOE) retains the right to curtail them when water supply forecasts drop below a critical level. Appropriators with priorities predating 1980 hold "uninterruptible rights", which means that the WDOE will not curtail them when instream flow targets are not met. As of 2011, there were 340 interruptible water rights issued on the Columbia River and 33 on the Snake River (WDOE 2011).

Rent seeking

Given that the bulk of appropriative rights pre-date 1980, Washington State does not offer instream flow rights much protection against traditional appropriative uses. This limited protection has been compromised by the rent-seeking activities of appropriators seeking to firm-up interruptible rights on the Columbia and Snake Rivers. For example, to settle a lawsuit brought by the Columbia Snake River Irrigators Association (CSRIA), Washington State directed the WDOE to develop rules allowing appropriators to convert interruptible to uninterruptible rights not subject to instream flow conditions (WDOE 2005). The WDOE also was directed by statute to develop "a new uninterruptible supply of water for the holders of interruptible water rights on the Columbia River mainstem…" (RCW 90.90.020(3)(c)).

In 2008, the WDOE entered into a "Voluntary Regional Agreement" (RCW 90.90.030) with the CSRIA to issue "drought permits to existing interruptible water right holders and new permanent water rights on the Columbia and Lower Snake Rivers under conditions of no adverse impact on instream flows during critical periods (WDOE 2011). WDOE also committed to "find, fund, and acquire mitigation water" to offset adverse effects on instream flows from converting interruptible to uninterruptible rights. The CSRIA committed to maintain compliance with irrigation Best Management Practices (BMPs), which "focus solely on irrigation water application [with emphasis] on an adoption of high efficiency water use practices…" (CSRIA, BMPs). The CSRIA would commit water savings from BMPs to mitigate negative impacts on instream flows. Relying on improved on-farm irrigation efficiency to generate water savings is troublesome. In the return-flow hydrologic systems characterizing the West, evidence is accumulating that improved on-farm irrigation efficiency fails to create additional water. In practice, it may unintentionally increase consumptive water use as water is applied to field crops in a more uniform and timely manner, and irrigators potentially increase irrigable acreage (Ward and Pulido-Velaquez 2008).

Rent-seeking pathologies

Washington State jeopardized the integrity of traditional prior-appropriative water rights by allowing some rights to be re-conditioned outside of doctrinal parameters. The seniority of instream flow rights was put at risk as the state (1) invested public resources to firm-up junior appropriative rights to make them immune to curtailment during critical flow periods; and (2) sanctioned mitigation measures of questionable hydrologic effectiveness. The public choice literature (Gwartney and Wagner 1988) identifies socially costly rent-seeking behaviors that are highly evident in the Columbia Basin.

First, the state created the expectation among junior appropriators that their water rights could be firmed up during droughts at public expense, and thus encouraged moral hazardous[1] investments in irrigated agriculture. For example, Yakima Valley farmers who established orchards and other perennial crops irrigated with junior water rights lobbied for the Black Rock Project to save their investments during drought (Columbia Institute for Water Policy 2007).

Second, rent seekers lobby for "fiscal discriminatory" projects that generate private benefits at public expense (Gwartney and Wagner 1988). The public bears the expense of water development for which they may not be the primary, or even the secondary, beneficiaries, and is "rationally ignorant" regarding the increased tax burden. Politicians favor these types of projects because they concentrate highly visible benefits on powerful special-interest groups and conceal the costs spread widely over voters.

Lobbying for fiscally-discriminatory water projects in the Columbia Basin diverts attention from regulatory causes of water scarcity; in particular, failure to enforce the strictures of the prior appropriation system. For example, the original appropriative rights in the CBP had a water duty in excess of five acre-feet of diversion per acre of appurtenant land based on the flood irrigation technology prevailing in the 1940s. Shifts to center-pivot irrigation and the large water duty spread irrigation to non-appurtenant land previously judged to be non-irrigable. Prior appropriative water law should have required water-spreading appropriators to apply for junior water rights to expand their use, which the state would have granted given the availability of unappropriated water. However, water spreading has occurred without challenge from state or federal regulators resulting in regulatory-induced water scarcity.

The Bureau of Reclamation plans to remedy water scarcity in the Odessa Subarea of the CBA by partially expanding the project toward the original congressionally-authorized size. The price tag is US$800 million (US$11,800 per irrigated acre). Expansion requires that Washington State contribute to the expense. The state declined to contribute to expansion in 1984 because a commissioned study demonstrated that project expansion costs far exceeded benefits. Expansion was jettisoned. Agricultural economists contend that current project expansion should be jettisoned for the same reasons (Steury 2013).

Water markets

Federal and state regulation has been ineffective in promoting a socially efficient water allocation among competitive appropriative and instream uses in the Columbia River Basin. Can water markets do better? Water markets are attractive because they do not require consensus to operate—just voluntary transactions that maximize social benefits by equalizing marginal water values across competitive uses. There are substantial potential benefits to be gained from water trading. For example, water transfers from irrigation to hydroelectric power in the Columbia Basin were estimated to generate potential benefits ten times greater than lost farm income (Hamilton et al. 1989).

Markets function best if property rights are well defined, freely transferable without unduly impairing the property rights of non-transacting parties, and well enforced. Unfortunately, water rights in the Columbia Basin deviate substantially from this ideal in practice. A major limitation, considered above, is that groups advocating for instream flow uses such as salmon preservation do not have standing to own water rights or directly engage in market transactions. They must rely on state water agencies—vulnerable to rent-seeking activities of competing user groups—to represent their interests. Water markets cannot equalize marginal water values across competitive uses by excluding a group of them from participation.

A second limitation is that western states place moderate to severe restrictions on transfers of appropriated water to prevent impairment to third-party rights caused by changes to the quantity, quality, and timing of irrigation return flows (Young 1986). Economists have recommended several specialized transfers designed to limit the extent and duration of impairment to third-party rights. These include "trial transfers" (transfers are modified or revoked given actual impairment), "one-time-temporary transfers" (transfers whose brevity reduces the duration of an injury), and "contingent transfers" (intermittent transfers triggered by some predetermined, often drought-related, contingency). To protect irrigation return flows that replenish surface and groundwater sources, the volume of transferred water should be limited to a seller's consumptive use (i.e., the volume of water consumed by crops and lost in evaporation) as opposed to the volume of water appropriated by diversion or pumping.

A third limitation is the reticence of Columbia Basin appropriators to participate in water markets because of seller and buyer uncertainty that transactions might fall askance of the "use it or lose it" requirement resulting in forfeiture of the transacted water right. This was the root cause of why Idaho potato growers continued irrigating in a poor market year rather than lease their water rights to hydroelectric power generators (*The Economist* 2001).

Finally, marketing in Columbia Basin water is limited because appropriative rights are not well enforced. For example, a Washington State court ruled that the WDOE is unauthorized to evaluate the priorities of water rights that have not undergone a formal judicial basin-wide adjudication procedure, or to issue regulatory orders to persons violating those rights (*Rettkowski* v. *Department of Ecology* 1983 (Sinking Creek)). A renowned water lawyer in Washington concluded: "Since most of the state's waters remain unadjudicated, the most obvious effect of Sinking Creek is that for most water users, priority—the keystone of Western water law—is now meaningless" (Dufford 1994).

Conclusion

The Columbia River Basin is blessed with abundant water resources that support an impressive array of profitable economic activities, but is cursed with the highly daunting task of balancing economic tradeoffs in allocating water among them. The activities are fierce competitors for Basin water because river system management compatible with one activity imposes large economic losses on another. The institutions tasked with reaching accommodation have significant limitations that substantially reduce their effectiveness. Federal agencies have struggled with resolving conflicts between federal and state laws regulating water use, and have been reticent to incur political fallout from imposing a coordinated water management plan on Basin water users. Water users have been unable to reach sufficient consensus to fill the vacuum with a plan on their own. State water agencies have their hands tied by (1) the inflexibility of the prior appropriation doctrine in allocating water to modern-day non-appropriative benefits; (2) the rent-seeking behavior of traditional appropriators that compromises instream flow rights and imposes social costs by encouraging moral hazardous water investment and recurrent proposals for fiscally-discriminating water projects ; and (3) judicial decisions that restrict regulatory authority that an agency requires to effectively police prior-appropriative water rights.

Columbia Basin states could break this cycle of ineffectiveness if they would recognize modern-day non-appropriative water activities as beneficial uses on par with traditional appropriative uses, protect the integrity of non-appropriative rights against the rent-seeking activities of traditional appropriators, and give holders of non-appropriative rights full standing to engage in water market transactions that would elicit the relative values of appropriative and non-appropriative uses more accurately than the present political allocation process. Third-party impairment caused by increased transactions could be mitigated with specialized water transfers designed to limit the extent and duration of harm.

Note

1　Moral hazard describes a situation in which economic agents engage in risky behavior that incurs costs borne by others (Bannock et al. 1998).

References

Associated Press (2001), "Farmers bypass canal head gate to water crops", *Spokane Spokesman-Review*, 16 July

Bannock, G., Baxter, R., and Davis, E. (1998), *Dictionary of economics*, 6th ed., Penguin Reference, England.

Bernard, J. (2001), "Trespass in irrigation protest", *Spokane Spokesman-Review*, 2 September.

Bonneville Power Administration, "About Us", http://www.bpa.gov/news/AboutUs, accessed 4 September 2013.

Brinckman, J. (1999), "Fisheries aide attacks salmon myths", *Oregonian*, 20 December.

California v. U.S., 438 U.S. 645 (1978)

Columbia Institute for Water Policy (2007), "History of Over-Appropriation in the Yakima River Basin", http://columbia-institute.org/blackrock/backrock/Overappropriation.html, accessed 4 September 2013

CSIRA (Columbia-Snake River Irrigators Association), "Who We Are", http://www.csria.org/who_we_are.php, accessed 4 September 2013

CSIRA (Columbia-Snake River Irrigators Association), "Best Management Practices", http://www.csria.org/bmp.php, accessed 4 September 2013

Dufford, W. (1994), "Water law after Sinking Creek", in: *Proceedings of the Sinking Creek decision: water rights in the 21st century*, University of Washington, Seattle, WA, pp. A2–A12

The Economist (2001), "You say potato, I say electricity", 12 July

FCRPS (Federal Columbia River Power System) (2013) Federal Columbia River Power System brochure, http://www.bpa.gov/power/pg/fcrps_brochure_17x11.pdf , accessed 4 September 2013

Gwartney, J. and Wagner, R. (1988), "Public choice and the conduct of representative government", in J. Gwartney and R. Wagner (eds) *Public Choice and Constitutional Government*, JAI Press, Inc, Greenwich, CN

Hamilton, J., Whittlesey, N., and Halverson, P. (1989), "Interruptible water markets in the Pacific Northwest", *Amer. J. of Agric. Econ.*, 71, pp. 63–73

Hanson, D. (2001) "Feds, Methow facing off in water dispute", *Spokesman Review*, 3 September

Hicks, L. (1999) "Ditch problems have no easy solutions", *Methow Valley News*, 2 September

Natural Resources Law Institute (1990), Anadromous Fish Law Memo 4 at p. 50

NOAA Fisheries, Northwest Regional Office (n.d.) Salmon and Steelhead Listings, http://www.nwr.noaa.gov/protected_species/salmon_steelhead/salmon_and_steelhead_listings/salmon_and_steelhead_listings.html, accessed 5 September 2013

NOAA Fisheries, Northwest Regional Office (n.d.) "What We Do", http://www.nwr.noaa.gov/whatwedo/overview/what_we_do_overview.html, accessed 5 September 2013

PCFFA (Pacific Coast Federation of Fisherman's Associations) (n.d.) "What is PCFFA?", http://www.pcffa.org/, accessed 4 September 2013

Pacific Northwest Waterways Association (PNWA) (2010) "Columbia Snake River System Facts", http://www.pnwa.net/new/Articles/CSRSFactSheet.pdf, accessed 4 September 2013

Rettkowski v. Department of Ecology ("Sinking Creek"), 858 P.2d 232 (1993)

Save Our Wild Salmon Coalition (n.d.) http://www.wildsalmon.org/about-us/save-our-wild-salmon-coalition.html, accessed 4 September 2013

Stevens, J. (1980) "The public trust: a sovereign's ancient prerogative becomes the people's environmental right", *University of California Davis Law Review*, 14, p. 195

Steury, T. (2013), "Water to the Promised Land", *Washington State Magazine*, 12(4), pp. 25–31

Tarlock, D. (1985) "The Endangered Species Act and western water rights", *Land Water Law Rev.*, 20(1), pp. 1–30

Ward, F. and Pulido-Velaquez, M. (2008), "Water conservation in irrigation can increase water use", *PNAS*, 10, pp. 18215–18220

Washington State Department of Ecology (WDOE) (2005), Columbia River Initiative (CRI) Archive 2002-2005, http://www.ecy.wa.gov/programs/wr/cwp/crihome_archive, accessed 4 September 2013

Washington State Department of Ecology (WDOE) (2011) "Voluntary Regional Agreements, 2011 Report to the Legislature", http://www.ecy.wa.gov/programs/wr/cwp/images/pdf/OCR_VRA_DRAFT-final.pdf

Washington State Department of Ecology (WDOE), "Instream Flows", http://www.ecy.wa.gov/programs/wr/instream-flows/isfhm.html, accessed 4 September 2013

Wilkinson, C. (1992), *Crossing the Next Meridian: land, water, and the future of the West*, Island Press, Washington, DC

Young, R. (1986), "Why are there so few transactions among water users?", *Amer. J. of Agric. Econ.*, 68, pp. 1143–1151

U.S. Army Corps of Engineers, "Mission Overview", http://www.usace.army.mil/Missions.aspx, accessed 4 September 2013.

U.S. Bureau of Reclamation (n.d.) Columbia Basin Project, http://www.usbr.gov/projects/, accessed 4 September 2013

U.S. Bureau of Reclamation, "About Us", http://www.usbr.gov/main/about/, accessed 4 September 2013.

20

WATER SALES, PECUNIARY EXTERNALITIES AND LOCAL DEVELOPMENT

Chinatown revisited

Gary D. Libecap

Introduction: transaction costs and the extent of water trade

Economists normally do not conclude that pecuniary externalities have important efficiency impacts. The notion is that small price changes and related changes in market exchange due to shifts in demand or supply have only distributional consequences on third parties that net out across relevant consumers and producers (Scitovsky 1954). With imperfect information and incomplete competition, however, pecuniary externalities could have productive implications because they impede the adjustment process. Under these circumstances, mitigating government tax and subsidy policies could be Pareto improving (Greenwald and Stiglitz 1986). This literature, however, does not make clear what factors impede a response to price change or develop how associated distributional outcomes could lead negatively-affected parties to devote real resources to block entry and exchange.

Transaction costs (Coase 1960, 1992; Demsetz 1968, 1969; Williamson 1975, 1979, 2010; Dahlman 1979; Barzel 1982) make these effects more transparent. Transaction costs are the resource costs of defining, enforcing, and exchanging property rights (Allen 2000). Significant transaction costs reduce beneficial trade and inhibit otherwise efficient resource allocation. They can arise from incomplete information, asset specificity, and bargaining strategies under imperfect competition that result in hold up and lack of contract enforcement. Transaction costs are the basis for Williamson's (1975, 1979) make-or-buy paradox facing firms.

In the case of water three related factors raise transaction costs and limit exchange: First, distributional concerns when water is moved from small, rural economies to urban ones leads to local resistance to water markets (Haddad 2000, xv; Hanak 2003, 123; Donohew 2008). About 70 percent of western US annual water consumption is in agriculture, and water transfers from agriculture to urban use could result in extensive land fallowing or other shifts in crop production that reduce demand for agricultural labor and other inputs. This, in turn, could lead to loss of population and other broader negative economic impacts in the source area. It may be very costly to agree upon and distribute sufficient compensation to the heterogeneous parties potentially affected to insure that no party is made worse off from the proposed exchange. The parties include farm labor, farm implement sellers, school districts impacted by lower property values, home owners, and so forth. Absent sufficient compensation, these parties have incentives to assign productive inputs to block otherwise general welfare-improving transactions. Fairness norms loom large in small, rural communities when water is traded to support new uses in different locations, generating major shifts in surpluses from source areas to new locations. The importance of distributional concerns and fairness in economic decision making is addressed more broadly, for example by Camerer and Thaler (1995) and Alesina and Angeletos (2005).

Second is the related issue of maneuvering by some agents to increase their share of the rents generated from reallocation, which also raises transaction costs, limiting water trade. The anticipated surpluses gained from moving water from low-valued marginal agricultural use to high-valued urban use in the western US often are so large that rent-seeking is stimulated. A dramatic example is where farmers in the Imperial Irrigation District of California (IID) paid US$13.50/acre-foot (AF, 326,000 gallons) in 2001, while a development near the South Rim of Grand Canyon National Park was prepared to pay US$20,000/AF to deliver the same Colorado River water (Brewer et al. 2008, 92). Indeed, efforts to move any water from the Imperial Irrigation District have encountered high transaction costs. Negotiations over selling water among IID, the Metropolitan Water District of Southern California, and the San Diego Water Authority began in 1984, but no lasting agreement could be reached until 2002, 18 years later, for a variety of reasons (Haddad 2000, 72–77; Hanak 2003, 73). There were opportunity costs for both sides in not reallocating water more rapidly and in securing more costly alternative sources for urban areas, as well as costs in the litigation used to halt preliminary agreements.

Third, the structures of surface water rights and water markets contribute to high transaction costs. Third-party technological externalities arising from reduced return flows for downstream are generated from upstream water trades out of basin. Maintaining water exchange within a basin reduces third-party impairment, but constrains the range of reallocation options and number of traders. As a result, water markets, especially informal ones, historically have been very localized and thin with few formal trades and limited price information. This setting leads to asymmetric information about water quantities, qualities, values, and objectives among water buyers and sellers. It raises the costs of defining compensation or other adjustments to mitigate third-party technological externalities. Further, small numbers of buyers and sellers in thin markets can result in bilateral monopoly, where outcomes are indeterminate and the bargaining parties use real resources to increase their share of the gains from trade.

Indeed, what is striking about western US water markets is how limited they are in the face of major disparities in the prices paid by urban and agricultural users. In the semi-arid parts of Australia and Chile 20 percent or more of annual water use is formally traded, compared to less than 3 percent in much of the US West (Brewer et al. 2008; Grafton et al. 2011). Price disparities exist in local water markets like Nevada's Truckee Basin, where the median price of 1,025 agriculture-to-urban water rights sales between 2002 and 2009 (2008 prices) was US$17,685/AF, whereas for 13 agriculture-to-agriculture water rights sales over the same period the median price was US$1,500/AF. In other markets, such as the South Platte, Colorado, the median price for 138 agriculture-to-urban sales between 2002 and 2008 was US$6,519/AF as compared to

US$5,309/AF for 110 agriculture-to-agriculture sales (Libecap 2011, 65). These prices are much closer, but they are not typical. Aggregating transactions across markets and time to gain an additional sense of differences in value across sectors reveals median prices for one-year leases across 12 western states between 1987 and 2008 to be US$74/AF for agriculture-to-urban leases (204 observations) and US$19/AF for agriculture-to-agriculture leases (207 observations) and median prices are US$295/AF for agriculture-to-urban sales (1,140 observations) as compared to US$144/AF for agriculture-to-agriculture sales (215 observations) (Libecap 2011, 65). The opportunities for arbitrage and mutually-beneficial trade appear to be unexploited and the questions are why and what role might transaction costs play?

This chapter analyzes the first major water rights exchange in the western US, the purchase of farms and associated water rights in Owens Valley, about 250 miles northeast of Los Angeles, and the urban water utility in Los Angeles, today the Los Angeles Department of Water and Power (LADWP). Negotiations largely took place between 1923 and 1934. The associated water rights acquired by the city raised Los Angeles' water supplies by over four times and made the city's growth possible. As late as 1998, Owens Valley and the Mono Basin to the north contributed over 70 percent of the city's water. Until 1941 and the arrival of Colorado River water, there were no other major sources of supply.

The Owens Valley was (and remains) a marginal agricultural area with a narrow growing season, limited access to markets, poor soil quality, and high water tables, creating alkaline deposits. In the 1920s, Owens Valley farms were very small, even by Great Basin standards, and there was a nationwide agricultural depression resulting in farm failure (Libecap 2007, 34–37). Even so, the negotiations were protracted and acrimonious, and have become famous for high costs and the inequities of the outcome—the so-called "theft" of Owens Valley water and the economic destruction of the local communities. The Owens Valley syndrome, as it is sometimes referred to, remains in the background of western water markets 90 years later:

> ...farmers remain suspicious of the "Owens valley syndrome"…The "theft" of its water…in the early 20th century has become the most notorious water grab by any city anywhere…the whole experience has poisoned subsequent attempts to persuade farmers to trade their water to thirsty cities.
>
> (*The Economist*, July 19, 2003, 15)

The Owens Valley story has become part of the popular media. The 1974 movie, *Chinatown*, starring Jack Nicholson and Faye Dunaway, dramatized conspiracies involving Owens Valley water and land speculation in Los Angeles. For these reasons, it is worthwhile examining the Owens Valley water exchange with Los Angeles to determine what is factual and to demonstrate how transaction costs can raise the costs of water market exchange.

Owens Valley water trade overview

The Owens Valley water transfer is an important case where the bargaining strategies of the parties to capture more of the gains from trade increased transaction costs and molded perceptions of fairness of the exchange held by both contemporary and subsequent observers of Owens Valley. Negotiated between 1905 and 1935, with most transitions between 1923 and 1934, this is one of the largest private land and water acquisitions by any local government in American history. 1,167 farms covering 262,102 acres were purchased for US$20,768,233 (US$274,659,881 in 2012 dollars) (Libecap 2007, 46, Table 3.3).

For some farms, sale negotiations between the owner and the LADWP were smooth and agreements were reached quickly, whereas for others bargaining was much more acrimonious, taking five to ten years to complete. To improve their bargaining position, farmers with the most water formed three sellers' pools along major ditches to bargain for their members. The pools coordinated bargaining positions. Pool members engaged in periodic violence, appealed to the state and national press, and called for intervention by politicians to bolster their demands for higher prices. Although the surplus was greatest from the early sale of farms with the most water to the city, farmers who engaged most effectively in collusion held back sales. The strategies of the antagonists to capture the surplus raised the costs of exchange, delayed property sales, raised the prices paid for some properties, and left a legacy of negative equity judgments, despite the fact that both parties benefited from the transactions.

In 1920, some 7,031 people farmed in Owens Valley or lived in five small towns. Its agriculture depended upon irrigation from the adjacent Owens River and feeder streams, cooperative mutual ditches, or groundwater pumping. There were 140,000 acres of farmland, of which about 40,000 were improved as pasture or in crops. The rest of the valley was semi-arid scrubland. All in all, the region had little agricultural potential. There was limited arable land; growing seasons were short; the soil was alkaline; and there were few outlets to markets. This information is important because it suggests that Los Angeles' offer prices, based on agricultural productivity (the assumed farmer reservation values) were low compared to perceived water values in Los Angeles, the actual reservation prices of many farmers. The surplus attained by reallocating water from each farm to Los Angeles and maneuvering to capture it are the basis for high bargaining costs and ensuing distributional concerns.

Although some water rights purchases took place as early as 1905, most properties were secured after 1922. When the LADWP could not reach agreement with one farmer, it would move on to others until it had acquired enough water-bearing land to meet its immediate objective. Accordingly, the LADWP kept returning to the Valley during the rest of the 1920s to re-open negotiations with those farmers where bargaining had reached an impasse and to buy other properties with less water. By 1934, the agency had acquired 95 percent of the agricultural acreage in the valley.

The unbroken line in Figure 20.1 details the gradual acquisition of water rights in Owens Valley by showing the cumulative percentage of total water acquired by the city as of each year. The figure covers the northern part of the valley where most of the farms were located, where most of the agriculture was found, and where protracted negotiations took place. The data are from 595 observations of all farm sales to Los Angeles between 1916 and 1934 as analyzed by Libecap (2007, 2008). Each observation in the data set includes information on the water available annually from a farm. The sum of these amounts represents the total water acquired by Los Angeles by 1934 and serves as the denominator of the fraction charted in the figure, while the numerator is the cumulative water acquired by the city as of each year.

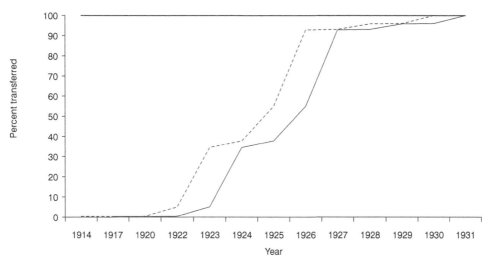

Figure 20.1 Percent of water transferred over time (source: adapted from Libecap 2008, 317).

Had there been no bargaining conflicts to drive up transaction costs and had sales been competitive, the pattern of water acquisitions shown in the figure would have shifted to the left between 1923 and 1934 as indicated by the dashed line. Accordingly, the welfare effects of higher transaction costs include the lost output of additional resources devoted to bargaining by both sides, any losses in surplus from delay in the acquisition of farms with the most water/acre and postponed reallocation of their water to the city, as well as the negative legacy of the bargaining conflict on the subsequent slow development of water markets and reduced movement of water to higher-valued uses in the western US. With the data available, it is not feasible to measure these welfare losses, but the sources of higher transaction costs can be identified in the bargaining history.

Figure 20.2 plots the distribution of the 595 Owens Valley farms by water/acre along with the mean. It is clear from the figure that farms varied considerably in their water holdings. The farms with above-mean water/acre were concentrated on ditches, whereas the below-mean farms were distributed throughout the valley, off major water sources.

The bargaining setting and transaction costs

Exchange requires locating the relevant parties; measuring the attributes of the asset to be traded; negotiating a sale price; contract drafting; and enforcement. The transaction costs literature emphasizes that each of these activities can be costly, affecting the timing, extent, and nature of trade (Coase 1960; Williamson 1979; Barzel 1982). In Owens Valley, bargaining for the most important water-bearing farms is characterized as bilateral monopoly between the LADWP and the sellers' pools. Bilateral monopolies have undefined pricing outcomes because they depend upon relative bargaining strength. Each party has incentive to misrepresent its position in order to extract a greater share of the gains of trade, and there is little competitive pressure to force more accurate information revelation. Accordingly, bilateral monopoly negotiations often break down and take a long time to complete (Williamson 1975, 238–247; Blair, Kaserman, and Romano 1989).

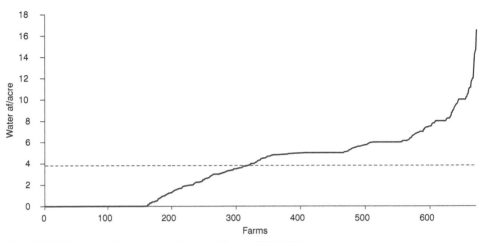

Figure 20.2 Water: acre-feet per acre (Source: Libecap 2008, 318).

On the one side was the LADWP, which essentially was the only buyer of farm land in Owens Valley. The agency built an aqueduct early in the twentieth century as it acquired initial water rights in Owens Valley, prior to the major acquisitions examined here. As such the aqueduct was a large, fixed, immobile investment whose value depended upon the flow of previously-purchased and additional Owens Valley water. The LADWP was made up of five members, appointed by the Mayor to staggered terms and confirmed by the City Council. As such, the agency was subject to citizen demands on the Mayor and City Council for a relatively constant, reliable flow of water per capita in the aqueduct and was under ratepayer oversight in the management of its funds. The LADWP's actions also were under scrutiny of other civic organizations, such as the Chamber of Commerce and Municipal League that were attempting to attract new residents and businesses to Los Angeles. Any negative publicity resulting from its dealings with farmers harmed the city's reputation. Sensitivity of the agency to negative publicity was a tool used by the farmers in their negotiations with it.

For these reasons, the LADWP sought to acquire water-bearing lands quickly to insure water supplies, smoothly to minimize transaction costs, and cheaply to stay within its water bond limits. In bargaining with farmers, the agency sought to buy farmlands and the water rights associated with them based on their agricultural values in Owens Valley, rather than Los Angeles' water values which were much higher. If the agency could obtain the farms at these prices, the total surplus from reallocation of water would go to the city's land owners, capitalized into the value of the land, and burdens on the LADWP's ratepayers would be minimized. Accordingly, the agency had incentive to maintain a competitive environment among sellers, blocking the formation of any collusive sellers' organization. If successful, the LADWP then could price discriminate in its purchases of farms based on the assessed value of agricultural productivity in Owens Valley. Because farms were small and no single one was essential for the city's water supply, a threat position of the agency was to leave an un-cooperating farmer isolated, surrounded by properties that were sold to the city. Indeed, some farmers complained that the LADWP engaged in a checker-boarding strategy, buying properties around holdouts who did not sell.

Because agricultural productivity was known best by the farmers and less well by agency personnel, to address this information asymmetry the agency established an appraisal committee to collect data on each farm's characteristics. These data were then compared to similar farms to arrive at an appraised productivity value. Offer prices were formed by multiplying appraised values times a fixed adjustment factor of 4.1 to gain an offer price. If adhered to, such a pricing rule potentially allowed for price discrimination by the LADWP as it moved along the supply curve of farm land. To meet its objectives of securing a reliable water supply in the face of a rapidly growing urban population and to minimize the transactions costs of negotiation, the agency was most interested in buying larger farms with more water than in negotiating with many smaller farmers, who had less. When the LADWP successfully reached agreement with a farmer, it acquired the right to the water associated with the land, either a riparian claim (less important) or appropriative claims to ditch water and/or groundwater beneath the surface property. It could then release the farm's water from the ditch and/or pump groundwater for flow down the Owens River to the Los Angeles Aqueduct intake.

On the other side were the farmers. Almost all farms were small, with each holding only a limited portion of the total water in Owens Valley. Hence, no one farmer had market power or was vital for meeting Los Angeles' water needs. To improve their bargaining power, farmers had incentive to negotiate collectively. Recall the agency had no other major sources of water supply and it had a potential stranded asset with the large aqueduct. Farmers would seek at least the present value of the agricultural productivity of their farms as their reservation prices, plus as much of the surplus value from trade as they could get. Farmers were well aware of how valuable Owens Valley water was in Los Angeles where land values were rising rapidly with the arrival of new water. Contemporary newspapers reported jumps in land values in the San Fernando Valley, where properties rose from a few dollars an acre to US$500 and more per acre as early as 1913 with completion of the aqueduct. Indeed, speculation in land with arrival of water was the backdrop for the movie *Chinatown*. To illustrate the effect of the arrival of Owens Valley water, taking US$300 as the average increase in land value in the non-urban areas of Los Angeles County alone, the gain was over US$37,000,000 for *new* agricultural lands added between 1910 and 1920 and possibly an additional US$227,395,500 for increases in the value of agricultural lands *existing* in Los Angeles Country in 1910 due to a more certain water source and greater urban land expansion. These values are in current 1910 prices.

Owens Valley farmers wanted their share of this surplus. Whether or not a farmer could extract more depended on being in an effective bargaining group that could give them some market power. Competitive farmers would reveal their reservation prices and be under pressure to accept offers that met them.

The sellers' pools

Three smaller sellers' pools were formed by owners of small clusters of adjacent, relatively homogeneous farms on two important ditches in 1923 and 1924. These farmers controlled about 17 percent of the valley's water. The organizations included the Keough pool on the Owens River Canal with 23 members, the Watterson pool of 20 members on Bishop Creek Ditch, and the Cashbaugh pool of 43 members also on Bishop Creek Ditch. The pools were organized and led by the largest land owner in each cluster, who acted as the bargaining agent for all pool members. By reducing competition among sellers, these farmers had more bargaining power. Pool members could threaten to hold out for higher prices until later when either Los Angeles became more desperate for the water and/or there were fewer remaining farms in the valley and fewer opportunities to purchase additional supplies. For example, in 1926 Keough Pool members refused a LADWP offer of US$1,025,000 for all of their properties, demanding instead US$2,100,000. The agency countered with US$1,250,000, followed by US$1,600,000 from the pool, which also was refused. Negotiations were not resolved with the last pool members until 1931. In holding out, however, members had to compare the expected returns from accepting the agency's offer with the option value of delay. Members suspected that Los Angeles would need more water, but there was uncertainty as to the amount required. Until late in the 1920s it was unclear to all parties just how much land and water Los Angeles would have to buy, given unexpected population growth and recurring droughts in the region. As a result, there was risk to holdouts of missing a sale if the LADWP determined that it had sufficient supplies and no longer needed to acquire additional properties in Owens Valley. This was an important problem that threatened to undermine the unity of each pool since some members might conclude that they would be better off by selling now rather than delaying.

The credibility of each pool's threat position depended on its size and cohesiveness. For the pools to be effective, they had to retain their members and avoid defection. If only a few small farmers left a pool, it could retain its effectiveness for those that remained. But the defection of a large farmer, especially the pool leader, was a serious blow. Among the three, the Keough was the most concentrated and tightly organized with a Herfindahl index (HHI, based on the size of farms in the pool) of 1,583. The Watterson pool had a Herfindahl index of 1,163, and the Cashbaugh, 410. Within the Keough Pool, there was some defection with 17 of the 23 members selling in 1926 and 1927, but these were very small farmers (14 of them had 10 acres each). The core of the pool, led by the largest land owner, Karl Keough with 4,482 acres (60 percent) of the 7,862 acres on the Owens River Canal and by far the most water of any other pool member, and five other farmers held out until 1931 for higher prices. Member George L. Wallace, for instance, offered his lands to the LADWP in 1926 for US$417 per acre, while the city countered with US$254 per acre. In 1931 (when farmland values everywhere else in the US were falling due to the Great Depression), he finally sold for US$466 per acre.

The other two pools suffered from early sales by their largest land owners. Within the Watterson pool, the leaders and biggest land owners, Wilfred and Mark Watterson, with 1,216 acres across three separate plots agreed to sell to the LADWP in 1926 at a slight premium over the agency's offer. They were quickly followed by all but 3 of the 20 pool members. The others sold in 1927. The Cashbaugh pool also had 20 of the 43 members selling in 1926 and the leader, William Cashbaugh with 596 acres, selling in 1927.

Pool leaders coordinated through joint organizations such as the Owens Valley Protective Association in efforts to pressure the LADWP to meet their price demands by appealing to state politicians and the press. In response, throughout the 1920s, the press was invariably critical of Los Angeles, portraying Owens Valley negotiations as ones of small farmers battling a large, wealthy city against uneven odds. Newspaper articles stressed the unfairness of Los Angeles' offers as it allegedly took the farmers' water and their livelihoods in order to fuel urban growth to the south. The allegations were repeated in other press coverage and subsequent evaluations of the Owens Valley transfer. Further, between 1924 and 1931 whenever negotiations stalled, the aqueduct and city wells were periodically dynamited, although the aqueduct was never seriously damaged.

Unorganized, competitive non-ditch farms

Non-pool farmers, who were not on key ditches, competed to sell their properties to the LADWP, often proposing their own offers to the LADWP. In 1925, agency Counsel W.B. Mathews commented on the "insistent demand" by some property owners for the city to buy their lands. Because these farms had smaller amounts of water the LADWP was less interested in them and often bought them later. There is no evidence that these farmers were involved in any of the bargaining conflicts in Owens Valley. Indeed, in these negotiations, the agency reported that "the prices paid, with few exceptions, have been entirely satisfactory to the seller." With no market power, these farmers had to accept offer prices from the LADWP once their reservation values were met or lose the sale. They would get little of the surplus from the transfer.

Empirical analysis of Owens Valley bargaining

The discussion has focused on how the bargaining environment in Owens Valley increased transaction costs and established long-term views of the distributional implications of water markets. It is not possible to directly test those claims. As indicated in Figure 20.1, had exchange proceeded at lower cost, the curve showing cumulative water acquisitions would have shifted to the left, with more water moving to Los Angeles sooner. Although it is difficult to measure the benefits of the counterfactual of more-rapid transfer of water to the city, it is possible to define the parameters underlying those hypothetical gains. They include any faster increase in land values due to fewer impediments in arrival of water; the use of those capital gains to support earlier investment in the city; and savings in bargaining costs over the surplus. In fact, of course, trading proceeding much more slowly and at higher transaction cost and these factors underlie the immediate welfare loss of the Owens Valley conflict over water exchange. The long-term welfare costs are the opportunity costs (or losses) of reduced water trading today due to rural community concerns attributable to the Owens Valley experience. Again, these effects are difficult to measure, but given the prominence of the Owens Valley example in contemporary rural-to-urban water exchanges, these losses may be large.

The data set allows for examination of the role of the sellers' pools in determining the timing of sale and price received by the farmers, relative to the unorganized farmer baseline. The actions of the sellers' pools were an integral part of the bargaining conflicts with the LADWP that characterize the history of Owens Valley. The data on farm properties purchased between 1916 and 1934 by the LADWP include 869 observations, which encompass almost all properties acquired by the agency during that time. Excluding properties of ten acres or less as not being farms, but town lots, as well as dropping incomplete entries leaves 595 observations. Of those, 367 farms were on

Table 20.1 Owens Valley farm property characteristics, mean values

Property Type	Price/Acre (current$)	Total Purchase Price (current $)	Year of Purchase	Size (acres)	Water Price/AF (current $)	Total Water AF/Farm	Water AF/ Cultivated Acre/Farm
All properties	198	23,425	1926	154	178	448	28
Farms not on a ditch	82	19,890	1927	207	473	261	14
Keough Pool	443	27,647	1928	79	77	366	69
Cashbaugh Pool	242	32,156	1927	126	69	544	33
Watterson Pool	237	33,983	1926	147	75	584	18
Non-pool on ditches	263	23,861	1926	122	112	581	30

Source: Libecap (2009, 326)

irrigation ditches and 228 were not on ditches, but spread throughout Owens Valley. Table 20.1 provides mean values for farm property owners in Owens Valley by various classifications.

As indicated by the mean values in the table, members of the most cohesive pool, Keough, on average sold two years later than the overall sample and commanded the highest price per acre of land. Members of the Cashbaugh and Watterson pools also did better on average in terms of price per acre and total purchase price than did competitive non-ditch properties that comprise the baseline. In general, farms on ditches sold for higher prices per acre and greater total prices than did those that were not on a ditch. The former had higher percentages of cultivated land; had more water; more water per acre of cultivated land; and their owners were more likely to be in a sellers' pool. A few non-pool farms that were on ditches and purchased preemptively to block the formation of broader sellers' pools brought more in total and per acre of land than did the non-ditch farms and slightly more on average than farms within the two weaker sellers' pools. All pool farms had sales prices considerably above the 1925 mean census farm values for four Great Basin counties Lyon, Douglas, Churchill in Nevada, and Lassen in California with similar farm environments—topography, precipitation, soil quality, growing season, access to markets. The per-acre land prices for pool properties were at least three times those of competitive farms not on ditches. These results are consistent with the notion that pool membership would increase bargaining power for farmers.

It is possible to calculate the implied price received for water. The price paid for water is obtained by dividing the sale price of the farm by the water acre-feet conveyed in its purchase. As shown in Table 20.1, although non-ditch farms sold for less in total and per acre of land, their owners earned more per water acre-foot than did farmers more favorably located on ditches. This outcome reflects the purchase of a bundled asset in the land market. While non-ditch farms had less water, the agency still had to pay at least their agricultural reservation values in order to secure sale. Because of limited arable land in Owens Valley, not all water on a farm translated directly into greater farm production. This was especially the case for those farms with the most water, where parts of farms were swampland. This condition also underscores the unfairness assessment of the results of bargaining over land in Owens Valley. When farm prices were based on agricultural productivity, as was desired by the LADWP, farmers with less water would receive more per unit of water than would their counterparts, who had greater water endowments but lower marginal agricultural values.

The mean values in Table 20.1 reveal that non-ditch competitive farms had an average sale price of US$19,890 or US$473/acre-foot of water. This total farm sale price is somewhat less than the mean 1925 census farm value for four comparable Great Basin counties noted above of US$21,167 (current prices), but these non-ditch farms were the least productive units in Owens Valley. This result is consistent with the notion that competitive farmers would receive their reservation values, the net present value of agricultural production. Even so, a sale value of nearly US$20,000 corresponded to six years of gross farm receipts for Owens Valley (Inyo County, California) farms during a time of agricultural depression. It is no wonder that these small competitive farmers sold whenever the LADWP offered to buy their farms. On the other hand, members of the Keough Pool who received the highest per-acre price of land in the sample, US$443/acre, gained far less per acre-foot of water, US$77/AF, than did the baseline competitive farm owners. Keough members, however, secured the highest water price per acre-foot devoted to cultivated acreage, US$69/AF as compared to US$14/AF for the baseline. This outcome also reflects payment by LADWP for the productive value of land, the small amount of cultivated acreage per farm, and the critical role of irrigation in cultivation. Mean values for the other ditch farms in pools do not generally follow those of Keough members.

Libecap (2007, 80–90; 2008, 325–335) presents econometric analysis of the determinants of the timing of sale as well as of observed land and water prices and compares water prices with what Los Angeles might have been willing to pay to get a sense of the distribution of the surplus. Controls include farm size, cultivated acreage, water/acre, riparian water rights to the Owens River, pool membership, non-pool ditch farms, Los Angeles aqueduct flow per urban population, lagged changes in the city's population, and the cumulative percent of total water purchased as of each sale to indicate supply constraints. The year of sale reflects both demand factors reflecting the interests of the LADWP and supply conditions reflecting the interests and efforts of farmers, especially those of pool members. Declines in the current flow of aqueduct water per capita and past growth in population encourage the agency to return to Owens Valley to negotiate for more properties to maintain water supplies. Farms with more water/acre and riparian water rights likely would be sought earlier to meet this demand. Further, all else equal, larger farms might have more water, and those with more cultivated acreage might have signaled access to water not captured in the other water variables, suggesting that these two variables would lead to earlier years of sale. Pool membership in general would make delay of sale more feasible whenever farmers believed that initial offer prices were too low. Collectively holding more water, pools were in a better position to reject LADWP offers and to demand higher prices at lower risk than they would have been if left isolated with unsold properties. The more cohesive the pool, the better it would be able to resist defection and hold out for higher prices. The non-pool ditch farms that were purchased by the LADWP to avoid joining any pool should also have earlier years of sale relative to the baseline of non-pool, competitive non-ditch farms.

The estimation reveals that as suggested in the mean values, members of the most cohesive Keough pool delayed sale by almost a year and a half longer than the competitive farmers not on ditches. For members of the less-cohesive Watterson and Cashbaugh pools, however, members on average sold between 11 and 15 months earlier than did non-ditch farmers. This result reflects the premature defection within those two pools. These farms had water and were comparatively more attractive to the LADWP than were the drier non-ditch farms. When the members defected, they sold. Those farmers who were on ditches but not in pools and who had their properties purchased to prevent their joining them on average sold about seven months earlier than the baseline farmers. Farms with more water, all else equal, were purchased earlier, reflecting the agency's desire to secure properties that brought more water to the aqueduct. An additional acre-foot of water/acre accelerated sale by over one month, all else equal. In terms of price paid for a farm, among the agricultural productivity variables, water endowments mattered the most, with an additional acre-foot of water/acre adding over US$37 per acre to the farm sales price. This contribution, however, grew at a declining rate. The fall off in the value of additional water/acre varied across the sample, with the farms at the center of the most contested negotiations having the largest negative effects. Those farmers with the most water had the greatest reason to resist efforts by the LADWP to purchase their water-bearing properties according to their agricultural productivity values. These generally were the farmers who colluded to secure higher per-acre prices. The fact that they were not able to secure more for their water through the land market contributed to the notion of "water theft."

Even though there was differential cohesion and ability of the sellers' pools to hold out, they exhibited market power in the land market. Members of the Keough pool earned about US$213 more per acre than did the 228 non-ditch property owners and US$145 more per acre than those farmers who were on ditches but defected from collusive efforts. Members of the Watterson and Cashbaugh pools earned approximately US$81 and US$52 more per acre respectively than the competitive baseline farmers. To keep some farmers from joining a pool, Los Angeles paid an additional US$68/acre for the non-pool ditch farms, an amount better than their owners would have earned in the Cashbaugh pool, but less than in the Watterson Pool.

It is possible to compare the implicit prices paid per water acre-foot with the price that the LADWP might have been willing to pay. In 1931 after most Owens Valley properties had been purchased, voters in the Metropolitan Water District, which included Los Angeles, approved bonding for US$220 million for construction of the Colorado River Aqueduct to bring 1.1 million acre-feet to the city annually. This translates approximately to US$220/AF for water from the Colorado River or US$9.50/acre-foot for an annual flow. Converting all implicit water prices for each Owens Valley farm into prices for an annual flow of water and plotting them in Figure 20.3 illustrates the position of the farmers relative to the Colorado River water price. This exercise does not include the added costs of pumping Colorado River water nor of treating it, since Colorado River water was more mineralized than was Owens Valley water. As shown, farmers generally received well below the maximum amount the agency might have been willing to pay, regardless of whether or not a farmer was part of a sellers' pool. Moreover, those farmers

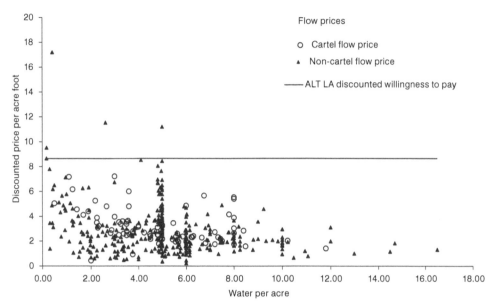

Figure 20.3 Water prices paid, sellers pools, non-pool members, and Los Angeles' willingness to pay
(source: Libecap 2008, 335)

with more water endowments per acre of land received among the lowest prices per acre-foot for their water.

Moreover, consider the total expenditures made by the Board relative to what it might have been willing to pay. The total outlay for Owens Valley farms by the LADWP between 1916 and 1934 in the data set used here was US$13,937,934 to secure 266,428 acre-feet of water for an average price per acre-foot of US$52.31 in current prices. Including the additional capital cost of the Los Angeles Aqueduct for carrying the purchased water results in a total outlay of US$23,076, 934. If the agency had paid US$220/AF for Owens Valley water, as it did for Colorado River water and related infrastructure, the total sales expenditures would have been US$58,614,375, or about two and a half times the actual outlay. It seems clear then that the LADWP paid less for Owens Valley water than it had to pay for Colorado River water and, hence, what it might have been prepared to pay. This is only a suggestive exercise, but the outcome reflects the relative bargaining power of the agency, and it underlines the distributional legacy of the transaction.

Concluding remarks: lessons of Owens Valley for understanding transaction costs and contemporary water transfers

Added transaction costs, driven by community concerns over distributional implications and pecuniary effects of water transfers, appear to limit contemporary water markets. Absent comprehensive data on water exchanges for econometric analysis, it is not possible at this time to quantify the size of this effect. It is feasible to examine the most famous water transfer, the Owens Valley-to-Los Angeles water exchange of the early twentieth century, to assess bargaining costs and how they may have impacted the timing and transaction costs of water trading. The analysis shows how efficiency-enhancing trade was tied up by distributional conflicts. There were intense disputes over property valuation and the sharing of the gains from re-allocating water to Los Angeles. The bargaining for water was complicated because it took place in a land market and the most contentious, lengthy negotiations took place between the LADWP and farmers who were part of collusive organizations. The efforts of the latter to delay sale in order to capture more of the gains of exchange slowed the valuable re-allocation of water from marginal agriculture to higher-valued urban uses in Los Angeles. Additional resources were expended by both parties in maneuvering to improve their positions. The LADWP had an inherent advantage in bargaining. It was a single entity, supported by urban property owners who had an informational advantage (relative to Owens Valley farmers) of the capital gains achievable with the acquisition of more water for the semi-arid city. The farmers, however, were competitive sellers, and had that condition persisted more of the surplus would have gone to the city. But many of the Owens Valley land owners formed sellers' pools to make the bargaining outcome somewhat more equal, although there were numerous pools and they suffered from defection. This finding supports the arguments of Fehr and Schmidt (1999) that competitive settings can lead to more unfair outcomes and, hence, bargaining parties have incentive to cooperate to improve their bargaining position. Had the farmers combined into a single unit, their efforts to achieve a more balanced outcome would have been far greater. As Fehr and Schmidt note, inequity aversion is a far less powerful constraint on the party advantaged relative to the others. Hence, the LADWP strove to block any unit or Owens Valley-wide collusion. The LADWP was serving the interests of its constituents. Inequity can, however, result in less production dynamically if bargaining and market exchange is discredited as a result and other potential water sellers resist trades and/or devote resources to improving their bargaining position.

This dynamic outcome appears at play today. Farmer appeals to the press about the inequities of the process created lasting perceptions of injustice in the transfer of water from Owens Valley to Los Angeles, where it generated dramatic increases in land values. Indeed, the ensuing bargaining disputes have lingered on as part of the historical legacy of Owens Valley, which is important for understanding the costs faced by water markets today. While increases in transaction costs arising from the bargaining strategies of the parties are not directly measured, it is possible to measure the outcomes. In general Owens Valley farmers did better through the sales of land to Los Angeles than if they had remained in agriculture. Colluding farmers earned more per acre of land than did their unorganized colleagues. Further, between 1910 and 1930, when Los Angeles was buying properties, farmland values in Inyo County (Owens Valley) as reported in the US census rose by 175 percent to US$143/acre, while in the four similar baseline Great Basin counties where no land/water purchases took place, they rose by 52 percent to an average of US$45.50/acre. Nevertheless, Owens Valley farmers did not do as well in the water market. Their sellers' pools captured more of the surplus, but these gains were small relative to the overall rents garnered by Los Angeles property owners.

Comparing US census data on the rise in value of agricultural land and buildings in Los Angeles County and Inyo County (Owens Valley) between 1900 and 1930 reveals a gain of US$11,568,000 in Inyo versus US$407,051,000 in Los Angeles. The Los Angeles values are about 40 times those of Inyo and, in part, they reflect access to a steady supply of Owens Valley water. Alternative sources from the Colorado River did not arrive until 1941. Even this is an understatement of Los Angeles' gain because the growth of urban land values is not reported in the census. If one uses property value data from the California State Board of Equalization a similar picture emerges. Between 1900 and 1930, the total value of all property in Inyo County rose by 917 percent, but in Los Angeles County the increase was 4,408 percent. It is no wonder that Owens Valley farmers wanted more of the benefits of the exchange.

The sense of inequity over the terms of trade also was driven by the nature of supply and demand for water. Urban users had relatively inelastic demand, whereas farmers competing for sale had comparatively elastic export supply. Hence, Los Angeles residents gained considerable consumer surplus from the transaction. When the gains from trade are very large, distributional outcomes move to the forefront as they did in Owens Valley negotiations. Generally, as Fehr and Schmidt imply, it may be the case that trades have lower transaction costs when the benefits are shared reasonably equally, as they would be in more competitive settings on both sides. Trading entails higher transaction costs when the distribution appears to be highly skewed towards one party. Assessing the size of potential third-party pecuniary effects and moderating distributional differences in the rents achieved from exchange through regulatory-created mitigation funds may smooth the path of water exchanges.

References

Alesina, Alberto and George-Marios Angeletos. (2005). Fairness and Redistribution. *American Economic Review* 95 (4): 960–80.

Allen, Douglas W. (2000). Transaction Costs. Boudewijn Bouckaert and Gerrit De Geest, eds, *Encyclopedia of Law and Economics*, Volume 1: Cheltenham, England: Edward Elgar Press: 893–926.

Barzel, Yoram. (1982). Measurement Cost and the Organization of Markets. *Journal of Law and Economics* 25(1): 27–48.

Blair, Roger D., David L. Kaserman, and Richard F. Romano. (1989). A Pedagogical Treatment of Bilateral Monopoly. *Southern Economic Journal* 55: 831–41.

Brewer, Jedidiah, Robert Glennon, Alan Ker, and Gary Libecap. (2008). Water Markets in the West: Prices, Trading, and Contractual Forms. *Economic Inquiry* 46(2): 91–112.

Camerer, Colin F. and Richard H. Thaler. (1995). Ultimatums, Dictators, and Manners. *Journal of Economic Perspectives* 9 (2): 209–19.

Coase, Ronald H. (1960). The Problem of Social Cost. *Journal of Law and Economics* 3: 1–44.

Coase, Ronald. (1992). The Institutional Structure of Production. *American Economic Review* 82 (4): 713–19.

Dahlman, Carl J. (1979). The Problem of Externality. *Journal of Law and Economics* 22: 141–62.

Demsetz, Harold. (1968). The Cost of Transacting. *Quarterly Journal of Economics* 82: 33–53.

Demsetz, Harold. (1969). Information and Efficiency: Another Viewpoint. *Journal of Law and Economics* 12 (1): 1–22.

Donohew, Zach. (2008). Property Rights and Western United States Water Markets. *Australian Journal of Agricultural and Resource Economics* 53, 85–103.

Fehr, Ernst and Klaus M. Schmidt. (1999). A Theory of Fairness, Competition, and Cooperation. *Quarterly Journal of Economics* 114 (3): 817–68.

Grafton, R. Quentin, Clay Landry, Gary Libecap, Sam McGlennon, and Bob O'Brien. (2011). An Integrated Assessment of Water Markets: Australia, Chile, China, South Africa and the USA. *Review of Environmental Economics and Policy* 5 (2): 219–39.

Greenwald, Bruce C. and Joseph E. Stiglitz. (1986). Externalities in Economies with Imperfect Information and Incomplete Markets. *Quarterly Journal of Economics* 101 (2): 229–64.

Haddad, Brent M. (2000). *Rivers of Gold: Designing Markets to Allocate Water in California*. Washington, DC: Island Press.

Hanak, Ellen. (2003). *Who Should Be Allowed to Sell Water in California? Third-Party Issues and the Water Market*, San Francisco, CA: Public Policy Institute of California.

Libecap, Gary D. (2007). *Owens Valley Revisited: A Reassessment of the West's First Great Water Transfer*. Palo Alto, CA: Stanford University Press.

Libecap, Gary D. (2008). *Chinatown Revisited:* Owens Valley and Los Angeles—Bargaining Costs and Fairness Perceptions in the First Major Water Rights Exchange. *Journal of Law, Economics, and Organization* 25 (2): 311–338.

Libecap, Gary D. (2011). Institutional Path Dependence in Adaptation to Climate: Coman's "Some Unsettled Problems of Irrigation". *American Economic Review* 101 (1): February, 64–80.

Scitovsky, Tibor. (1954). Two Concepts of External Economies. *Journal of Political Economy* 62 (2): 143–51

Williamson, Oliver E. (1975). *Markets and Hierarchies: Analysis and Antitrust Implications*, New York: Free Press.

Williamson, Oliver E. (1979). Transaction Cost Economics: The Governance of Contractual Relations. *Journal of Law and Economics* 22 (2): 233–61.

Williamson, Oliver E. (2010). Transaction Cost Economics: The Natural Progress. *The American Economic Review* 100 (3): 673–90.

21

AGRICULTURAL WATER MANAGEMENT AT THE VILLAGE LEVEL IN NORTHERN CHINA

Qiuqiong Huang, Jinxia Wang, Siwa Msangi,
Scott Rozelle, and Jikun Huang

Introduction

China's water resource availability is among the lowest worldwide. The most recent estimate of the annual renewable internal freshwater resources per capita is 2,093 m³ (FAO, AQUASTAT), which is far below the estimated world average of 8,349 m³ (ESCAP, 2010). In addition, water resources are not evenly distributed. Northern China has only 21 percent of the country's water endowment (Ministry of Water Resources of China, 2011).[1] Northern China, however, remains an important region, with about 35 percent of the total population, 38 percent of the nation's gross domestic product (GDP) and almost half of the grain production (NBSC, 2013).

There is a consensus that China is facing increasing water shortages (Jiang, 2009; Wang et al., 2009a; Yang et al., 2003). The available water supply is shrinking. The total estimated water resources of China dropped by 16.5 percent from 1997 to 2011 (Figure 21.1). The demand for water continues to increase steadily over time, especially in the industrial and residential sectors. The total water use of all sectors has increased by 9.7 percent from 1997 to 2011 (Figure 21.1). The increasing water scarcity is particularly challenging to the agricultural sector. The government has decided that agricultural use will not be given priority for any additional future allocations of water (China, 2002). The share of agricultural water use in total water use has declined from 70.4 percent in 1997 to 61.3 percent in 2011 (Ministry of Water Resources of China, 1997, 2011). At the same time, the government is intent on keeping its food security target of food self-sufficiency at 95 percent or more of domestic grain supply (National Development and Reform Commission of China, 2008). How to maintain agricultural production in an environment of increasing water scarcity is the key challenge in agricultural water management.

After trying out other policies such as promoting the use of water-saving technologies and water trading, China's leaders have started to focus on village-level irrigation management reform as a key part of their strategy to combat China's agricultural water problems.[2] The primary state agency charged with managing the state's water is the Ministry of Water Resources (MWR) and its provincial counterparts. Irrigation districts

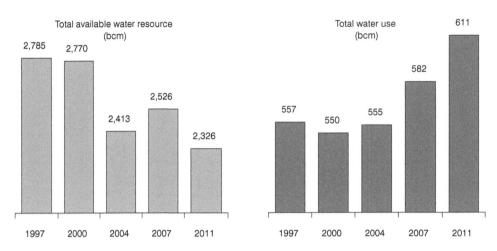

Figure 21.1 Changes in available water resource and water use over time in China

(IDs) and local water resource bureaus (WRBs) often manage the upper levels of irrigation systems (the main canals and branch canals) that transfer water out of major rivers (e.g., the Yellow River) or reservoirs and channel it to lower levels. Local irrigation systems (tertiary canals and below) are administered by county, township governments and village committees (Xie, 2007). The canal network in the village, then, is completely maintained by the village. In the surface water sector, officials have mostly focused on promoting Water User Associations (WUAs) and contracting. With the assistance of international agencies, such as the World Bank (Reidinger, 2002), WUAs were created to take the place of village committees in the management of village-level irrigation systems. Unlike other policy efforts such as promoting water-saving technologies and trading of water rights, the reform has spread rapidly nationwide. Since the first WUAs were established in south China in 1995 under the World Bank "Yangtze Basin Water Resources Project" (World Bank, 2003a), by 2006 there were more than 30,000 WUAs in China (Ministry of Water Resources of China, 2007). Contracting is also encouraged, which transfers the management responsibilities of the village-level irrigation system to an individual contractor. The reform in the groundwater sector is characterized by the rise of private wells and the concurrent emergence of groundwater markets. With these changes, the management of groundwater has also been shifted from village committees to individual farmers.

Despite the fact that agriculture is still the largest user of water and the rapid expansion of WUAs nationwide and contracting regionally, surprisingly not many studies have examined the impact of water management reform in northern China. One major reason is the lack of data. The objective of this chapter is to provide an overview and synthesis of the irrigation management reform that has reshaped the institutions that govern local irrigation systems in northern China. The following section focuses on surface water institutions. The third section focuses on groundwater institutions. In both of these sections, we first provide the specific historic context that has motivated the irrigation reform in the surface water and groundwater sectors; we then document the development of institutional arrangements over time; we also describe the characteristics of governance under different forms of water management; we then evaluate the impact of institutional reforms on water allocation and agricultural production. Most empirical analyses discussed in the second and third sections are based on the village-level and household-level data sets collected by the authors of this chapter in northern China: the China Water Institutions and Management survey (CWIM) and North China Water Resource Survey (NCWRS).[3] The final section concludes by discussing the challenges and the potential of these institutional arrangements to facilitate further reforms such as water trading schemes and water pricing policies.

Surface water institutions

This section focuses on the irrigation reform that took place in the surface water sector. It begins by describing the problems in the surface water sector and the changes in surface water management as a result of the reform. The second subsection describes the time trend of the development of surface water institutions. The third compares several key features of governance under different surface water institutions such as farmer participation, information transparency and monetary water saving incentive. The final subsection evaluates the impact of the water management reform on crop water use, productivity of water and irrigation system performance (maintenance expenditure, rate of water fee collection and timeliness of water delivery).

Background

The most common form of irrigation water in China is surface water (Huang et al., 2006). Before 1980s, the village committee (i.e., village leaders) took direct responsibility for the allocation of irrigation water, operation and maintenance (O&M) tasks and fee collection (Huang et al., 2009). While the village leaders made the major decisions, a water manager was hired for a small stipend to carry out daily irrigation-related activities. In larger villages, the manager may have interacted with a group of subordinate water managers representing subgroups of households in the village. The water manager coordinated with the ID on the timing of water deliveries and informed farmers of it. Traditionally surface water fee collection at the farm level was managed by the village committee. This was because up until the early 2000s, water fees were bundled with other village taxes and fees. Increasingly, water managers are also responsible for collecting water fees from farmers. In addition to water fees, farmers were required to contribute obligatory labor to community activities including building and maintaining irrigation infrastructure (Lohmar et al., 2008). This corvée labor contribution is a carry-over from the collective period when most rural infrastructure was constructed using teams of collectively managed labor under the communes.[4] When the village committee is in charge, we consider the village's irrigation system to be run under collective management, and we refer to this as the traditional system.

A series of reforms starting in the 1990s have continuously weakened the ability of the village committee to invest in and maintain irrigation infrastructure (Lohmar, et al., 2003). The 1994 fiscal reform reduced tax revenues local governments (provincial and below) could access by reassigning them to the central government (Wang, 1997). The 2003 tax-for-fee reform stripped the village committee of the authority to levy fees and surcharges on households (Oi et al., 2012). In 2006, agricultural tax was completely eliminated nationwide (Oi et al., 2012). In the 2000s, provincial governments started to gradually remove the requirement of corvée labor (Oi et al., 2012). For example, Hebei province started to limit village leaders' ability to mobilize corvée labor in 2002 and Henan province in 2004. These reforms have restricted or removed the two means (collective fiscal resources from tax and fees and corvée labor) village leaders previously had to invest in irrigation infrastructure. Villages now have to rely on fiscal transfers from upper-level governments (e.g., provincial, prefecture and county governments) to provide public goods in the village including investment in irrigation infrastructure. As a result, tertiary canals and on-farm structures became dysfunctional in many areas (Xie, 2007).

In addition to problems within villages, there was also a lack of coordination between the village and IDs. Farmers and sometimes village water managers often could neither choose when their irrigation water was delivered nor how much they received (Lohmar et al., 2008). Although IDs delivered water to villages, IDs only received payment for water after water fees had been routed through various levels of bureaucracy (from farmers to village; village to township irrigation stations; towns to counties; and finally from counties to IDs; Lohmar et al., 2008). In general, each level of bureaucracy charged a fee for the service of handling these payments, further reducing the amount ultimately remitted to the ID.

The more fundamental problem is the low water price. Surface water is priced between 30 percent and 50 percent of the cost of supply (Zheng, 2002). Farmers using surface water supplied by the ID pay for water on a per unit of land basis. In addition to discouraging water savings, low prices provided IDs with few incentives to deliver water in a timely manner, partly because revenue generated from water fees were not sufficient to maintain delivery infrastructure (Lin, 2003). The poor delivery services then resulted in farmers refusing to pay their water fees. Farmers were rarely punished for forfeiting their water fees. The poor rate of water fee collection in turn exacerbated the IDs' insufficient funds (Lohmar et al., 2008).

The establishment of WUAs is a major movement to improve irrigation management in China (Huang et al., 2009; Xie et al., 2009). In principle, a WUA is a farmer-based, participatory organization in which farmers come together to elect a board to manage the village's irrigation system (World Bank, 2003a). The argument is that with more active participation, farmers have a greater stake in the system, therefore will be more willing to remit water fees in return for improved irrigation services. In addition, by attending regular WUA meetings, farmers may be more aware of their water use and the cost of water, which may induce water conservation behavior (Lohmar et al., 2003). In addition to providing more timely irrigation deliveries to farmers, WUAs are also expected to maintain the village's irrigation infrastructure and collect water fees. Most WUAs bypass the traditional village–township–county channels and directly purchase water from the ID on a volumetric basis. It is hoped that this more direct way can increase the actual amount received by the ID and lead to lower fees for farmers (Lohmar et al., 2003; Xie, 2007). As villages are the basic hydrologic units within which irrigation is managed, most WUAs were established at the village level. Some WUAs are also established at the level of branch canal and govern several villages.

Although not formally initiated by the effort of upper-level governments, contracting also emerged as a way to manage a village's irrigation system (Huang et al., 2009).

Contracting is a commonly observed form of management in many different contexts of China's reforms (e.g., in the management of township and village enterprises; grain enterprises; extension system agencies etc.; Park and Rozelle, 1998). It is similar to WUAs in that they turn irrigation management over from village officials to a specified manager, but instead of a whole village's irrigation infrastructure being turned over, just a lateral canal, which may service only part of a village, is turned over to the manager. We consider WUAs and contracting as *reform-oriented management systems* (in contrast to the traditional collective management).

Development of institutions over time

Using both CWIM and NCWRS data, Huang et al. (2009) documented the trend of surface water management reforms between 1995 and 2004 in seven provinces in northern China (Inner Mongolia, Hebei, Henan, Liaoning, Ningxia, Shaanxi and Shanxi, see Figure 21.2). The data clearly show that collective management is on the decline (Figure 21.3, Panel A). The share of collective management declined from 90 percent in 1995 to 73 percent in 2004. WUAs and contracting have developed at about the same pace. By 2004, 10 percent of villages managed their surface water through WUAs

Figure 21.2 Sample provinces of survey data referred to in the chapter
Note: CWIM sample provinces 1. Hebei; 2. Henan; 3. Ningxia. NCWRS sample provinces: 1. Hebei; 2. Henan; 4. Inner Mongolia; 5. Liaoning; 6. Shanxi; 7. Shaanxi. Bank Survey sample provinces: 8. Gansu; 9. Hubei; 10. Hunan.

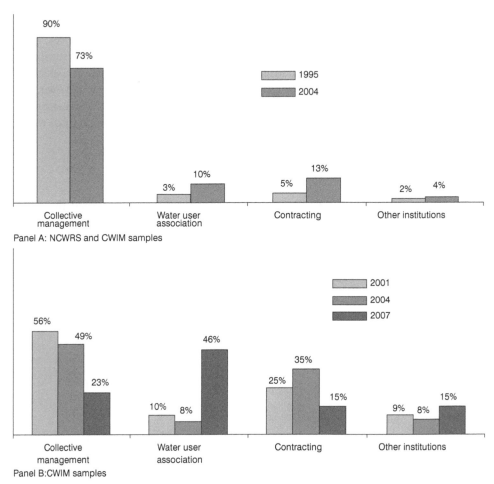

Figure 21.3 Changes in surface water management institutions

and 13 percent through contracting. The mixed systems also rose from 2 percent to 4 percent between 1995 and 2004. While collective management was still the dominant form of management, 27 percent of villages in northern China had been affected by water management reform by 2004.

Huang et al. (2010a) update the development in the CWIM sample provinces by 2007 (Figure 21.3, Panel B). Even by 2004, the share of villages under collective management in Hebei, Henan and Ningxia provinces was 49 percent, lower than the rate of 73 percent in the combined NCWRS and CWIM sample (Figure 20.3, Panels A and B). The share of collective management continued to drop to 23 percent in 2007. This drop is mostly driven by the changes in Ningxia province. By 2007, none of the sample villages in Ningxia province remained under collective management. WUAs were managing canals in about 72 percent of the villages and were jointly managing canals with contractors in another 22 percent of the villages.

Both the NCWRS and the CWIM samples have shown that water management reform varies significantly across the sample provinces (Huang et al., 2009; Huang et al., 2010a). The reform efforts in Ningxia province clearly focused on creating WUAs.

In contrast, in Hebei and Henan, most villages replaced collective management with contracting, instead of WUAs. Clearly the reforms are far from universal, which is what we would expect in China, a nation that often allows local governments considerable freedom in making their own decisions on the exact form and timing of reform. Huang et al. (2008) showed that the choice of managerial form in different villages can be modeled as a function of village leaders' ability relative to that of potential managers, the conditions of the canal system in the village and the characteristics of the water resources. Some sample villages either discontinued or partially discontinued WUAs by 2004 after having created them in 2001. Village leaders and canal managers in these villages explained that they only nominally adopted WUAs and the WUAs were not at all active in irrigation management. There are also several sample villages that reverted back to collective management by 2004 after adopting contracting in 2001, because nobody wanted to contract the canals. These shifts into and out of WUAs and contracting may indicate that water management reform is not universally successful.

Characteristics of governance

Participation

The international literature emphasizes the importance of farmer participation in the promotion of successful WUAs (World Bank, 2003b). Huang et al. (2009) and Huang et al. (2010a) assessed the level of participation. To avoid reporting bias that could arise if responses from the village leaders were used, only information from the farm household survey in the CWIM sample is used. Under collective management, most decisions were made with little consultation or participation of farmers (Huang et al., 2009; Wang et al., 2005a). Similarly, there was little participation by farmers in contracting villages (Huang et al., 2009; Wang et al., 2005a). Village leaders usually decided whether to contract out canals or not. Farmers rarely played any role in the transition. Only in a few villages did farmers participate in electing the contractor. Once the contractor was selected, there was no formal role for farmers in supervising, coordinating or in any other governance aspects.

Even when the reforms that led to the creation of WUAs explicitly attempted to encourage farmer participation, practice often varied from principle. Evidence suggests that WUAs are not reaching the expected level of "broad-based participation" envisioned by advocates of participatory programs. Although farmers still participate in irrigation activities, mostly in the form of their cash and labor contributions, their involvement in management has been minimal, or at most an input supplementary to WUA management in providing irrigation services (Huang et al., 2010a). Farmers were not always consulted before the establishment of WUAs. For example, in 2007 only about 32 percent of WUAs in the CWIM sample involved farmers in the decision on their establishment. In less than half of the WUA villages, farmers had power in appointing the chair or the board members of WUAs. In most villages, these positions were appointed by either the ID or the village committee. Partly because of this, WUA leadership had close ties to the village leadership. In more than 60 percent of the villages either the head of the WUA board or one more board members were also village leaders. Farmers were not active in other aspects of water management either. Although the share of the WUAs that invited farmers to attend their regular meetings rose sharply to 75 percent by 2007, the median percentage of farmers that actually attended was only 15 percent.

The limited participation of farmers may be partly explained by the small farm size, large number of households and high off-farm employment participation in rural China (Huang et al., 2010a). On average, farm size in the CWIM sample areas is about 0.6 ha per household and the average number of households within the command areas of canals is 400. Given the small farm size and the large number of households, the cost of participation that accrues to each farmer likely exceeds the benefit. In villages where farmers are busy with wage-earning jobs or other off-farm employment, the opportunity cost of attending meetings is high. Some sample farmers in the CWIM survey reported "busy" as their reason for not participating (Huang et al., 2010a). Other farmers also pointed to "the size of the village" (in larger villages, farmers tend to think they benefit less from participation) and "level of education" (illiterate farmers are less likely to attend meetings; Huang et al., 2010a).

Transparency

Management under WUAs is more transparent than under collective management or contracting. In 2007 almost 90 percent of the WUAs in the CWIM sample shared all three key types of information about irrigation management with farmers: a) the total amount of water fees collected; b) the volume of water actually delivered by the ID to the village; and c) the actual area irrigated (Huang et al., 2010a). In contrast, more than 30 percent of the contractors chose to share no information with farmers. Most villages under collective management shared only one of the three key types of information.

Incentive

One unique characteristic of China's water management reform is that in some IDs' water managers are provided with monetary incentives tied to water savings (Wang et al., 2005a). The use of incentives is not new in the context of China's overall economic reform effort (Naughton, 1995). For example, the core component of the household responsibility system (HRS) in the 1980s is that farmers can earn profit from the land allocated to them, which has induced them to exert more effort, allocate resources more efficiently and enter into new economic activities (Lin, 1992). The water-saving incentive mechanism works because of the way surface water is measured and priced (Lohmar et al., 2008; Wang et al., 2005a). Most IDs measure water flows at the branch or lateral canal that delivers water to villages. This allows IDs to charge villages a volumetric price for the amount of water delivered. Within the village, there are no water meters along the tertiary and lower canals that deliver water to farmers' fields. So farmers pay for water on per unit of land basis. The level of the water fee farmers pay is determined prior to the irrigation season and is related to the historical level of water use. If a water manager could generate water savings in an irrigation season and thus need to pay the ID less than the amount of the water fees collected from farmers, the manager could keep the difference as his own profit.

There are sharp differences in the implementation of the incentive part of the reform packages across villages. In none of the collectively managed villages were village leaders provided with incentives, perhaps because any profits from saving water would be counted as village fiscal revenue (Huang et al., 2009; Huang et al., 2010a). In 2004, in 73 percent of the contracting villages and 32 percent of the WUA villages, managers were provided with incentives (Huang et al., 2009). Clearly, the provision of incentives distinguishes contracting significantly from collective management or WUAs. However, by 2007, the trend had changed. In 2007, incentives were provided to more than 80 percent of the WUAs in the CWIM sample provinces (Huang et al., 2010a). Again this was driven by the change in Ningxia province, where policy makers are active advocates of water-saving incentives. The use of incentives in contracting dropped to 31.3 percent in 2007 (Huang et al., 2010a). Although the reasons are not yet clear, the fall in the use of incentives from 2004 to 2007 might have been due to the declining effectiveness of incentives. Over time, the room for saving water would shrink, thus making profits harder to earn.

Impact of water management reform

In one of the few studies that evaluate the performance of water management reform, Wang et al. (2005a) argued that when there was only a nominal shift of institution, not many changes in water management were observed. Using the 2001 panel of the CWIM survey data, Wang et al. (2005a) show that in villages where no water-saving incentives were offered, contractors and WUA managers acted much like village leaders in collectively managed systems. There was little effort in any of these villages to save water. In contrast, in villages in which managers were provided with incentives to save water, water use declined by about 40 percent. The research also showed that water savings were achieved without negatively affecting crop yields or cropping income (Wang et al., 2005a). In some villages, as an inducement for farmers to accept reduced but timelier water deliveries, WUA managers pass some savings on to the farmers by charging lower water fees (Lohmar et al., 2008).

Using the 2007 panel of the CWIM survey, Huang et al. (2010a) showed that WUAs improved other aspects of irrigation services relative to that under either traditional collective management or contracting. For example, villages with WUAs spent 27.5 yuan/m on maintaining canals (1 USD = 7.5 yuan in 2007), while villages under traditional collective management or contracting spent less than 10 yuan/m. Water was delivered on time in villages with WUAs more than 92 percent of the time. By contrast, water was delivered on time less than 60 percent of the time in collectively managed villages, and less than 80 percent of the time in villages under contracting. The improvement in irrigation services increased farmers' willingness to pay water fees. Improving the transparency of management by sharing more irrigation information with farmers helped too. Huang et al. (2010a) show that the proportion of water fee collected, defined as the actual water fees collected from farmers as a portion of the total fees payable by farmers, was 93.5 percent under WUAs and 92 percent under contracting but only 72 percent under collective management. The improved water fee collection in turn reinforces the improvement in irrigation service. This is because with sufficient water fees from villages, local IDs are able to supply water to the village on time (Lohmar et al., 2008).

Wang et al. (2010) focus on evaluating the performance of WUAs set up by the World Bank versus villages nearby including both non-Bank WUAs and collective managed villages. Using the Bank survey that covers 60 villages in Gansu, Hubei and Hunan provinces,[5] Wang et al. (2010) found that in 2005 water use in the Bank villages on all major crops was between 16 and 24 percent lower than that in collectively managed villages. For example, farmers in the Bank villages used 439 m^3/mu to irrigate rice while those in the collectively managed villages used 543 m^3/mu (1 hectare = 15 mu). Farmers in the Bank villages also used 18 percent and 16 percent less water on wheat and maize, respectively, compared to farmers in the collective villages. The multivariate analysis also found that water use was lower in the Bank villages and was statistically different. In addition, the multivariate analysis also showed that difference in crop yields between the Bank villages and the collectively managed villages was not significant. This can be interpreted as the Bank villages improving the productivity of water in that they used less water but obtained the same yields. No statistical differences were found between non-Bank WUAs and collectively managed villages in terms of water use or crop yield. The explanation was that when setting up and operating WUAs, the Bank WUAs followed the principles laid out by the World Bank (World Bank, 2003b) more closely than non-bank WUAs.[6]

The focus group with farmers conducted in the Bank survey showed that farmers' perception about water management has improved after WUAs were established in the Bank villages (Wang et al., 2010). All the Bank villages said that overall water was better managed by WUAs. Farmers also stated that there was more effort centered on saving water (93 percent); more timely delivery of water (96 percent); a reduction in water charges to farmers—in either an absolute or relative sense (89 percent); and less conflict when water fees were being collected (81 percent). The comparable percentage figures were much lower for the non-Bank villages. In one Bank WUA village in Hubei province, farmers reported it was easier for WUA managers to collect water fees from farmers, because WUA managers were elected by and thus were trusted by farmers (Wang et al., 2010). In addition, if some farmers really had no money to pay water fees before the irrigation season (but could pay later), WUA managers were able to secure alternative funds to make up for the shortage in water fees in order to pay the ID in time. In another village in Hubei province, a non-bank WUA village, farmers also linked the improvement in irrigation services to more adequate water fee collection as well as better canal maintenance and investment in canal lining (Wang et al., 2010).

Several reasons explain why WUAs were able to improve water allocation. First, establishing WUAs improved the scheduling of irrigation services. Instead of going through layers of bureaucracies such as township and county government, WUAs interacted directly with IDs. WUAs could communicate irrigation demands easily to IDs. WUAs and IDs coordinated easily to schedule the delivery of water at the times needed. Second, with the water saving incentive, managers were more motivated to provide satisfactory irrigation services so that farmers would not refuse to pay the irrigation fees. WUA managers were more likely to make investments such as lining canals that reduce water loss and improve water deliveries to fields. Often they applied funds from upper-level governments or raised funds from farmers to make such investments. Third, compared to village leaders, WUAs could more effectively manage water allocation operations. WUAs made more effort in supervising water deliveries to make sure that each farmer got their share of water and that the sluice gates along the canals were opened and closed in time so that water would not flow into the fields when not needed. Since water could be allocated to farmers more efficiently, they received a better irrigation service even though less water may be drawn from the system. In contrast, village leaders were constrained in their time to spend on supervision activities because they were burdened with a myriad of other administrative responsibilities such as running village enterprises and maintaining local schools and health facilities.

WUAs, however, have not succeeded in all aspects. Wang et al. (2010) did not find evidence that WUA increased crop income, the aspect that China's leaders often focus on. This is perhaps because farmers still pay for water on per unit of land basis. Even if water use were reduced, the water fee would remain the same. Huang et al. (2010a) also show that in many villages, either WUAs still shared responsibility with the village committee on tasks such as conflict resolution and canal maintenance or the village committee was still in charge of these tasks. This is probably because WUAs also lack the authority of village leaders that can facilitate water management. In the focus group discussion in the Bank survey (Wang et al., 2010), farmers reported that their WUA did not have enough power to coordinate water allocation or resolve water conflicts inside the village. If they encountered any difficulty in the management of water, they still needed help from the village leaders.

Groundwater institutions

This section focuses on the irrigation reform that took place in the groundwater sector. The first subsection describes the importance of groundwater to agriculture in northern China and how groundwater was managed before the 1980s. The rise of private wells and the emergence of groundwater markets are documented in the two middle subsections. The final subsection evaluates the impact of private well ownership and groundwater markets on crop water use, crop production and rural income.

Background

Groundwater resources are playing an increasingly important role in the economy of northern China. In 2011, on average, 35.5 percent of the total water supply (industrial, residential and agricultural sectors) came from groundwater (Ministry of Water Resources of China, 2011). Agriculture relies even more heavily on groundwater. As public investment in canal systems waned and deliveries became more unreliable, farmers in northern China began to rely more on small irrigation systems fed by groundwater. The rapid expansion of groundwater irrigation has stimulated the growth of agriculture in northern China. In the NCWRS survey sample villages, in 2004, with the exception of rice, at least 70 percent of the area sown to grains and other staple crops were irrigated by groundwater (e.g., 72 percent of wheat and 70 percent of maize; Wang et al., 2009a). Groundwater also irrigates most cash crops (e.g., 70 percent of cotton, 62 percent of oil seed crops and 67 percent of vegetables).

In most rural areas in northern China, central and regional governments have little control over groundwater use. China's National Water Law (China, 2002) stipulates that all property rights over groundwater resources belong to the national government, including the right to use, sell and/or charge for water. In practice, however, villages that lie above the aquifers have the de facto rights to groundwater resources. Unlike the US, water rights are not associated with land ownership or historic use. Often they are associated with the ownership of wells. Despite the plethora of laws and policy measures created by government officials, there has been a lack of enforcement (Wang et al., 2009a). One of the reasons is the difficulty in regulating millions of small, water using farmers. For example, despite the nearly universal regulation that requires the use of a permit for drilling a well, less than 10 percent of the sample well owners in the 2004 NCWRS survey obtained one before drilling. Only 5 percent of sample villages gave any consideration to well spacing (Wang et al., 2009a).

With the lack of control from upper level governments, groundwater use is organized and managed at the village level. Before the rural reforms in the 1980s, wells in almost all rural villages were collectively owned and financed primarily by collective retained earnings and additional funding from township governments. Village leaders were largely responsible for arranging for the water resource bureau-run well drilling companies to sink tubewells. Pumps all came from either the water resource bureau pump supply company or the state-run local agricultural inputs corporation. Village leaders made decisions on all aspects of water management: when and where to sink the wells, how many wells to sink, and, importantly, how much water to extract during each season. Village leaders often hired a well operator to pump water and deliver to households under their instruction. In most villages individual farmers at most contributed their labor for tubewell construction and maintenance.

Rise of private wells

The series of reforms described in the earlier section (e.g., the 1994 tax reform, the 2003 tax-for-fee reform) have also weakened village leaders' ability to invest in and maintain wells. During this period, a number of collective tubewells became inoperable because the groundwater table fell below the depths of the wells. The village collective lacked the fiscal resources to either replace the collective wells or to maintain pumps, engines or other equipment. In response, farmers began to take the place of the collective to re-establish wells and groundwater irrigation in many regions where farmers had come to rely on this resource (Wang et al., 2006). The transition of well ownership from collective ownership to private ownership took place in the macro environment in which policy makers started to gradually relax the constraints on private activities. In particular, the HRS reform has shifted income and control rights of land from the collective to the individual household.

The shift of ownership started in the 1980s and accelerated in the 1990s. The survey conducted by Wang (2000) in Hebei province showed that in the early 1980s collective ownership accounted for 93 percent of all tubewells but diminished throughout the late 1980s and 1990s. During this period, the share of private tubewells increased from 7 percent to 64 percent. This is consistent with findings from the NCWRS and the CWIM surveys (Wang et al., 2006). In 1995, 58 percent of wells in groundwater-using villages were still under collective ownership (Figure 21.4, Panel A). By 2004, the share of privately owned wells rose to 70 percent, shifting a large part of groundwater management into the hands of individual farmers. Tubewell ownership in the CWIM study area also shifted sharply from collective to private (Figure 21.4, Panel B). In 1990 collective ownership in the Hebei and Henan sample counties accounted for 51 percent of all tubewells. Between 1990 and 2004, however, the collective ownership of tubewells

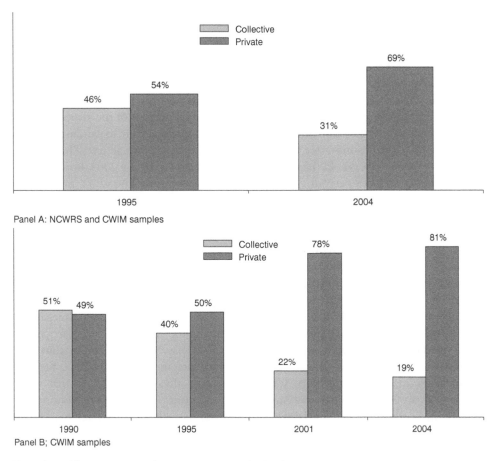

Panel A: NCWRS and CWIM samples

Panel B; CWIM samples

Figure 21.4 Changes in groundwater management institutions

diminished to 19 percent; during the same period the share of private tubewells increased from 49 percent to 81 percent. The shift in ownership is more significant in villages that face more severe water shortage problems (Wang et al., 2006). Government programs such as fiscal subsidies and loans may have also influenced the pattern of tubewell ownership (Wang et al., 2006).

The shift of tubewell ownership is the result of the establishment of new tubewells rather than ownership transfers of collective tubewells although the absolute number of collective tubewells has declined (Wang et al., 2006). In both the NCWRS and the CWIM sample, the number of private wells sunk by farmers (either an individual farmer or a group of farmers jointly) has increased significantly. The NCWRS data show that, except for Hebei province, individual ownership is the main form of private ownership (Wang et al., 2006). Shareholding is an institutional arrangement that is mostly found in Hebei province. This is probably because most wells in Hebei province are deeper than those in other provinces. Since it is more expensive to drill deeper wells and purchase the pumps and engines needed for those wells, farmers are more likely to pool their capital.

Emergence of groundwater markets

Concurrent with the trend of increasing privatization of wells is the development of groundwater markets. Although almost nonexistent before 1980, by 1995, groundwater markets were present in 9 percent of the NCWRS villages that used groundwater and had private wells (Zhang et al., 2008). Groundwater markets spread at a much faster rate over the next 10 years. By 2004, tubewell operators in 44 percent of the sample villages were selling water. At the same time that groundwater markets were expanding spatially, in villages that had groundwater markets, markets became more active (Zhang et al., 2008). In 1995 water was sold from only 5 percent of tubewells; by 2004, however, this number increased to 18 percent. Tubewell owners that were selling water were selling a large portion of the water they pumped out. In the CWIM sample area (Hebei and Henan provinces), about 80 percent of the water pumped from these wells was sold on the groundwater market (Zhang et al., 2008).

Zhang et al. (2008) summarized four characteristics of groundwater markets in rural China. First, almost all groundwater markets in China are informal in that farmers buy and sell water without a contract and their oral commitments are only enforced by social norm. The informal nature is consistent with the general environment in China where the rule of law is still weak. It significantly reduces transaction costs (such as legal fees to draw up a formal contract and the cost of enforcing the contract) which participants in the markets would have to incur otherwise. This may be one of the reasons why groundwater markets were able to grow at a fast rate in northern China. Second, groundwater markets in northern China are almost always localized in that water transactions are mostly limited to households in the same village. Third, groundwater markets in northern China are largely unregulated. Fourth, groundwater markets in northern China are largely impersonal in that most sellers do not charge different prices for different types of buyers. Generally the fee charged to irrigators is based on the electricity used rather than the volume of water, but electricity used is highly correlated with the volume of groundwater pumped and the depth of the well.

Impact on private well ownership and groundwater market

Partly because the shift to private well management during 1990s coincided with the rapid decline of water levels in aquifers, some scholars have blamed private well management for the accelerated decline in groundwater levels in northern China (Zhang and Zhao, 2003). When wells are managed by the collective, the authority associated with village leaders entails the presence of some governance structure in the groundwater sector, which is often missing in most groundwater economies including India (which is now the largest groundwater economy worldwide; Shah, 2009). Village leaders, as the custodian of the village's assets including water resources, may have an incentive to conserve groundwater for future use.[7] In contrast, when wells are controlled by farmers, the incentive of a well owner to conserve water may be limited. Given the typically large number of wells in groundwater-using villages, the incentive diminishes rapidly as the number of competitors increases. Even if the well owner wants to regulate water use, he does not have the same authority as village leaders and thus would be less effective in influencing his fellow villagers' water uses. As a result, it is entirely plausible that unregulated pumping by well owners could result in the tragedy of the commons.

Empirical analysis, however, shows that the differences in water use between farmers that depend on collective wells and farmers that depend on private wells (either as buyers or sellers) are not statistically significant. Using the 2004 panel of the CWIM data, Zhang et al. (2008) show both farmers who bought groundwater to irrigate wheat (3,241 m³/hectare) and farmers who pumped water from their own tubewells (3,571 m³/hectare) used less water than farmers relying on water from collective tubewells (3,660 m³/hectare). When regression analysis was used to control for the characteristics of villages, households and plots, the difference in water usage between farmers that used water from their own wells and farmers that relied on collective wells disappeared. There is also no evidence to support that collective well management and private well management differ in their effects on groundwater. Wang et al. (2005b) show that groundwater levels were not lower in the villages in which wells were managed by private owners.

When trying to explain this result, Huang et al. (2013) show that the hydrology of the aquifers plays a key role. If water in an aquifer is accessible not only to the village on top of the aquifer but also to neighboring villages, the water level in one village may be affected by the pumping of neighboring villages (or nearby cities) and vice versa. If this is the case, then the aquifers underlying the different villages are connected. In villages pumping from connected aquifers (called *connected villages*), users are not assured that water not used in this period is available in future periods because their neighboring villages may pump it out. In such cases even village leaders do not have an incentive to conserve water. Huang et al. (2013) test this hypothesis by including a dummy variable that equals one if a plot is in a connected village in the regression with plot level water use as the dependent variable. The regression results show that only in villages that are hydrologically isolated from other villages do we observe a higher level of water use by farmers that depend on private wells for irrigation. Farmers that pumped from private wells used 80 percent more water than those that pumped from collective wells and the difference was statistically significant at 1 percent. Given that a large share of the villages (more than 60 percent) is connected, it is not surprising to find no difference between collective well management and private well management.

The privatization of tubewells was found to induce the adjustment of crop patterns. Both descriptive analysis and multivariate regressions have found that in villages with higher shares of private wells, farmers grew less water-intensive but low-valued crops such as wheat (which requires up to five irrigations per season because its growing season does not overlap with the rainy season) and allocated a higher share of their sown areas to cash crops such as cotton, fruits and vegetables (Wang et al., 2009b). The area devoted to cash crops (mostly horticulture crops) rose from 6 to 10 percentage points (Wang et al., 2006). The shift to private tubewell ownership has facilitated the expansion of high-valued crops by augmenting the quantity of irrigation water as well as improving the reliability of water supply. The relationship between ownership and yields is less clear. The coefficient of the tubewell ownership variable is not significant in either the wheat or the maize yield equation (Wang et al., 2009b).

There is also evidence that farmers' income increases with the shift to private tubewell ownership (Wang et al., 2009b). One possible reason for the positive income effect is the shift from lower-valued wheat to higher-valued cash crops enabled by the privatization of wells. In addition, the presence of groundwater markets may have also contributed to the positive income effect. Groundwater markets provided poorer farmers with access to groundwater and thus opportunities to earn more crop income (Zhang et al., 2008). Households in the sample that buy water from groundwater markets are poorer than water-selling households. The per capita crop income of water buyers is 902 yuan, which is only 61 percent of that of tubewell owners (1,482 yuan) and 77 percent of that of farmers getting irrigation from collective tubewells (1,168 yuan).

The research on groundwater markets also sheds some light on the potential of a water pricing policy in China. Both descriptive analysis and multivariate analysis show that when farmers buy water from groundwater markets, they use less water than those that have their own tubewells or that use collective wells (Zhang et al., 2008). Water price may play a role here. In groundwater-using areas, farmers pay for water according to the number of hours that pumps are operated to irrigate their crop. Therefore, the cost of groundwater is closely related to the energy cost of lifting water out of wells (either electricity or diesel) and the depth to water in wells. Practically, groundwater is volumetrically priced (Webber et al., 2008), which makes groundwater pricing policies different from surface water policies. Farmers that buy water have higher outlays for their water than farmers that pump from their own tubewells or depend on water delivered from collective tubewells. On average water buyers paid 0.39 yuan/m^3 to irrigate wheat, which was more than two times the pumping cost well owners incurred (Zhang et al., 2008). Despite the lower level of water use, crop yields of water buyers were not negatively affected (Zhang et al., 2008). Field observations suggest that this is because those that buy water may be working harder at not wasting water. So when faced with higher water prices, farmers would exert more effort to save water and agricultural productivity may not be negatively affected.

Conclusion

During the past decade, reform-oriented institutions, such as WUAs and contracting, have largely replaced the traditional institution of collective management in village-level surface water irrigation systems. A feature unique to China is that WUAs and contractors are provided with monetary incentives to save water. WUAs have not yet achieved the broad-based participation of farmers that some advocates consider a primary goal for forming the associations. Many village leaders also serve as the leaders of WUAs, thus possibly reducing opportunities for receiving operational input and policy direction from farmers. However, we observe improved performance of irrigation systems managed by WUAs, relative to collective management, in terms of water savings, maintenance expenditures, the timeliness of water deliveries, and the rates of fee collection. Performance has also improved in systems managed by contractors, although not as substantially as in the case of WUAs.

There is still substantial room for more reform efforts. Current policy reforms serve primarily to trim the bureaucracy in irrigation water management, clarify incentives to improve service and streamline the path of fee remittances to improve the capacity for local IDs to be self-sufficient. Most surface water systems in China still suffer from a lack of volumetric pricing mechanisms to achieve efficient allocation and enable IDs to be financially self-sufficient. Research has shown that WUA managers are motivated by monetary incentives to save water, which is enabled by volumetrically measuring water at the village level. WUAs could play a role in extending volumetric pricing to more places. WUAs could also play a role in other mechanisms to improve water allocation efficiency. In water trading between agriculture and other sectors in Spain, WUAs, instead of individual farmers, are the sellers of water (Garrido et al., 2012). Chinese WUAs can also take on the role of representing farmers in the market, which will surely lower the transaction cost significantly relative to the case of individual farmers being the market participants.

In the groundwater sector, farmers have also taken over the management of groundwater from village leaders by sinking their own wells. The privatization of the wells does not seem to accelerate the decline in groundwater levels relative to collective management. The increased access to groundwater also has a positive income effect. However, this does not mean the government should leave complete control of groundwater to farmers. The development of tubewell irrigation should take into consideration the long-run effect. In particular, to reduce the threat to the sustainability of regional growth if groundwater resources are diminished, governments should put in more effort to monitor the changes in groundwater resources.

Although groundwater is practically volumetrically measured, the price does not include the scarcity value. Groundwater resources in rural China are still treated as a common property resource. Huang et al. (2010b) show that even within the agricultural sector, the cost of groundwater is below the value of the marginal product of groundwater in agricultural production. Unless large price increments are used in pricing reform, farmers are not likely to be responsive to price incentives to conserve water. Groundwater markets do play a role in increasing the allocative efficiency of groundwater because groundwater can be considered as volumetrically priced. However, so far most groundwater markets are localized and trading only occurs within the same village. Scaling up groundwater markets to be cross-village and even between rural villages and industrial sectors would further increase the allocative efficiency of groundwater use.

Notes

1 In this chapter, northern China includes north China (*huabei*, Beijing and Tianjing municipalities, Hebei, Shanxi and Inner Mongolia provinces), north-east China (*dongbei*, Liaoning, Jiling and Heilongjiang provinces), north-west China (*xibei*, Shaanxi, Gansu, Qinghai, Ningxia and Xinjiang provinces) and part of Henan province.

2 After several decades of water policies that focused on increasing water supply by constructing more canals and larger reservoirs (Ross, 1983), China's leaders have started to recognize the need to stem the rising demand for water. In the 1990s, the national government embarked on a series of water policy reforms to encourage water conservation. Since agriculture is still by far the largest user of water, these reforms generally targeted agriculture. The efforts to promote water-saving technologies did not result in a widespread adoption. For example, the proportion of sown area on which farm households adopted sprinkler irrigation was only 3 percent in northern China in 2004 (Blanke et al., 2007). One of the major reasons is the absence of economic incentives facing water users (Lohmar et al., 2003; Yang et al., 2003). Similar to the situation in many places around the world, the price of irrigation water is low in China and surface water is not volumetrically priced.

Water pricing reform is on the agenda of policy makers (e.g., General Office of the State Council of China, 2004), but progress has been slow. The prices of water in the industrial and residential sectors are on the rise. The rates of water price increments, however, have been modest. For example, the price of residential water in Beijing was increased from 3.7 yuan/m³ in 2009 to 4 yuan/m³ in 2010, which was only a 3.8 percent increment after adjusting for inflation (an 8.1 percent increase, Beijing Development and Reform Commission, 2009). In addition, the scheduled price increments of 0.3 yuan/m³ in 2011 and 2012 were not implemented due to political considerations. Moreover, the price increase is not implemented in the agricultural sector. This is because there will be strong resistance against any policy that may result in lower rural incomes. China has made remarkable progress in alleviating poverty in its rural areas in the past and the leaders are definitely intent on continuing to alleviate poverty in rural China (Rozelle et al., 2003). In recent years, many irrigation districts were encouraged by the national government to reduce the price of surface water charged to farmers through improving irrigation infrastructure and better water management (Ministry of Water Resources of China, 2005).

A variety of projects that examine ways to allocate water rights and promote water conservation are also being carried out in China (Speed, 2009; Xie et al., 2009). One such case is the trading between the cities of Dongyang and Yiwu in Zhejiang Province (Gao, 2006). Dongyang signed China's first intercity trading agreement that permanently transfers a total of 50 million m³ of water each year to the nearby fast-growing Yiwu at a price of 200 million yuan. So far water transfers between agriculture and other sectors are restricted to the "saved" water (Speed, 2009). For example, in the Yellow River Basin, the Inner Mongolia Water Resources Department initiated pilot projects in which water saved through channel lining is transferred to downstream industries, with the costs of lining paid by industrial beneficiaries (Speed, 2009; Xie et al., 2009). Farmers also benefited from reduced irrigation fees as they no longer had to pay for water that was lost during conveyance. In other projects, officials experimented with policies that granted farmers the use rights to water at a low price and then allowed them to sell these rights to other users or back to the water management authorities at a higher price (Jia and Duan, 2006). These projects were less successful, mostly because the high transaction cost easily outweighed the benefit accrued to individual farmers. Although water trading in China has the potential to increase the efficiency of water allocation, significant institutional barriers still need to be overcome before it can be scaled up from just pilot projects to a nationwide effort. One such challenge is that there is no consensus over how to determine or allocate water rights (Jia and Duan, 2006; Liu, 2003; Speed, 2009; WET, 2006).

3 The CWIM survey was a panel survey that tracked a sample of randomly selected villages in 2001, 2004 and 2007. Enumerators interviewed village leaders, surface water irrigation managers, well operators and farmers in 80 villages in Hebei, Ningxia and Henan provinces (Figure 21.2). Separate survey questionnaires were used for each type of respondent. The NCWRS survey was conducted in January 2005. A stratified random sampling strategy was used to select 401 villages from six provinces: Inner Mongolia, Hebei, Henan, Liaoning, Shaanxi and Shanxi (Figure 21.2). The strata were defined using indices of water scarcity. Only village leaders were interviewed in the NCWRS survey due to limited time and budget. An extended version of the CWIM village leader questionnaire was used. Sample provinces in the CWIM survey and the NCWRS survey cover all or part of four major river basins in northern China: the Hai River basin (Hebei), the Yellow River basin (Ningxia, Inner Mongolia, Shanxi, Shaanxi, Henan), the northern bank of the Huai River basin (Henan) and the Songliao River basin in the north-east (Liaoning).

4 Corvée is a term that refers to an obligation imposed on inhabitants of a district to perform services, such as the repair of roads, for little or no pay.

5 The Bank survey was also conducted by the authors of this chapter in 2006. Three provinces were selected: Gansu, Hubei and Hunan (Figure 20.3). In each province, 10 villages were randomly selected from a list of the Bank WUA candidate villages. These are called the Bank WUAs or Bank villages. A set of non-Bank WUA villages and a set of collectively-managed villages were also selected from within the same physical proximity as the Bank WUAs. The "same physical proximity" was defined as 10 kilometers or the closest set of 10 villages (if there were not any within a 10 kilometer radius). No contracting villages were found on these study sties. In total the sample includes 30 Bank WUAs, 15 non-Bank WUAs and

15 collectively-managed villages. In each village, village leaders and the head of the WUA board were interviewed. A focus group with farmers was also conducted with groups of five farmers.

6 The World Bank has identified five principles: adequate and reliable water supply; legal status and participation; WUAs organized within hydraulic boundaries; water deliveries can be measured volumetrically; WUA equitably collects water charges from members (Wang et al., 2010).

7 Several studies (Kung et al., 2009; Zhang, 2007) have quantified empirically that the decision making of village leaders is more consistent with agents that care as much about status and other intangible objectives as they do about maximizing profits and implementing policies of upper level governments; leaders care about acting as the village head, and protecting and furthering the interests of the village, which leads to higher status and the prestige that comes with such a role. In fact, village leaders often act like traditional village heads or elders of the village that rise to their positions because they care about villagers and are able to help the village solve its problems. Zhang (2007) further shows econometrically that village leaders did not gain from their village cadre status in terms of income, assets, consumption, access to off-farm employment for their families or profits for their own business. Often they are better off because of their human and physical capital, family background, and other unobservable heterogeneities. Of course there were always the evil, self-serving village leaders that were corrupt and exercised power arbitrarily for their own benefit. These stories, however, according to Shue (1988) and Oi (1989), are the exception rather than the norm and the ones the media tends to pick up.

There are also reasons to believe that village leaders may consider the long-run use of the resource (Huang et al., 2013). First, leaders are members of the village and they will be living in the village even after they step down as leaders. Their children will live here too (mobility in China traditionally has been very low). Close kinship and friendship ties could make them take a long-run view. Second, many village leaders—even those facing elections—have been in office for many years. In the villages in the CWIM sample in Hebei and Henan, the average tenure of a village party secretary was 11.4 years; the village committee head had an average tenure of 10.4 years (Huang et al., 2013). Many of the leaders told the enumerators that they expect to be in office for a long time.

References

Beijing Development and Reform Commission. (2009) No. 2555: Notice on the adjustment of water resource fee and waste water treatment fee of residential water use in Beijing (In Chinese), December 21, 2009

Blanke, A., Rozelle, S., Lohmar, B., Wang, J. and Huang, J. (2007) 'Water saving technology and saving water in China', *Agricultural Water Management*, vol 87, pp. 139–150.

China. (2002) Water Law of the People's Republic of China. China.

Economic and Social Survey of Asia and the Pacific (ESCAP) (2010) *Statistical Yearbook for Asia and the Pacific 2009*. United Nations Publication. Bangkok.

Food and Agriculture Organization (FAO). AQUASTAT data, http://data.worldbank.org/indicator/ER.H2O.INTR.PC. Accessed April 19, 2013.

Gao, E. (2006) *Water Rights System Development in China* (in Chinese), Beijing: China Water and Hydropower Publishing.

Garrido A., Rey, D. and Calatrava J. (2012) 'Water trading in Spain', in de Stefano, L. and Llamas, M. R. (eds) *Water, Agriculture and the Environment in Spain: can we square the circle?* Leiden, The Netherlands: CRC Press/Balkema, Taylor and Francis. pp. 205–216.

General Office of the State Council of China (2004) Notice on pushing the reform of water prices to promote water savings and water resources protection. No. 36. Beijing: General Office of the State Council of China

Huang, Q., Wang, J., Polasky, S., Rozelle, S. and Liu, Y. (2013) 'The effects of well management and the nature of the aquifer on groundwater resources', *American Journal of Agricultural Economics*, vol 95, no 1, pp. 94–116.

Huang, Q., Wang, J., Easter, K. W. and Rozelle, S. (2010a) 'Empirical assessment of water management institutions in northern China', *Agricultural Water Management*, vol 98, no 2, pp. 361–369.

Huang, Q., Rozelle, S., Howitt, R., Wang, J. and Huang, J. (2010b) 'Irrigation water demand and implications for water pricing policy in rural China', *Environment and Development Economics*, vol 15, no 3, pp. 293–319.

Huang, Q., Rozelle, S., Wang, J. and Huang, J. (2009) 'Water management institutional reform: a representative look at Northern China', *Agricultural Water Management*, vol 96, no 2, pp. 215–225.

Huang, Q., Rozelle, S., Huang, J., Wang, J. and Msangi, S. (2008) 'Water management reform and the choice of the contractual form in rural China', *Environment and Development Economics*, vol 13, no 2, pp. 171–200.

Huang, Q., Rozelle, S., Lohmar, B., Huang, J. and Wang, J. (2006) 'Irrigation, agricultural performance and poverty reduction in China', *Food Policy*, vol 31, no 1, pp. 30–52.

Jia, Y. and Duan, J. (2006) 'Impact analysis of the pilot projects of water saving society at Zhanye City', *Water Resources Planning and Design*, vol 2, no 1, 5–10 (in Chinese)

Jiang, Y. (2009) 'China's water scarcity', *Journal of Environmental Management*, vol 90, no 11, pp. 3185–3196.

Kung, J., Cai, Y. S. and Sun, X. L. (2009) 'Rural cadres and governance in China: incentive, institution and accountability', *China Journal*, vol 62, pp. 61–77.

Lin, J. Y. (1992) 'Rural reforms and agricultural growth in China', *American Economic Review*, vol 82, pp. 34–51.

Lin, Z. (2003) 'Participatory Management by Farmers – Local Incentives for Self-financing Irrigation and Drainage Districts in China', Environment and Social Development, East Asia and Pacific Region Discussion Paper, World Bank, Washington, D.C.

Liu, B. (2003) 'Water Rights in China', paper prepared for the International Working Conference on Water Rights: Institutional Options for Improving Water Allocation, Hanoi, Vietnam, February, 12–15, 2003.

Lohmar, B., Huang, Q., Lei, B. and Gao, Z. (2008) 'Chapter 12. Water pricing policies and recent reforms in China: the conflict between conservation and other policy goals', in Molle, F. and Berkoff, J. (eds) *Irrigation Water Pricing: The Gap Between Theory and Practice*, Wallingford: CAB International.

Lohmar, B., Wang, J., Rozelle, S., Huang, J. and Dawe, D. (2003) 'China's agricultural water policy reforms: increasing investment, resolving conflicts, and revising incentives', Agriculture Information Bulletin No. 782, Market and Trade Economics Division, Economic Research Service, U.S. Department of Agriculture, Washington, DC.

Ministry of Water Resources of China. (2011) Water Resources Bulletin 2011.

Ministry of Water Resources of China. (2007) Water Resources Bulletin 2007.

Ministry of Water Resources of China. (2005) No. 2247: Management of pilot projects on reconstructing tertiary canal network for supplying agricultural water use (in Chinese), August 31, 2005

Ministry of Water Resources of China. (2004) Water Resources Bulletin 2004.

Ministry of Water Resources of China. (2000) Water Resources Bulletin 2000.

Ministry of Water Resources of China. (1997) Water Resources Bulletin 1997.

NBSC (National Bureau of Statistics of China) 2013. National Data. 2013. http://data.stats.gov.cn/workspace/index?m=hgnd. Accessed February 16, 2014.

National Development and Reform Commission of China. (2008). China's Food Security Medium and Long-term Planning Outline (2008-2020). Document No. [2008]24.

Naughton, B. (1995) *Growing Out of the Plan: Chinese economic reform, 1978-1993*, New York: Cambridge University Press.

Oi, J. C. (1989) *State and Peasant in Contemporary China: the political economy of village government*, Berkeley, CA: University of California Press.

Oi, J. C., Babiarz, K. S., Zhang, L., Luo, R. and Rozelle, S. (2012) 'Shifting fiscal control to limit cadre power in China's townships and villages', *The China Quarterly*, vol 211, pp. 649–675.

Park, A. and Rozelle, S. (1998) 'Reforming state-market relations in rural China', *Economics of Transition,* vol 6, no 2, pp. 461–480.

Reidinger, R. (2002) Participatory irrigation management reform: Self-managing irrigation and drainage districts in China. Sixth International Seminar on Participatory Irrigation Management, Beijing, China.

Ross, L. (1983) 'Changes in water policy in the People's Republic of China', *Water Resources Bulletin,* vol 19, pp. 69–72.

Rozelle, S., Zhang, L. and Huang, J. (2003) 'China's war on poverty', in Hope, N. C., Yang, D. T. and Li, M. Y. (eds) *How Far across the River? Chinese Policy Reform at the Millennium,* Stanford, CA: Stanford University Press.

Shah, T. (2009) *Taming the Anarchy: Groundwater governance in South Asia,* Washington DC: Resources for the Future.

Shue, V. (1988) *The Reach of the State: Sketches of the Chinese body politic,* Stanford, CA: Stanford University Press.

Speed, R. (2009) 'Transferring and trading water rights in the People's Republic of China', *International Journal of Water Resources Development,* vol 25, no 2, pp. 269–281.

Wang, J., Huang, J., Zhang, L., Huang, Q. and Rozelle, S. (2010) 'Water governance and water use efficiency: the five principles of WUA management and performance in China', *Journal of the American Water Resources Association,* vol 46, no 4, pp. 665–685.

Wang, J., Huang, J., Rozelle, S., Huang, Q. and Zhang, L. (2009a) 'Understanding the water crisis in northern China: what government and farmers are doing', *International Journal of Water Resources Development,* vol 25 no 1, pp. 141–158.

Wang, J., Huang, J., Huang, Q., Rozelle, S. and Farnsworth, H.F. (2009b) 'The evolution of groundwater governance: productivity, equity and changes in the level of China's aquifers', *Quarterly Journal of Engineering Geology and Hydrogeology,* vol 42, no 3, pp. 267–280.

Wang, J., Huang, J., Huang, Q. and Rozelle, S. (2006) 'Privatization of tubewells in North China: Determinants and impacts on irrigated area, productivity and the water table', *Hydrogeology Journal,* vol 14, no 3, pp. 275–285.

Wang, J., Xu, Z., Huang, J. and Rozelle, S. (2005a) 'Incentives in water management reform: assessing the effect on water use, production, and poverty in the Yellow River Basin', *Environment and Development Economics,* vol 10, pp. 769–799.

Wang, J., Huang, J. and Rozelle, S. (2005b) 'Evolution of tubewell ownership and production in the North China Plain', *Australian Journal of Agricultural and Resource Economics,* vol 49, no 2, pp.177–195.

Wang, J. (2000) 'Innovation in groundwater irrigation system property rights, efficiency and policies', Ph.D. dissertation, Center for Chinese Agricultural Policy, Academy of Chinese Agricultural Sciences.

Wang, S. (1997) 'China's 1994 fiscal reform: an initial assessment', *Asian Survey,* vol 37, pp. 801–817.

Webber, M., Barnett, J., Finlayson, B. and Wang, M. (2008) 'Pricing China's irrigation water', *Global Environmental Change,* vol 18, no 4, pp. 617–625.

WET (2006) Water Entitlements and Trading Project (WET Phase 1) Final Report November 2006 (in English and Chinese), Beijing, Ministry of Water Resources, People's Republic of China and Canberra, Department of Agriculture, Fisheries and Forestry, Australian Government). Available at: http://www.environment.gov.au/water/action/international/wet1.html. Accessed September 30, 2013.

World Bank. (2003a) China – Yangtze Basin Water Resources Project: Implementation Completion and Results. The World Bank, No. 25903.

World Bank. (2003b) 'Water User Association Development in China: Participatory Management Practice under Bank-Supported Projects and Beyond', no 83.

Xie, J., Liebenthal, A., Warford, J. J., Dixon, J. A., Wang, M., Gao, S., Wang, S., Jiang, Y. and Ma, Z. (2009) 'Addressing China's water scarcity; recommendations for selected water resource management issues', World Bank, Washington, DC.

Xie, M. (2007) 'Global Development of Farmer Water User Associations (WUAs): Lessons from South-East Asia', Proceedings of the Regional Workshop on WUAs Development: Water Users' Associations Development in Southeastern European Countries – Bucharest, Romania.

Yang, H., Zhang, X. and Zehnder, A. J. B. (2003) 'Water scarcity, pricing mechanism and institutional reform in northern China irrigated agriculture', *Agricultural Water Management*, vol 61, no 2, pp. 143–161.

Zhang, J. (2007) 'Make rural China run: three essays on entrepreneurs, regulators and cadres', Ph.D. dissertation, University of California, Davis.

Zhang, L., Wang, J., Huang, J. and Rozelle, S. (2008) 'Development of groundwater markets in China: a glimpse into progress to date', *World Development*, vol 36, no 4, pp. 706–726.

Zhang, X. and Zhao, Y. (2003) 'Existing problems and counter measures on exploitation and utilization of groundwater in Hengshui City', *Groundwater*, vol 25, no 2, pp. 87–89.

Zheng, T. (2002) 'Several issues on the reform of the pricing of water supplied in irrigation projects', Presented on the China Water Pricing Reform Workshop, Haikou, Hainan Province, China.

22

IMPLEMENTING THE EUROPEAN WATER FRAMEWORK DIRECTIVE IN GREECE

An integrated socio-economic approach and remaining obstacles

Phoebe Koundouri and Osiel González Dávila

Introduction

The purpose of the EU Water Framework Directive (WFD) is to establish a common framework for the protection of inland surface waters, transitional waters, coastal waters and groundwater in order to prevent further deterioration and to enhance the status of aquatic ecosystems. Hence, achieving "good status" for all waters by 2015 is one of its key objectives (EC 2000). Each Member State is responsible for the implementation of this directive (adoption of implementing measures before a specified deadline, conformity and correct application) within its own legal system. Three main steps for the implementation of the WFD can be identified: 1) setting of ecological standards, 2) identification of anthropogenic pressures and 3) adoption of corrective measures. However, the implementation of the WFD in Greece has faced several challenges that range from technical problems to complex institutional issues. The Commission of the European Communities is responsible for ensuring that EU law is properly applied. Thus, if a Member State fails to comply with EU law, the Commission has powers of its own to try to bring the infringement to an end and, where necessary, may refer the case to the European Court of Justice. The European Commission has referred Greece to the EU Court of Justice a number of times for failing to take measures to guarantee that different water-related directives are correctly applied. In order to understand the requirements and challenges in implementation of the WFD in Greece the evolution of the different water-related directives and regulations will be analysed in this chapter. Then, we will move on to describe the situation of the water sector in Greece. In the context of a legal action of "Non-Conformity" started by the EC against Greece (case C-264/07), the reasons for not catching up fast enough with the implementation of the WFD will be outlined. The following issues will also be discussed: a) a methodology that enables rapid assessment of the status quo water situation in each Greek catchment and

b) the implementation of this methodology on each of the fourteen Greek river basin districts (RBDs) and related outcomes. As will be explained in the following sections, it is evident that reforms in the current pricing policy are required in order for the water bodies to reach good ecological status and to ensure full recovery of the cost of water services. In general, it can be expected that Southern European countries with weak institutional structure and limited financial resources will face further difficulties during the implementation of the WFD.

The evolution of European legislation on water resources

In this section, we analyse how the European legislation on water resources has evolved in the last fifty years. This is important to understand the requirements and challenges in implementation of the Water Framework Directive in Greece. In the literature it is widely acknowledged that the evolution of European legislation and policies for water resources have gone through three different waves (see Kallis and Nijkamp 1999; CEC 2000; Kaika 2003; Dworak et al. 2006; REC 2008). Table 22.1 shows the directives passed in each of the three waves of European legislation on water resources. In the following sections each of the three waves will be presented and analysed.

Early European environmental legislation and first wave of European legislation on water resources

The first wave of legislation started with the launch of the First Environmental Action Programme (EAP) 1973–1976 (OJ C 112, 20.12.73). This early European environmental legislation established the objectives and principles of the environmental policies of the European Commission (EC). The programme was mainly concerned with water protection and waste emissions and comprised a special reference to agriculture and spatial planning (Hey 2005; Dworak et al. 2006). In consequence, a number of directives were enacted in order to reduce and prevent water pollution based primarily on a regulatory approach. For example, the Directive on Water Pollution by Discharges of Certain Dangerous Substances (76/464/EEC) was one of the first water-related directives to be decreed. Its aim was the regulation of impending water pollution by chemicals produced in Europe at that time. The concepts of list I and list II substances were introduced for the first time in the annex of this directive.

Pollutants included in list I include substances regarded as particularly dangerous on the basis of their persistence, toxicity and bioaccumulation. Pollutants in list II included substances that are considered less dangerous but which, nevertheless, can have a deleterious effect on the aquatic environment (Horth et al. 2003). The directive covered discharges to surface waters, coastal waters and groundwater. The protection of groundwater was removed from the directive (76/464/EEC) and a new directive on the protection of groundwater against pollution caused by certain dangerous substances (80/68/EEC) was established (CEC 2000). In general, two different approaches to tackle water pollution have been used: the water quality objective (WQO) approach which defines the minimum quality requirements of water to limit the cumulative impact of emissions, both from point sources and diffuse sources. This approach was mainly used in the first wave of water directives. On the other hand, the emission limit value approach

Table 22.1 The three waves of European legislation on water resources

Early European water policy	
1973	First Environmental Action Programme (EAP) 1973–1976 (OJ C 112, 20.12.73)

First wave of legislation	
Focus on water quality objectives (WQO)	
1975	Surface Water Directive (75/440/EEC)
1976	Bathing Water Directive (76/160/EEC)
1976	Directive on Water Pollution by Discharges of Certain Dangerous Substances (76/464/EEC)
1978	Fish Water Directive (78/659/EEC)
1979	Shellfish Water Directive (79/923/EEC)
1980	Directive on the protection of groundwater against pollution caused by certain dangerous substances (80/68/EEC)
1980	Drinking Water Directive (80/778/EEC)

Second wave of legislation	
Focus on emission limit value approach (ELV)	
1991/1998	Urban Waste Water Management Directive (91/271/EEC and 98/15/EEC)
1991	Nitrates Directive (91/676/EEC)
1996	Directive for Integrated Pollution and Prevention Control (96/61/EC)

Third wave of legislation	
Integrated approach	
1996	Commission's Communication on European Water Policy (COM(96)0059 – C4-0144/96)
1997	Commission's Proposal for a Water Framework Directive (COM(97) 49)
1997	Commission's amended proposal following consultation (COM(97) 614)
1998	Commission's further amendment of proposal following consultation (COM(98) 76)
2000	Water Framework Directive (WFD) (2000/60/EC)
2006	New Bathing Water Directive (2006/7/EC)
2006	Groundwater Directive (2006/118/EC)
2007	Floods Directive (2007/60/EC)
2008	Marine Strategy Framework Directive (MSFD) (2008/56/EC)
2008	Directive on Environmental Quality Standards (2008/105/EC)
2009	Directive on the Sustainable Use of Pesticides (2009/128/EC)
2010	Industrial Emissions Directive (2010/75/EU)

Source: Adapted from Kaika (2003), Dworak et al. (2006) and EC (2012b).

(ELV) was mainly used in the second wave of water legislation during the 1990s and focuses on the maximum allowed quantities of pollutants that may be discharged from a particular source into the aquatic environment (REC 2008).

Second wave of European legislation on water resources

In the first half of the 1990s, the increasing eutrophication of the ocean and a deteriorating state of water resources were considered the major water-related problems within the EU. The second wave of European legislation started in 1991 and occurred after a review of the existing regulations and the identification of gaps and required improvements. In this second wave the directives defined quality standards, which had to be achieved through certain measures for controlling emission levels (Kaika 2003; Dworak et al. 2006). Two new directives were adopted and set stringent rules on the treatment of waste water and the use of nitrates in agriculture: the Urban Waste Water Management Directive (91/271/EEC and 98/15/EEC) established legally binding measures at the community level in order to regulate the collection and treatment of urban waste water and the discharge of industrial waste water from the agro-food industry. On the other hand, the Nitrates Directive (91/676/EEC) complements the Urban Waste Water Directive by reducing and preventing the nitrates pollution of water from agricultural sources (chemical fertilizer and livestock manure), to protect drinking water supplies and to prevent eutrophication (Dworak et al. 2006). Other relevant amendments were done in the Drinking Water and Bathing Water Directives (in 1994 and 1995 respectively), in order to update them. Further, in 1994 a proposal for an ecological quality of water directive was drafted (REC 2008).

Third wave of European legislation on water resources

The third wave of European water legislation started in 1996. A communication of the European Commission on water policy of the Community (COM(96)0059 – C4-0144/96) was published in February that year. This document concluded that in order to improve the efficiency of water protection legislation a Water Framework Directive was required (CEC 1996). After four years of analysis and negotiations, the WFD was passed by the EU's Parliament in September 2000 and came into force in December 2000 (Kaika 2003). The aim of this directive is to achieve good ecological status of all waters by 2015. In contrast to the previous waves of legislation, the area covered by this directive extends to all aquatic systems, surface waters (rivers and lakes), groundwater and coastal waters. The WFD meant a radical shift to measure quality of all waters using a range of biological communities rather than the more limited aspects of chemical quality (REC 2008; Moss 2008). The directive (2007/60/EC) on the assessment and management of flood risks entered into force on 26 November 2007. Its aim is to reduce and manage the risks that floods pose to human health, the environment, cultural heritage and economic activity. It requires that Member States prepare flood risk maps in order to assess the extent and assets and humans at risk in relevant areas and to take adequate and coordinated measures to reduce this risk. The Marine Strategy Framework Directive (2008/56/EC) was published in the Official Journal of the European Union on 17 June 2008. It states that pressure on natural marine resources and the demand for marine ecological services are too high. Thus, its main goal is to achieve good environmental status of the EU's marine waters by 2020 and to protect the resource base upon which marine-related economic and social activities depend. The Groundwater Directive (2006/118/EC) is a WFD daughter since groundwater protection should be tackled separately (see WFD Article 16). It was passed by the European parliament on 12 December 2006 and entered into force on 16 January 2007. The directive aims to prevent the pollution of groundwater from agricultural residues such as pesticides and other harmful substances (REC 2008).

The Water Framework Directive

As established in its first article, the purpose of the WFD is to establish a common framework for the protection of inland surface waters, transitional waters, coastal waters and groundwater in order to prevent further deterioration and to enhance the status of aquatic ecosystems and to contribute to mitigating the effects of floods and droughts (EC 2000). One of the key objectives of the directive is to achieve good ecological status for all waters by 2015, and preserving such status where it already exists. The directive classifies ecological status of aquatic habitats on a scale from high (effectively pristine) to bad and includes intermediate steps of good, moderate and poor. The concept "good ecological status" implies that water can be used as long as the ecological function of the water body is not significantly impaired (Moss 2008; REC 2008). The WFD recognizes the need to approach the interaction of anthropogenic activities and water resources in an integrated approach in order to attain sustainable water resources management. It should be noted that the WFD combines the two approaches to pollution control previously discussed: ELVs and WQOs. Both are used to mutually reinforce each other and, in any particular situation, the more rigorous approach is applied (Kaika 2003; REC 2008). In particular, the WFD establishes in its Article 10 that all Member States shall ensure the establishment and/or implementation of emission controls based on best available techniques, or relevant emission limit values (ELV), or in the case of diffuse impacts the controls including, as appropriate, best environmental practices described in the water-related directives (e.g. the Urban Waste Water Management Directive (91/271/EEC and 98/15/EEC) and the Nitrates Directive (91/676/EEC)) and any other relevant Community legislation. Thus, the WFD provides an integrated framework for water resources management and protection in Europe, both in terms of quality and quantity to achieve the objective of good ecological status (EC 2000). Further, the role of economics in reaching environmental and ecological objectives is explicitly acknowledged in this directive and requires the application of economic principles, approaches, tools and instruments at river basin district (RBD) level. Public participation in decisions is also required. The economic elements of the WFD are discussed in the following articles:

- Article 5. Characteristics of the river basin district, review of environmental impact of human activity and economic analysis of water use
- Article 9. Recovery of costs for water services
- Article 11. Program of measures
- Annex III. Economic analysis.

Table 22.2 Total economic cost of water

Financial cost	Costs of providing and administering water services: capital cost, operation cost, maintenance cost and administrative cost.
Environmental cost	The environmental cost represents the costs of damage that water uses impose on the environment and ecosystems and those who use the environment (e.g. a reduction in the ecological quality of aquatic ecosystems or the salinization and degradation of productive soils).
Resource cost	Resource cost represents the costs of foregone opportunities which other uses suffer due to the depletion of the resource beyond its natural rate of recharge or recovery (e.g. linked to the over-abstraction of groundwater).

Source: Koundouri, Kountouris and Remoundou (2010).

In addition, since the 2015 deadline to achieve good water status in all water bodies of the EU is be unrealistic, the WFD contemplates the use of disproportionate cost analysis in terms of extended deadlines and less stringent environmental objectives (Esteban, Le Quesne and Strosser 2006). Article 4.4 establishes that an extension to the 2015 deadline is permitted if "completing the improvements within the timescale would be disproportionately expensive" and Article 4.5 states that "Member States may aim to achieve less stringent environmental objectives for specific bodies of water when they are so affected by human activity… or their natural condition is such that the achievement of these objectives would be infeasible or disproportionately expensive" (EC 2000:10). On the other hand, it should be noted that Article 9 states that Member States "shall take account of the principle of recovery of the costs of water services, including environmental and resource costs, having regard to the economic analysis conducted according to Annex III, and in accordance in particular with the polluter pays principle" (EC 2000:12). The environmental cost reflects social welfare losses associated with water quality deterioration, caused by the water uses, while the resource cost represents additional costs required to cover water demand under water deficits due to the overexploitation of available water resources. Furthermore, the WFD also states that the cost recovery of water services should be analysed for different water uses, which should be at least disaggregated into households, industry and agriculture. Table 22.2 shows the disaggregation of the total cost of water services.

Challenges in the implementation of the Water Framework Directive

The implementation of the WFD has faced several challenges. Almost all EU Member States have spent substantial resources to develop tools, to gain the required data and to prepare River Basin Management Plans (RBMPs) by 2009 as required under Article 13 (Moss 2008). According to the Commission's Compliance Report (EC 2007), one of the main deficiencies in the WFD implementation is the economic assessment. According to Annex III of the directive, the economic analysis reports should contain adequate information on the major drivers and pressures in each RBD and on the contribution of water uses in the recovery of costs consistent with the polluter pays principle, to enable the selection of the program of measures on a cost-effectiveness basis (EC 2000). Even though all EU Member States sent country reports in accordance to Article 5, half of them did not supply information at all on cost recovery. This reflects the informational and methodological difficulties that Member States face when implementing the economic elements of the WFD. The European Commission has taken Belgium, Denmark, Greece and Portugal to court over their failure to comply with EU water legislation and submit plans for managing their river basins. These plans should have been adopted by 22 December 2009 at the latest. Besides, according to the Report from the Commission to the European Parliament on the implementation of the WFD, the assessment of the RMBPs

> shows the poor quality of the assessment of costs and benefits. A strong improvement in this area and the definition of a shared methodology for the calculation of costs (including environmental and resource costs) and benefits (including ecosystem services) is necessary. Otherwise, it will be possible neither to ensure the implementation of effective pricing policies nor to avoid disproportionate and inadequate measures.
>
> (EC 2012a)

Finally, according to Article 9 of the directive, by 2010 Member States should have introduced pricing policies and economic instruments with the element of cost-recovery for the environment's benefit. However, "there are very few Member States that have implemented a transparent recovery of environmental and resource costs. Cost recovery is implemented, to a greater or lesser extent, in households and industry. For agriculture, in many areas, water is charged only to a limited extent" (EC 2012a). When available, cost-recovery levels vary significantly. Member States that have provided information on households have indicated a cost-recovery rate of services for households between 70 and 100 percent. For industry, the Member States providing information reported a cost-recovery rate between 40 and 100 percent. For agriculture the cost-recovery rate is reported to vary between 1 and 100 percent (WWF 2006). Therefore, the EC (2012a) report suggests the EU Member States improve cost–benefit assessments to ensure cost recovery and to ensure the transparency and fairness of water pricing policies. Some recent reviews on the implementation of water pricing in Europe reach similar conclusions. Despite the guidelines provided by the WFD and the general consensus about the importance of cost recovery and incentive pricing for an effective water policy, tariff designs and price levels between EU Members and even between RBDs remain largely heterogeneous (EC 2012c; EEA 2013). In the agricultural sector (and despite its importance as water user) a significant portion of water abstractions is not priced and the recovery of the operation and maintenance costs of water services remains limited: "In more than one third of the Member States, farmers do not pay for their water abstractions" (EC 2012c:45). This phenomenon is prevalent in water stressed Southern European Member States.

The water situation in Greece and the implementation of the EU WFD

The political and administrative structures in Greece can be described as hierarchical and centralized with institutions dependent mainly on the state and the political parties. There is little participation of the civil society and lack of agreements among political elites. The central government determines the planning and the allocation of resources, considerably limiting the participation of other actors to provide wider legitimization. In consequence, the introduction of participatory arrangements required for the implementation of different European policies (including the WFD) has been very difficult (Demetropoulou et al. 2010). In the late 1980s emerged the first attempts to decentralize and to include integrated and participatory forms of water governance at water basin level with Law 1739/87 "for the Management of Water Resources." Its main goal is the reservation of adequate water supply to satisfy the present and future demand for different water uses and defined and established procedures and structures that permitted water management on a national and a regional scale (Sofios et al. 2008; Demetropoulou et al. 2010). Greece has an area of 131.957 km^2 and was divided into 45 river basins (RBs) (see Figure 22.1) and grouped into 14 RBDs (see Figure 22.2). The Law 3199/2003 integrated the WFD into the national legislation of Greece (MoEPPW 2003). In Greece, water supply, waste water collection, treatment and disposal are public services. However, the top-down approach of policy-making and implementation is prevalent and the coordination between the different governmental levels has remained incoherent (Demetropoulou et al. 2010). In Athens and Thessaloniki, water companies controlled by the Ministry of Environment Physical Planning and Public Works own and operate the treatment plants on a not-for-profit basis. In other cities, water supply is managed by municipal companies operating as private enterprises DEYA (Municipal Enterprise for Water Supply and Sewerage) but owned by the municipalities. Each DEYA determines its pricing policy on the basis of their cost and is approved by the municipal council. The mean price per cubic meter of water in Greece is estimated at €1.27. Irrigation water prices are set based on private cost criteria. Forty percent of Greece's abstraction and distribution of irrigation water is provided by local irrigation companies and the mean price per irrigated thousand square meters is €13.73 (Koundouri et al. 2014).

Table 22.3 shows the most significant water uses in each RBD. The demand of water for residential use was calculated taking into account the resident population (200 liters person/day) in each RBD and all the nights tourists spent in each region (300 liters person/night since the flow of tourists is higher in summer) according to the 2001 census. The highest demand for water for residential use was found in Attica. Not surprisingly, this RBD also has the biggest population. The demand for water for irrigation was calculated taking into account crop needs for water. Then, based on the cultivation area of each crop species in each RBD, the water demand was estimated. The highest demand of water for irrigation was found in Thessaly. The water requirements for industrial use were calculated according to data on water consumption in industrial areas. The highest demand of water for the industry was found in Central Macedonia.

The need for a "quick appraisal"

In November 2007, the European Commission initiated legal action of "Non-Conformity" against Greece (case C-264/07). In this action it is stated that

> by failing to draw up by 22 December 2004 for each river basin district falling within its territory an analysis of its characteristics, a review of the impact of

Figure 22.1 Greek river basins (source: http://www.minenv.gr/nera/ WFD Article 3 report – Greek maps)

Figure 22.2 Greek river basin districts (source: http://www.minenv.gr/nera/ WFD Article 3 report – Greek maps)
01: West Peloponnese; 02: North Peloponnese; 03: East Peloponnese; 04: West Sterea Ellada; 05: Epirus; 06: Attica; 07: East Sterea Ellada; 08: Thessaly; 09: West Macedonia; 10: Central Macedonia; 11: East Macedonia; 12: Thrace; 13: Crete; 14: Aegean Islands

Table 22.3 Economic analysis of water uses and pressures in each RBD

River basin district	Population (2001)	Area (km²)	Demand for residential use (hm³/year)	Demand for irrigation (hm³/year)	Demand for industry (hm³/year)
1. West Peloponnese	331 180	7 301	23.0	201.0	3.000
2. North Peloponnese	615 288	7 310	36.7	395.3	3.000
3. East Peloponnese	288 285	8 477	22.1	324.9	0.030
4. West Sterea Ellada	312 516	10 199	22.4	366.5	0.350
5. Epirus	464 093	10 026	33.9	127.4	1.000
6. Attica	3 737 959	3 207	400.0	99.0	1.500
7. East Sterea Ellada	577 955	12 341	41.6	773.7	12.600
8. Thessaly	750 445	13 377	69.0	1,550.0	0.054
9. West Macedonia	596 891	13 440	43.7	609.4	30.000
10. Central Macedonia	1 362 190	10 389	99.8	527.6	80.000
11. East Macedonia	412 732	7 280	32.0	627.0	0.321
12. Thrace	404 182	11 177	27.9	825.2	11.000
13. Crete	601 131	8 335	42.33	320.0	4.100
14. Aegean Islands	508 807	9 103	37.19	80.2	1.240

Source: MoEPPW (2008).

human activity on the status of surface waters and on groundwater and an economic analysis of water use, in accordance with the technical specifications set out in Annexes II and III, the Hellenic Republic has failed to fulfil its obligations under Article 5(1) of Directive 2000/60/EC establishing a framework for Community action in the field of water policy, while, by failing to submit summary reports of the analyses required under that article, it has also failed to fulfil its obligations under Article 15(2) of that directive.

In response, the Ministry of Environment, Physical Planning and Public Works[1] (MoEPPW) financed and supervised a series of studies required for the implementation of the WFD. A rapid-appraisal approach was required because of the severe information deficiencies and a time frame limit of two months defined by the ministry. The outcome was the Greek Report on the Implementation of Article 5 of the WFD (MoEPPW 2008). One of major challenges when producing this report was the significant lack of information. The preliminary analysis of water uses, pressures and impacts, under the first step of the implementation procedure, aiming to inform and guide the subsequent economic analysis was fragmentary. The only available source of information with regard to water uses in each RBD was a master plan study conducted by the Ministry of Environment that only contained general socio-economic information. Financial data from the drinking and irrigation water companies were not always available since their legal status does not oblige them to report their economic elements. At the time of the study, MoEPPW classified water quality in each RB in one of three categories:

Table 22.4 Overall condition of water quality

River basin district	Concentration									Total Condition
	NO$_3$			P			NH$_4$			
	Low	Moderate	High	Low	Moderate	High	Low	Moderate	High	
1. West Peloponnese	2	0	1	2	0	0	0	0	0	Good
2. North Peloponnese	1	0	0	1	0	0	0	1	0	Good
3. East Peloponnese	1	1	0	2	0	0	1	0	0	Good
4. West Sterea Ellada	11	0	0	10	2	1	0	9	0	Good
5. Epirus	7	0	0	8	0	1	3	4	0	Good
6. Attica	1	0	0	2	0	0	1	0	0	Good
7. East Sterea Ellada	3	0	2	1	2	0	0	3	0	Moderate
8. Thessaly	3	1	0	2	2	0	0	4	0	Moderate
9. West Macedonia	12	1	1	10	3	10	0	11	5	Bad
10. Central Macedonia	7	0	0	2	1	7	0	7	0	Moderate
11. East Macedonia	7	0	0	3	3	3	0	7	0	Moderate
12. Thrace	8	9	1	0	5	12	0	13	4	Bad
13. Crete	4	0	0	4	0	0	4	0	0	Good
14. Aegean Islands	–	–	–	–	–	–	–	–	–	Good

Source: MoEPPW (2008)

good, moderate and bad. This categorization was done in terms of the concentrations of nitrates (NO_3), phosphorus (P) and (ammonium) NH_4 in each RBD. The concentration for each pollutant is characterized as low, moderate or high according to the levels of the pollutant factor.[2] Table 22.4 presents the total number of the available measurements per pollutant and concentrations as well as the final condition of the RBD. West Macedonia and Thrace RBDs showed a bad quality condition under this classification.

Although Greece participated in the WFD intercalibration exercise (EC 2009), the implementation of the Article 4, which defines the environmental objectives per RBD, was not completed at the time of the study and thus the environmental quality assessment was based on approximations from other existing studies. The agricultural census was not organized per RBD and the information with respect to cultivations and water demand were approximated. In the following section, we briefly discuss the methodologies[3] used to calculate the total economic cost of water resources in Greece for the Report on the Implementation of Article 5 of the WFD (MoEPPW 2008).

Methodologies used for the calculation of the total economic cost of water resources

Financial cost

The financial costs faced by water companies providing supply, sewerage and irrigation services include expenditures for cost of capital, operation and maintenance cost of the network, administrative costs, depreciation and other financial costs. The data were collected from the enterprises' annual published financial reports. However, financial data were not available for all the companies in all the RBDs. Therefore, the total financial cost was approximated assuming for the remaining enterprises the Greek mean financial cost per enterprise and aggregating over all operating enterprises. Table 22.5 presents the financial costs for domestic and agricultural water supply in each RBD.

Environmental cost

The cost associated with the reduction of water quality and the subsequent limitation of water resources' capacity to provide goods and services is the environmental cost. In an optimal scenario, original valuation studies are carried out in each water body of each RBD addressing the particular environmental problem in the area in order to obtain accurate welfare loss estimations. However, such studies are expensive and were not available. In this case a benefits transfer approach can be used in order to calculate the environmental cost. This approach allows values from existing studies to be transferred to policy sites of interest after correcting for certain parameters (Desvousges et al., 1992; Kirchhoff et al., 1997). Therefore, a single value transfer was applied in the report and the environmental cost was estimated using existing valuation studies after proper adjustment. Welfare estimates in the studies considered are reported as willingness to pay for improvements in water quality (from bad or moderate to good) per individual and they were then aggregated over the population of each RBD. The zeros in the table indicate that the RBD has a good environmental status (recall Table 22.4). The results of the environmental cost calculations are reported in Table 22.6. The largest

Table 22.5 Financial cost per RBD

River basin district	Financial cost (€)	
	Domestic (€/hm³)	Irrigation (€/ha)
1. West Peloponnese	4 108 662	27.3
2. North Peloponnese	4 612 819	14.6
3. East Peloponnese	6 895 954	253.0
4. West Sterea Ellada	4 762 739	334.4
5. Epirus	5 684 518	319.2
6. Attica	833 711	13.0
7. East Sterea Ellada	3 378 763	10.1
8. Thessaly	6 850 916	63.9
9. West Macedonia	3 934 249	33.5
10. Central Macedonia	2 091 853	53.0
11. East Macedonia	5 193 781	95.7
12. Thrace	2 746 149	28.6
13. Crete	5 258 926	33.8
14. Aegean Islands	9 530 520	10.3

Source: MoEPPW (2008)

Table 22.6 Environmental cost per RBD

River basin district	Environmental cost (€)
1. West Peloponnese	0
2. North Peloponnese	0
3. East Peloponnese	0
4. West Sterea Ellada	0
5. Epirus	0
6. Attica	0
7. East Sterea Ellada	7 037 232
8. Thessaly	9 137 486
9. West Macedonia	14 535 598
10. Central Macedonia	16 586 149
11. East Macedonia	5 025 462
12. Thrace	9 842 713
13. Crete	0
14. Aegean Islands	0

Source: MoEPPW (2008).

environmental cost was found in northern Greece and specifically in Macedonia. Water quality in these regions is under severe stress since most of the industrial activity in the country is concentrated in these regions.

Resource Cost

The resource cost results from scarcity arising due to overexploitation of water resources beyond their rate of replenishment. The resource cost was calculated for the water districts of Aegean Islands, East Sterea Ellada, Thessaly and East Peloponnesos, where water demand surpasses supply as indicated by their water balance. The resource cost is approximated by the cost of backstop technology to cover excess demand (Koundouri 2004). In several Aegean islands desalination plants are installed. Therefore, the price of this backstop technology was used to estimate the resource cost. The exploitation of other non-conventional water sources such as recycled water was the backstop technology relevant for East Peloponnese and East Sterea Ellada. Finally, the excessive water demand in Thessaly is covered by the diversion of the river Acheloos. The resource cost was estimated by the product of the excess demand times the backstop technology cost per cubic meter of water which is €1.5m^3 for desalination plants and €0.5/m^3 for recycled water (WDD 2005) and €0.818m^3 for the Acheloos diversion (personal communication MoEPPW). Table 22.7 reports the resource cost in each RBD. Thessaly has the largest resource cost since it is the most important agricultural region in Greece. The resource cost is also high in the Aegean Islands due to water shortages.

Table 22.7 Resource cost per RBD

River basin district	Resource cost (€)
1. West Peloponnese	0
2. North Peloponnese	0
3. East Peloponnese	3 510 184
4. West Sterea Ellada	0
5. Epirus	0
6. Attica	0
7. East Sterea Ellada	20 515 680
8. Thessaly	89 356 467
9. West Macedonia	0
10. Central Macedonia	0
11. East Macedonia	0
12. Thrace	0
13. Crete	0
14. Aegean Islands	26 792 100

Source: MoEPPW (2008).

Assessment of cost-recovery level

Recovery from charges to users

An estimation of the cost-recovery level is possible once the total cost of water services is determined and the revenues of water companies are calculated. The cost recovery for irrigation companies was obtained by multiplying the irrigation requirements times the mean irrigation water price per RBD. The cost recovery of water companies in the domestic sector comes from potable water pricing, sewerage connection and waste water treatment pricing. The latter is calculated as a surcharge of 80 percent to the value of water consumption. Revenues from water consumption were obtained by multiplying the consumed cubic meters of water with the mean water price in each RBD whereas the sewerage expenses were inferred given the number of households in each RBD and the relevant fees set by the water companies.

Estimating the cost-recovery level

The cost-recovery level was calculated using the following equation:

$$\text{Cost recovery level} = \frac{\text{Recovery}}{\text{Total economic cost}}$$

Table 22.8 presents the results of the analysis of the cost-recovery level. The mean cost-recovery level per RBD in Greece was found to be 59.18 percent. In general, the revenues of water and sewerage services providers are not sufficient for financial cost recovery and the cost-recovery level of irrigation water is even lower.

Table 22.8 Cost-recovery level in each RBD

	Cost-recovery level (%)		
River basin district	*Domestic*	*Irrigation*	*Total*
1. West Peloponnese	62.21	11.44	50.54
2. North Peloponnese	77.31	19.41	68.22
3. East Peloponnese	37.89	15.66	34.18
4. West Sterea Ellada	61.29	14.28	46.19
5. Epirus	71.00	22.44	68.11
6. Attica	108.14	21.30	106.13
7. East Sterea Ellada	75.10	15.98	57.61
8. Thessaly	33.66	6.38	29.82
9. West Macedonia	53.55	41.05	51.71
10. Central Macedonia	86.58	12.04	78.27
11. East Macedonia	79.39	27.38	70.74
12. Thrace	103.29	11.05	78.28
13. Crete	49.67	56.25	50.91
14. Aegean Islands	42.94	1.78	37.84

Source: MoEPPW (2008)

These estimates are broad estimates of the true recovery level. This approach is based on benefit transfers and reasonable assumptions, and allows for valuable conclusions to be reached regarding the limitations of the pricing policies that fail to reflect the true value of the resource and efficiently allocate it to competing demands. The methodology followed in the report can assist future efforts to fully comply with the EU WFD reporting requirements.

Conclusions

In this chapter the evolution of different water-related directives and regulations were presented in order to understand the requirements and challenges in implementation of the WFD in Greece. Then, the situation of the water sector in Greece was analysed in the context of a legal action of "Non-Conformity" started by the EC against Greece (case C-264/07). A number of reasons explain the slow pace in the implementation of the WFD in Greece. For example, the hierarchical and centralized political and administrative structures in Greece have hindered the participation of civil society in policy development for the management of water resources. There is little evidence of coherent interaction between the different levels of government in policy implementation, monitoring and enforcing. Overlapping responsibilities between competent authorities pose a serious constraint in efficient and sustainable water resources management. In consequence, the introduction of participatory arrangements required by the WFD has been extremely difficult. Therefore, difficulties can be expected in the implementation of the WFD in countries with weak institutional structure and limited financial resources. The Greek Report on the Implementation of Article 5 of the WFD (MoEPPW 2008) provides a good example of the challenges and limitations encountered during the implementation of the directive. A complete analysis of all pressures and impacts and a detailed assessment of the cost recovery in each RB were extremely difficult due to the lack of information and the constrained time frame. Nevertheless, the results clearly highlight the need for reforms. The pricing policy in Greece requires changes to achieve good ecological status in all water bodies and to ensure full recovery of the cost of water services.

Notes

1 In 2009 the name of the ministry changed to Ministry of Environment, Energy and Climate Change.
2 Low concentration: $P<0.17mg/l$, $NO_3\text{-}N<5$ mg/l, $NH_4\text{-}N<0.04$ mg/l; moderate concentration: 0.17 mg/l$<P<0.31$ mg/l, 5 mg/l$<NO_3\text{-}N<11$ mg/l, 0.04 mg/l$<NH_4\text{-}N<1$; high concentration: $P>0.31$ mg/l, $NO_3\text{-}N>11$, $NH_4\text{-}N>1$
3 For a detailed explanation of the different methodologies using an ecosystem services approach see Koundouri and Dávila (2014).

References

Commission of the European Communities (CEC) (1996). European Community Water Policy. Communication from the Commission to the Council and the European Parliament. COM (96) 59 final, 21.02.1996.
Commission of the European Communities (CEC) (2000). Guide to the Approximation of European Union Environmental Legislation. Part 2: Overview of EU environmental

legislation. http://ec.europa.eu/environment/archives/guide/guidfin.pdf accessed 26 August 2014.

Demetropoulou, L., Nikolaidis, N., Papadoulakis, V., Tsakiris, K., Koussouris, T., Kalogerakis, N., ... and Theodoropoulos, K. (2010). Water Framework Directive implementation in Greece: introducing participation in water governance–the case of the Evrotas River Basin management plan. *Environmental Policy and Governance, 20*(5), 336–349.

Desvousges, W.H., Naughton, M.C. and Parsons, G.R., (1992), Benefit transfer: conceptual problems in estimating water quality benefits using existing studies. *Water Resources Research*, 28 (3), pp. 675–683.

Dworak, T., Kampa, E., de Roo, C., Alvarez, C., Bäck, S. and Benito, P. (2006). Simplification of European Water Policies (IP/A/ENVI/FWC/2006-172/Lot 1/C1/SC5). Policy Department Economic and Scientific Policy. European Parliament, Brussels.

Esteban, A., Le Quesne, T. and Strosser, P. (2006). *Economics and the Water Framework Directive: A User's Manual*, WWF/RSPB Publication. Available at: http://rackspace-web1.rspb.org.uk/Images/wfdguide_tcm9-137442.pdf accessed 26 August 2014.

European Commission (EC) (2000). Directive 2000/60/EC of the European Parliament and of the Council of 23rd October 2000 establishing a framework for Community action in the field of water policy, *Official Journal of the European Communities*, L327, 22.12.2000, pp. 1–72. European Commission, Brussels.

European Commission (EC) (2007). Towards Sustainable Water Management in the European Union. First stage in the implementation of the Water Framework Directive 2000/60/EC. Accompanying document to the Communication from the Commission to the European Parliament and the Council. COM (2007) 128 final, European Commission, Brussels.

European Commission (EC) (2009). Water Framework Directive Intercalibration Technical Report. EUR 23838 EN/1 – 2009, European Commission, Brussels.

European Commission (EC) (2012a). Report from the Commission to the European Parliament and the Council on the Implementation of the Water Framework Directive (2000/60/EC) River Basin Management Plans. SWD (2012) 379 final, European Commission, Brussels.

European Commission (EC) (2012b). Communication from the Commission to the European Parliament, the council, the European Economic and Social Committee and the Committee of the Regions: A Blueprint to Safeguard Europe's Water Resources. COM(2012) 673 final, European Commission, Brussels.

European Commission (EC) (2012c). The role of water pricing and water allocation in agriculture in delivering sustainable water use in Europe – Final report. European Commission, Brussels.

European Environment Agency (EEA) (2013) Assessment of cost recovery through pricing of water. EEA Technical report No 16/2013. http://www.eea.europa.eu/publications/assessment-of-full-cost-recovery accessed 26 August 2014.

Hey, C. (2005). EU environmental policies: a short history of the policy strategies. Chapter 3, 18–30. In S. Scheuer (ed.), *EU Environmental Policy Handbook. A Critical Analysis of EU Environmental Legislation*, European Environmental Bureau, Brussels.

Horth, H., France, S. and Zabel, T. (2003). Pollution Reduction Programmes in Europe: Updated report on the Assessment of Programmes under Article 7 of Directive 76/464/EEC, conducted by Water Research Centre (WRc) in the framework of a project on "Transitional Provisions for Council Directive 76/464/EEC and related Directives to the WFD 2000/60/EC". http://ec.europa.eu/environment/water/pdf/report1.pdf accessed 26 August 2014.

Kaika, M. (2003). The Water Framework Directive: a new directive for a changing social, political and economic European framework. *European Planning Studies, 11*(3), 299–316. DOI: 10.1080/09654310303640

Kallis, G., and Nijkamp, P. (1999). Evolution of EU water policy: a critical assessment and a hopeful perspective. *Research Memorandum, 1999*, 27.

Kirchhoff, S., Colby, B.G. and LaFrance, J.T. (1997). Evaluating the performance of benefit transfer: an empirical inquiry, *Journal of Environmental Economics and Management, 33*, 75–93.

Koundouri, P. (2004). Current issues in the economics of groundwater resource management. *Journal of Economic Surveys, 18*(5), 703–740.

Koundouri, P., and Dávila, O.G. (2014). The use of ecosystem services approach in guiding water valuation and management: inland and coastal waters. In A. Dinar, and K. Schwabe (eds.), *The Handbook of Water Economics*. Cheltenham: Edward Elgar Publishers.

Koundouri, P., Kountouris, Y. and Remoundou, K. (2010), A Note on the Implementation of the Economics of the EU Water Framework Directive under Data Limitations: A rapid appraisal approach. DEOS Working Papers, Athens University of Economics and Business.

Koundouri, P., Papandreou, N., Remoundou, K., and Kountouris, Y. (2014). A bird's eye view of the Greek water situation: the potential for the implementation of the EU WFD. In P. Koundori and N. Papandreou (eds) *Water Resources Management Sustaining Socio-Economic Welfare* (pp. 1–24). Springer Netherlands.

Ministry of Environment Physical Planning and Public Works (MoEPPW) (2003). The law for the management of water resources of Greece, Law N.3199/03. Athens, Greece.

Ministry of Environment Physical Planning and Public Works (MoEPPW) (2008). Report on the Implementation of Article 5 of the WFD. Athens. Available at: http://www.aueb.gr/users/koundouri/resees/en/aswposprojen.html (in Greek).

Moss, B. (2008). The Water Framework Directive: total environment or political compromise? *Science of the Total Environment*, *400*(1), 32–41.

Regional Environmental Center (REC) (2008). Handbook for Implementation of EU Environmental Legislation. http://ec.europa.eu/environment/archives/enlarg/handbook/handbook.pdf accessed 26 August 2014.

Sofios, S., Arabatzis, G., and Baltas, E. (2008). Policy for management of water resources in Greece. *The Environmentalist*, 28(3), 185–194.

Water Development Department (WDD), Republic of Cyprus (2005). EU summary report articles 5&6. Available at: http://www.cyprus.gov.cy/moa/wdd/Wdd.nsf/All/B8D7262CBFCC 9AF8C225711E00303F5A/$file/Page1-20.pdf (last accessed February 2014).

WWF (2006). EU Water Policy: Making economics work for the environment: Survey of the economic elements of the Article 5 report of the EU Water Framework Directive, available at: http://assets.panda.org/downloads/eu_water_policy___may_2006.pdf

23

WATER CONSERVATION AND TRADING

Policy challenges in Alberta, Canada

Henning Bjornlund and K.K. Klein

Introduction

Water scarcity is becoming a major issue in the southern part of Alberta, Canada. The environmental impacts of the current level of extraction are being felt with 22 out of 33 main stem river reaches rated as moderately impacted, 5 as heavily impacted and 3 as degraded (AE, 2005). The Alberta Government has responded by ceasing to accept applications for new licensed water allocations in the South Saskatchewan River Basin (SSRB). To secure a continued supply of sufficient volumes of water of adequate quality for human and economic uses, the Alberta Government has introduced a number of new institutions, legislation and two policy strategies: the Water for Life strategy in 2003 and the Land Use Framework in 2009 to integrate land and water management. The Water for Life strategy relies on improving water use 'efficiency' and 'productivity' by 30 percent, without any specification of what these terms mean, to reach its objective. Economic instruments, voluntary reallocations of water, and public partnerships in water planning are the main tools suggested to be used to move water to meet new demand as well as meet and secure conservation objectives. Since 97 percent of all allocated water resources in Alberta is from surface water (AE, 2014), and 75 percent of that water is allocated to irrigators (AE, 2002) the focus of these water policies is very much on the irrigation sector; the focus of this chapter is therefore also on surface water and the irrigation sector.

It is increasingly acknowledged that current water allocation and management institutions are an impediment to achieving efficient allocation outcomes (Taylor, 2009; Bjornlund, 2010). However, recent attempts to develop a new system have not resulted in any major changes and Alberta is currently going through another public review process. Reflecting the inadequacies of current institutions, there is widespread opposition from both irrigators and other stakeholder groups to expand the use of markets for water. As a result there has been widespread community opposition to the few attempts to reallocate or share water that is held by existing license holders, especially the irrigation districts, and trading has led to only minor reallocations to date.

While most of the river systems in the SSRB are over-allocated and some rated as environmentally degraded due to the current level of extraction, the full environmental impacts of the over-allocation have not yet been experienced as license holders on average use only 60 percent of their allocated water (Alberta Government, 2013). Many fear that water markets will activate the unused portions of existing water allocations and thus exacerbate already suffering river systems. The drive for efficiency improvements also raises concerns over the net impact of this on the environment.

The second section of this chapter provides an overview of the development of water policy in Alberta to address growing water scarcity and identifies the problems associated with water sharing under current water allocation and management institutions. The third section discusses the use and potential impacts of water markets. The fourth section examines the policy focus of increasing water use efficiency and productivity as a way of conserving water that can be shared with other users. The fifth section investigates the potential policy options for sharing water that currently are being considered in Alberta in light of public perception and acceptability. The final section provides some conclusions and policy recommendations.

Alberta's policy response to scarcity

Historical context

Alberta is one of Canada's three Prairie Provinces, which is part of the region that drains into the Hudson Bay. It was granted by Charles II as a Royal Charter to the Hudson Bay Company in 1670 (Knafla, 2005; Harris, 2008) to administer as long as it did not contravene British Law, of which the Riparian Doctrine is part. In 1870 the territory was transferred to the Dominion of Canada and the Prairie Provinces became part of the Northwest Territory, and the Riparian Doctrine became formal law.

As the railroad spread across Canada there was a great desire to settle the prairie. For farming to thrive in this semiarid environment, large scale irrigation was needed. The Riparian Doctrine proved to be an impediment as it limited water use to riparian land (Percy, 1977). In this semiarid region, water needed to be conveyed to land distant from the source in order to support large scale irrigation settlements. As a consequence, local visionaries looked at alternative water allocation mechanisms. They studied the prior appropriation doctrine that underpinned large scale irrigation in the western United States and the Victoria (Australia)Water Act, 1884, which rejected prior appropriation as unworkable (Percy, 1996). Alberta settled for a compromise. As in the Victoria Act, the North West Irrigation Act (NWIA), 1894, vested ownership of water in the crown allowing the government to grant licensed water allocations to access and use water in pursuit of their policy objectives. Such licenses were perpetual in nature but non-transferable as they were tied to the land to which they were granted. Although in this respect the NWIA was, in large part, a copy of the Victoria Act, it rejected the Victorian approach of proportional sharing under scarcity. Instead it adopted the United States system of First-In-Time, First-In-Right (FITFIR) and introduced a prior allocation system that allocates water during scarcity according to the seniority of the license. Subsequent water legislation in Alberta retained this system and attempts to change it have been met with fierce opposition from the largest and most senior license holders: irrigation farmers and hydropower corporations (Bjornlund, 2010).

Irrigation in Alberta started in 1879 when Mormon farmers came north from Utah with expertise in irrigation. Spurred by the NWIA of 1894, Alberta got its first irrigation Act in 1905 when it became a province (although the federal government retained control over water resources until 1930, after which Alberta legislated a Water Resources Act in 1931) and Irrigation Districts Act in 1915 (Percy, 2005). The Irrigation Districts Act enabled irrigation districts to obtain provincially guaranteed loans to construct infrastructure and to levy a tax on irrigated land as a means to pay for administration and maintenance. Until the 1920s, irrigation development was private business with the government providing technical advice only. These early attempts failed financially or ended with large debt. After 1920 the government relieved the irrigation districts of their debts and introduced water charges according to their ability to pay rather than the cost of supply. The 13 irrigation districts in Southern Alberta were formed by 1968 and in 1970 the province started to assist irrigation districts in rehabilitation of their capital works by paying 86 percent of the cost. This was reduced to 80 percent in 1969 and 75 percent in 1995, which is the current rate. Total irrigated area increased from 19,223 hectares in 1911 to 495,786 hectares in 2008 with major increases taking place after the 1970s (AIPA, 2002).

Current generation of water policies

As the demand for surface water increased faster than supply and as water quality increased in importance, concerns started to emerge over the potential increase in and impact of water allocations. The government's first response to this concern came in 1991 when it capped the amount of water that could be allocated for irrigation (AE, 2003a) and commenced a review of water policies and legislation. After several years of public consultations and consideration, and following cancellation of licenses for about one million acre-feet, the Water Act was passed in 1996 and came into force on January 1, 1999. That Act confirmed the FITFIR principle and protects all existing licenses in good standing. It introduced three important institutions for management of water under scarcity: water trading, public participation and water planning.

Three methods of water trading were introduced: transfers and leases of licensed allocations and assignments of the right to use water during a season. The latter was allowed only between existing license holders and thus did not allow new users to get established. Also under section 33(3) of the Water Act, irrigators can purchase assignments to use only the volume specified in their license; hence their main function is to allow irrigators with low priority licenses to secure supply from more senior licenses during periods of scarcity. Under the Act, the Minister must initiate water planning processes, identify water conservation objectives (WCOs) and secure community involvement in these processes. Upon the approval of water plans the WCOs will be protected by licenses, with the seniority being the date the plan is approved. Consequently, they would be the most junior licenses and the last to receive water during periods of shortage. While the Act provided for licenses to be issued to protect WCOs, it also stated that only the government can hold licenses for in-stream purposes, thereby effectively preventing non-government entities from using their own funds to achieve environmental or recreational objectives, as has been done both in the United States and Australia (Lane Miller et al., 2013). In February 2013 the Water Conservation Trust challenged this by applying for a license for habitat enhancement and recreation, fish and wildlife, and water management, based on a donation of a licensed water allocation from a private company. After this application was denied, the Trust appealed the decision to the Alberta Environmental Appeal Board. In September 2013, the Board denied the appeal (Calgary Herald, 2013). The water plans also set out the conditions under which trading can take place. The local director can approve transfers under these rules subject to the 'Administrative Guidelines for Transferring Water Allocations' (AE, 2003b). The director must ensure that transfers will not cause harm to the environment or third parties and has the authority to withhold 10 percent of the transferred volume to meet WCOs.

District irrigators' ability to trade water is set out in the Irrigation Districts Act, 2000. Trade between irrigators within the same district is relatively unrestricted. However, trading water to outside the district requires the approval of a majority of irrigators in a plebiscite. Under this Act, irrigation districts have another option to share their water: they can enter into a water supply agreement with a third party. Such contracts can be for longer or shorter terms. To do this, their license has to make explicit provision for this and specify how much water can be used for non-irrigation purposes.

During 2001/02, the province experienced a severe drought and irrigators in the southern part of the SSRB were subject to cuts to their allocations. In response, senior licensees agreed to suspend their seniority temporarily so that water was shared proportionally among license holders. Failing to have done so, an important agricultural processor with a junior license would have had to close due to lack of water, which would have hurt local irrigators indirectly. This period also saw trading in water assignments. These two sharing arrangements assisted irrigators and processors to manage the drought (Nicol and Klein, 2006).

Following this drought, the Alberta Government placed a moratorium on the issuing of new licensed allocations within the southern tributaries of the Oldman River (AE, 2003a) and commenced a process of public consultations to develop a new strategy for the sustainable management of Alberta's water resources. This led to release of the Water for Life (WFL), Alberta's strategy for sustainability, in November 2003. The strategy document acknowledges that water within the SSRB is fully or overcommitted and demand for water is likely to increase due to population and economic growth, while demands for in-stream flows are increasing. The objectives of WFL are to secure a continued supply of adequate water of sufficient quality to support continued economic prosperity, human wellbeing and healthy ecosystems. The major method to achieve this is a 30 percent increase in productivity and efficiency of water use (without defining what this means). It is hoped that this will conserve water to meet increased demands and that voluntary reallocations are the means by which this can take place. The strategy also places great emphasis on community engagement in water planning processes.

Following the Water Act, 1999, and the WFL strategy in 2003, three major community based governance institutions were established to develop water plans for the watersheds in Alberta. Watershed Planning and Advisory Councils (WPAC) are engaged in developing draft water management plans, Water Stewardship Groups are involved in implementing on-the-ground water actions and the Alberta Water Council works with the Alberta Government and the WPACs to resolve water planning issues. The draft Water Management Plan for the SSRB was published in 2005. It identified widespread over-allocation, environmental degradation and growing demand throughout the basin and recommended that the government cease issuing new licenses (AE, 2005). In response, the minister in 2005 closed the SSRB (except the Red Deer Sub-basin) so that no new applications for water licenses would be accepted after that date.

In recognition of the growing problems associated with Alberta's water allocation system, the minister in 2008 announced a process to develop a new water management and allocation policy. As a result, major reports were released by: i) the Minister's Advisory Group on a New Water Management and Allocation Framework (MAG, 2009); ii) the Alberta Water Council (AWC, 2009); iii) the Alberta Water Research Institute (AWRI, 2009); and iv) the CD Howe Institute (Bjornlund, 2010). In these reports, several alternative methods were discussed to improve water management. However, the process seems to have stalled and no further actions have been taken.

In 2008, the Land Use Framework (LUF) was introduced to integrate regional planning and management of land, water and other natural resources and given legislative support with the Alberta Land Stewardship Act (ALSA) in 2009 (Alberta Government, 2008, 2009). As a result, major regional planning exercises commenced in northern and southern Alberta in 2009 (Alberta Government, 2009). However, the geographic boundaries for the regional plan do not match those of the SSRB. It is unclear what the role of the WPACs and the water plans they had developed, or were in the process of developing, will be in the new regional plans. According to the LUF and ALSA, the regional plans should set out how to achieve the best outcomes from use of a region's natural resources. ALSA is the overarching act and overrules the Water Act, or any other Alberta Act, in case of inconsistency (Alberta Government, 2009). It has, however, been stated that the regional plans will build on the work of the WPACs and the water management plans will be an integral part of the regional plans.

Water reallocation or water sharing

The uptake of water trading among irrigators has been very slow and originally was viewed with suspicion. A survey of managers and directors of the 13 irrigation districts in 2005 indicated that only 8 percent agreed with trading licensed allocations and 15 percent with short term trades (Bjornlund et al., 2007). During the first five years of trading in licensed allocations, 23 applications were registered but only 6 were arm's length market transfers. Three were approved by 2004 and another three by 2008 (Nicol et al., 2008). Some trades of assignments took place during the 2001/02 drought (Nicol and Klein, 2006). Surveys in 2006, 2007 and 2012 of district and private irrigators suggest that very few trades occurred between individual irrigators, with only 1–3 percent having sold or leased water over the previous five years as a result of high transaction costs and inadequate flow of market information about supply, demand and prices (Nicol et al., 2008). A higher proportion of irrigators reported purchasing either a license or a lease of water: 28 percent of district irrigators reported they had purchased a license and 17 percent had purchased a lease. These, however, do not represent market transactions as most of the buyers are irrigators who purchased additional irrigated land from district expansions. Irrigators' intentions to trade over the next five years (2012–2016) were significantly higher: the intention to buy was highest among district irrigators and the intention to sell was highest among private irrigators. This likely reflects the continued expansion of irrigation districts and the fact that private irrigators have a greater ability to trade under current legislation (Bjornlund et al., 2014a). Current institutions for water trading clearly are constraining trade.

The closure of the SSRB in 2005 should have generated more activity in water markets. However, since irrigators, except in the most extreme years, use only 60 percent of their licensed water (Alberta Government, 2013), most are not short of water and have capacity to expand. However, the closure generated a greater interest from rural towns and developers, many of which do not have adequate water to meet increased demands from growing populations and industrial developments. There have been a few transfers to meet this type of demand. In some instances, irrigation districts have supplied neighboring communities or new industries, especially food processors that provided new markets for irrigated commodities.

The largest and most controversial water transfer took place in 2007. A developer wanted to construct a major shopping mall, casino and horse racetrack at Balzac, a small community north of Calgary. The development was just outside the boundary of the City of Calgary so the city was not obliged to supply water to the development. Calgary has a large water license, enough to supply three times its population, and its supply infrastructure stretched to the door of the proposed development. Calgary tried to use its water power to expand its jurisdiction by offering to supply the development if the land were transferred to the City of Calgary. The Municipality of Rocky View refused. The municipality then tried to obtain a license from the Red Deer River where licenses are still being issued. That would involve an inter-basin transfer, which requires ministerial approval, and was denied due to the political sensitivity of the issue and opposition from environmental groups. Finally the developer negotiated with the Western Irrigation District (WID) to purchase 2,500 dam^3 for the Municipality of Rocky View at a price of C\$6,000 per dam^3 (one dam^3 = 1,000 m^3). This was a record breaking price not only in Alberta but in most jurisdictions, including Australia and the United States (Saliba and Bush, 1987; Bjornlund et al., 2013a). The developer would then construct a new supply pipeline, basically duplicating the existing City of Calgary infrastructure: a very expensive solution.

The transfer had to be approved by the irrigators in a plebiscite. The WID management tried to sell the idea to irrigators by suggesting that the proceeds from the sale could be used to replace an old leaking canal with a pipeline that would 'save' more water than was sold. Management also argued that the government wanted the transfer to be approved and if they failed to do so, the water might simply be taken away. The transfer should be a clear example of what WFL calls for: a voluntary transfer, a win–win proposal. The proposal, however, caused heated debate among irrigators and the plebiscite narrowly approved the transfer. Subsequent research identified four underlying issues that influenced how irrigators voted; two influenced a yes vote: i) they feared the intervention of the government if they voted no; and ii) the money was invested in such a way that the irrigation district would be left with at least the same net supply; and two influenced a no vote: i) the transfer was necessary only because Calgary refused to supply the water, which was considered wasteful; and ii) they did not agree with the way the money was to be spent. Irrigators also expressed a general reluctance to permanently relinquish control over any part of their licensed allocation (Lafreniere et al., 2013). The failure of management to address irrigators' concerns in the lead-up to the plebiscite resulted in the narrow yes vote. The government's approval of the transfer also proved controversial; several stakeholders protested and tried to prevent the approval and the Calgary Water Authority tried to block it through the courts. However, the case eventually was dismissed.

Since the irrigation districts control the largest and most senior licenses and have large volumes of unused water, agriculture's contribution to future water sharing is essential if the WFL is to achieve its objective. A survey of irrigators across all 13 irrigation districts was conducted in 2012 to better understand how irrigators would vote in future plebiscites under different transfer scenarios. The findings confirm some of the findings of Lafreniere et al. (2013) but also provided further insights. Irrigators' voting intentions were no longer influenced by the government's position. The debates that occurred during and after the WID plebiscite apparently alleviated their concerns. The survey identified four major factors that would influence a yes vote: i) 25 percent would be influenced mainly by the way the proceeds would be spent, i.e., to increase efficiency so that all the water sold is saved; ii) 34 percent would be influenced mainly by the purpose for which the water was transferred (17 percent if the water was for environmental purposes and 17 percent if it was for municipal or irrigation purposes); iii) 14 percent would be influenced mainly by who would benefit from the transfer and the price paid, i.e., they would personally gain from the way the proceeds were spent and that the price would be a market record; and iv) 8 percent would be influenced mainly by the efficiency of the transfer, i.e., if this was the best way to meet the new demand. Fourteen percent would not vote yes under any scenario (Lafreniere et al., in press). These findings stress the importance of the message sent to irrigators when water sharing options are proposed to deal with shortages.

Considering the general opposition among irrigators to sell licensed water allocations and public pressure on the industry to share its water, industry leaders have taken two steps. First, they announced a 'People First' policy whereby basic human needs will always be met before irrigation. Second, districts are pursuing an alternative water sharing option under the Irrigation Districts Act, using water supply agreements to supply commercial, municipal and other users. This will not require districts to relinquish control of their licenses, only to commit them to supply a certain volume of water over a specified period. To do this, their licenses must specify the volume of water that can be used for non-irrigation purposes. One irrigation district got its license amended to facilitate this in 2003. It was argued that such supply agreements circumvent the rigorous environmental impact assessment required for transfers of licensed water allocations (Banks and Kwasniak, 2005). This is of concern given the level of over-allocation and environmental degradation as a consequence of the current level of extraction as such arrangements could activate unused water and increase total extraction.

Following the WID plebiscite in 2007, there was a wave of new applications for license amendments. Alberta Environment ceased processing applications when a number of individuals, mainly members of environmental groups, challenged the first application in the courts. In 2011, the Environment Appeals Board ruled that public interest groups have no standing to challenge approvals of such amendments, as only people directly affected have this right. Following this decision, Alberta Environment approved all pending amendments. This decision was tested by the Queens's Bench of Alberta as a number of environmental groups sought a judicial review. The final verdict, handed down in January 2013, confirmed that there is no public standing at the Alberta Environment Appeal Board (Fluker, 2013). These conflicts over transfers of licensed allocations and amendments of districts' licenses are a product of a policy failure to properly address environmental issues prior to introducing institutions to deal with shortages and have impeded their ability to deliver the anticipated outcomes.

While there has been a lot of debate regarding alternative water allocation mechanisms, there have been relatively few economic studies of their expected impacts. A recent study that compared the current prior allocation system with three proportional allocation systems under various levels of water shortages found that proportional reductions based on each district's past five year average diversions produced superior outcomes in terms of reallocation of water and maximizing economic rents (He et al., 2012). Even better outcomes could be achieved if the districts could participate in water trading providing the transaction costs of trading are not too high (which has been identified as one of the main impediments to more widespread trade (Nicol et al., 2008)). The sources of the economic gains from proportional reductions and the use of water markets are the quite diverse marginal values of water in each irrigation district.

Proportional reductions based on past diversions create a potentially inefficient incentive for irrigators to increase current water diversion in water-surplus years, in order to increase their relative share in times of shortage. Also, proportional sharing would create winners and losers during years of shortage through redistribution of income and wealth. Despite these concerns, He et al. (2012) provided clear support for the superiority of proportional sharing over the current seniority-based system from an overall economic efficiency perspective.

Improving productivity and efficiency

Reallocating water away from agriculture would come at a cost. The incremental increase in gross margins of irrigated crops in Alberta over the same crops grown on dry land has been estimated to average C$244 million (in 2008 dollars) per year. On a per hectare basis, the annual average incremental increase in gross margin was C$495, and per 1000m^3 of water was C$191 (Klein et al., 2012a). The WFL strategy implies that if water use efficiency (WUE) could be improved, it would be possible to maintain or even increase production with the use of less water. Irrigators therefore face increasing pressure to improve WUE so that more water will be available for industrial and municipal uses as well as for water conservation objectives. An important issue is how to measure WUE as soil, crop and irrigation scientists all have their own definitions.

Effective management of water within a river basin requires knowledge of aggregate water use efficiencies (Haie and Keller, 2008; AE, 2008). The Organisation for Economic Co-operation and Development (OECD, 2001) proposed two indicators of WUE at the aggregated level: i) Water Use Technical Efficiency (WUTE), which is the physical mass of agricultural production per unit of water used; and ii) Water Use Economic Efficiency (WUEE), which is the monetary value of agricultural production per unit of water used. WUTE places emphasis on bulky crops and disregards irrigators' financial results. WUEE provides a base for evaluating the impacts of changes in input and output prices, government policies and other factors that can change year by year. WUTE assumes that all crops have the same impact on the wellbeing of irrigators. WUEE removes the major limitation of WUTE by including prices and costs of each crop. A further issue is how to define the total monetary value; should it be defined as gross revenue, net revenue, or incremental increase in revenue?

Both technical and economic aggregated measures of WUE were estimated for the four river sub-basins encompassing the 13 irrigation districts in southern Alberta over the five year period, 2004–2008 (Klein et al., 2012b). The average level of WUTE varied from 3.5–6.2 Mt/dam³. The gross economic value of crop production varied from C\$345–C\$592/dam³. The net economic value of crop production varied from C\$163–C\$268/dam³. The incremental increase in net value of crop production varied from C\$130–C\$199/dam³. The correlations between WUTE and WUEE were relatively low and statistically insignificant. Although the three measurements of WUEE were positively correlated, they presented different pictures of changes in WUEE across river sub-basins and across the five year time dimension. These findings highlight the need for an adequate and agreed definition of WUE to measure improvements over time and space.

All indicators that compare WUE over time have two main weaknesses. First, moisture to produce a crop comes from two sources: natural rainfall and irrigation. In most instances, natural precipitation is ignored. Since some years have above normal rainfall and others may be drought-like, they can produce misleading and contradictory results as producers adjust the level of irrigation depending on weather conditions. For example, 2005 was a year of above average rainfall in southern Alberta; as a result net diversions of water for irrigation were much lower than in the other years. Since the net diversions are the divisors in the calculations, this high rainfall year had the highest levels of WUTE and WUEE (Klein et al., 2012b). Second, the measures are not comparable across regions due to differences in crop mix, soil type, weather patterns and other factors. Bulky crops, such as forages, generally yield higher tonnage than cereals or oilseeds and since the numerator in WUTE includes the total amount of biomass produced for all crops (not the monetary value), the WUTE results can be misleading. For forages, the numerator is the total harvestable biomass whereas for grains and oilseeds, the numerator includes only the weight of the seeds. These results demonstrate that measurements of WUE are not straightforward and interpretations of WUEs need to consider factors such as crops grown, output prices, input prices, methods of irrigation and amount of net water used.

In a first attempt to establish a baseline for measuring improvements in water use efficiency, Ali and Klein (2012) estimated total factor productivity for 12 of the irrigation districts (all but the smallest) over the decade 2000–09. Total factor productivity can be decomposed into two components: technical efficiency and technical change. Their results showed that total factor productivity has been increasing at a geometric average annual rate of 7.1 percent over the decade but this was made up of an average annual increase of 10.6 percent due to improvements in technology combined with an average annual decrease in technical efficiency of 3.2 percent. This is consistent with findings by AIPA (2002) and provincial water research (Wood, e-mail communication, 28 November 2013) that show improvements in water application technologies in Southern Alberta from flood to wheel-move to high pressure center pivot and to low pressure center pivot have led to improvements in on-farm irrigation efficiency of 34 percent in 1965 to 77.5 percent in 2012 (Figure 23.1). Follow-up studies should be

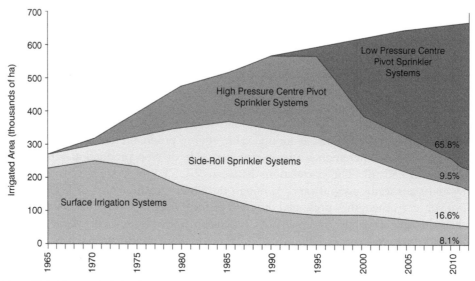

Figure 23.1 Changes in irrigation system use in Alberta from 1965 to 2012 (source: Alberta Agriculture and Rural Development, 2013)

conducted to further refine estimates of technical and economic efficiencies of water use in Alberta so valid comparisons can be made over time and appropriate, evidence-based judgments can be made regarding water use and allocation. The above discussion illustrates the importance of terminology and measures when setting policy objectives. However, regardless of the measure of efficiency used, improving the aggregate level of WUE depends on irrigators' decisions to adopt new technologies and management practices, which have the potential to increase on-farm irrigation efficiency. Surveys of irrigators in 2006/07 and again in 2012 suggest that the rate of adoption is slowing but also show that there is plenty of room for improvement (Bjornlund et al., 2008; Wang et al., 2014). The major driver of technology adoption seems to be yield and crop quality with 60 percent of district irrigators and 30 percent of private irrigators rating that as the most important factor, followed by saving energy cost (13 percent for district irrigators and 24 percent for private irrigators) and saving labor (16 percent for district irrigators and 27 percent for private irrigators). The major reasons for not adopting are that irrigators already perceive that they use the most efficient technology (28 percent of district irrigators and 15 percent of private irrigators) and financial constraints (10–28 percent depending on district and 14 percent for private irrigators). For private irrigators, physical field condition was most important for 28 percent while 16 percent found that while the adoption would save them money and increase yield they did not think the benefits would outweigh the costs (Bjornlund et al., 2009; Nicol et al., 2010).

Recent research indicates that to encourage a higher rate of adoption as well as intensity of adoption (percentage of irrigated land on which adoption has occurred) of new and potentially more efficient technologies, policy makers should encourage, facilitate and support: i) the supply of relevant information about new technologies and their potential benefits and costs through extension officers and farmer organizations; ii) more effective support services after an adoption decision has been made to assist farmers in its implementation; iii) establishment of farmer peer groups, farmer meetings and field days to increase the exchange of experiences and dissemination of knowledge; iv) an increase in extension services; v) farmers in developing the most efficient business structure for their farm business and farm succession process; and vi) programs that particularly target and accommodate small scale farmers (Wang et al., 2014).

Public perception of policy options

The WFL strategy focuses on water reallocation and efficiency and productivity improvements to meet future demand from both consumptive and non-consumptive uses. The debate in the province over potentially changing the current allocation mechanism with or without further use of water markets and other water sharing mechanisms has been rather heated, with various stakeholders expressing strong and conflicting views. Legislative constraints on who can hold water licenses for in-stream flows also are impediments to improve in-stream flows. The capacity to further improve on-farm irrigation efficiency might be declining under current physical, technological and financial constraints. Further, the early willingness of industry to share water saved though efficiency improvements with other users was low (Bjornlund et al., 2007). Also, there are disputes about the extent to which water saved through improved efficiency really represents system savings or simply reduces in-stream flows and thereby downstream supplies. These findings create doubts about the ability of the current processes and institutions to achieve their economic, environmental and social objectives.

The institutions and policy instruments currently pursued in Alberta could be implemented in different ways. Water sharing could be achieved through voluntary or compulsory means with or without compensation. Improving water use efficiency could happen by regulatory requirements or voluntary actions. Either way, the associated costs could be left to the water users or partially subsidized. Importantly, the water saved could either be redirected to new uses (compulsorily) or left to the current license holders' discretion. The policy choice taken will influence the likely success of implementation. The political willingness to introduce effective policies and institutions depends on politicians and their advisors' perceptions of the level of public support for such policies.

Recent surveys (2009–12) of more than 2,600 households across southern Alberta who reside in communities with different levels of exposure to water scarcity, environmental degradation and dependence on irrigation for their economic prosperity aimed to elicit the level of support for various policies and instruments that currently are being considered in Alberta and what influences their level of support. The nine policies and instruments tested reflect three different policy orientations. One would vest strong powers in the government in water allocation and management, another would focus on providing strong support for the environment and the third would provide strong support for existing license holders (Bjornlund et al., 2013a, 2014b). Support for the nine policy statements varied significantly among all four regions and irrigators (Table 23.1). Among non-irrigators, there is a high level of support for policies that grant strong powers to the government in water management and allocation and the setting of minimum flows, as well as giving non-government entities the right to hold licenses for in-stream flows. These are interesting findings as current government policies minimize the government's role and instead rely on public participation and market mechanisms. The difference between irrigators and non-irrigators is especially significant when the policy option has

Table 23.1 Percentage who agree or strongly agree with policy statements

	Most to least resource dependent				
	Irrig[1]	RTMS[2]	Strat[3]	Leth[4]	Calg[5]
The government, rather than market forces, should determine who uses Alberta's water.	44	57	57	64	65
If an irrigation district/private license holder isn't using all water allocation, the government should be able to take water for environmental purposes, without compensation.	11	48	41	53	51
If water is to be traded among irrigation districts and/or municipalities, the government should set the price.	17	44	44	55	40
Minimum flows of water should be set for all rivers and streams, and only the water above those minimum flows should be used for economic purposes.	62	65	72	76	77
Private groups and individuals should be able to hold water licenses for environmental purposes.	39	40	43	42	49
Public funds should be used to improve irrigation systems only if the water that is saved is left in rivers.	40	50	56	54	59
The government should buy water from current license holders, like irrigation districts, so that more water can be left in the rivers.	21	31	46	34	41
The seniority of a water license must be honored under all circumstances.	78	35	26	24	20
Water that is saved through improved water use efficiency should be used to increase economic activity.	73	45	41	39	35

Notes
1 Irrigators; 2 Raymond, Taber, McGrath, Sterling, four small irrigation communities around Lethbridge; 3 Strathmore, town 50 km east of Calgary, head office of Western Irrigation District; 4 Lethbridge, regional center in Southern Alberta, in the center of 65 percent of all irrigation in Canada; 5 Calgary, the largest City in Alberta, financial and oil industry center.
RMTS and Lethbridge are in the most water scarce and most degraded southern region of the SSRB; Calgary and Strathmore in the least scarce and impacted northern region.

Source: Based on Bjornlund et al., 2013b and 2014b.

specific impacts on irrigators' rights and profitability such as accepting the seniority of all licenses, how saved water should be used, and the government's powers to interfere in the rights of irrigation districts to trade water. These findings suggest that irrigators are becoming politically marginalized and might explain why industry leaders increasingly are very public in their attempts to find ways of sharing their water.

The results provide important insights into how various sectors of the community perceive that water should be shared among existing and new users, including the environment. There is strong evidence to suggest that the level of resource dependence and the exposure to water scarcity and environmental degradation influence people's policy preferences. While people living in rural areas align more closely with their irrigator neighbors than do city dwellers, their preferences are more closely aligned with city dwellers than with irrigators. This reflects other findings in the literature that the rural–urban gap in environmental values and behavior has been narrowing in recent years (Huddart-Kennedy et al., 2009; Sharp and Adua, 2009) and further supports the observation that irrigators are becoming increasingly marginalized politically. Once location has been controlled for, socio-demographic characteristics such as education, age, income and occupation also matter. These findings refute other findings in the literature that urban–rural differences represent differences in socio-demographic characteristics such as income and education (e.g. Salka, 2001). Further, people's involvement in water and land issues either through work, recreation or membership in environmental or natural resources organizations such as water planning and advisory councils or water stewardship groups also influence their policy preferences for water sharing. However, demographic characteristics, involvement and participation influence policy preferences differently in different locations and whether the person is an irrigator or non-irrigator; this further stresses the importance of context and geography (Bjornlund et al., 2013b, 2014b).

Conclusions and policy recommendations

Alberta has been active in introducing new institutions to deal with scarcity, improve water use efficiency and water sharing. The main emphasis has been on improving water use efficiency and productivity to generate water savings as a source of supply to meet new demands and rely on voluntary reallocations to move the saved water to new users. Public participation, which includes defining water conservation objectives and setting the framework for water trading, has been seen as an integral part of water planning processes. The process has not been smooth. The prior allocation mechanism that has been ingrained in Alberta's water laws since 1894, and the historical pattern of issuing licensed allocations, has consolidated control of the largest and most senior licenses in a very small group of users that currently have an excess supply during normal weather conditions. Recent research in Alberta has indicated that water use efficiency and economic gains to society could be improved by changing the way water is shared during periods of water shortages. Also, there is clear evidence that finding ways to lower transaction costs and promote trade in water allocations would lead to increased benefits to society.

Despite a low level of utilization, senior license holders initially showed limited willingness to share their water with other sectors, even if the water was 'saved' through improved water use efficiency. Due to political pressure, some of those license holders recently have shown an increased willingness to share their water. However, as current levels of water use are starting to show evidence of environmental degradation, special interest groups are very vocally and strongly opposing recent attempts to share water held by irrigation districts as they see this as potentially activating unused water, thereby increasing total extraction and environmental degradation. The Alberta case illustrates the consequences of a policy failure to revise the historical prior allocation doctrine, address the issues of unused or underused licensed allocations, provision of water for the environment and the means by which this can be achieved, as part of the process of introducing new institutions to deal with water scarcity.

The findings discussed in this chapter provide insights for policy makers who are tasked with designing and implementing water management and allocation institutions to deal with potential water shortages. The overall assessment is that context and geography matter and that policy solutions therefore need to be context specific to ensure their acceptance within the target community. Taking into account stakeholders' varied concerns will reduce conflicts and opposition among stakeholders, increase the chance of acceptance and improve the likelihood of achieving policy objectives. While no policy can be designed to satisfy all interests, it is important to be aware of the likely different responses from different stakeholders so that their concerns can be acknowledged in the implementation phase and an effort made to introduce instruments that allow each stakeholder to manage the new policy or institution in ways that best accommodate their concerns.

Acknowledgements

The authors are thankful of the funding provided by Alberta Innovates: Energy and Environment Solutions, the Social Sciences and Humanities Research Council of Canada and the Canadian Water Network, which have made the research underlying the discussion in this chapter possible.

References

AE, Alberta Environment (2002) 'Water for Life – Facts and Information on Water in Alberta', Publication #I/930. AE, Alberta Environment, Edmonton, Alberta.

AE, Alberta Environment (2003a) 'Summary, SSRB Background Studies', AE, Alberta Environment, Edmonton, Alberta.

AE, Alberta Environment (2003b) 'Administrative Guidelines for Transferring Water Allocations. Publication # I/949', AE, Alberta Environment, Edmonton, Alberta.

AE, Alberta Environment (2005) 'Background Information for Public Consultation on the South Saskatchewan River Basin's Draft Water Management', AE, Alberta Environment, Edmonton, Alberta.

AE, Alberta Environment (2008) 'Water for Life: A Renewal', AE, Alberta Environment, Edmonton, Alberta.

AE, Alberta Environment (2014) 'Water Allocation', http://environment.alberta.ca/03134.html (accessed on 27 March 2014).

Alberta Agriculture and Rural Development (2013) 'Alberta Irrigation Information 2012,' Irrigation and Farm Water Division. Lethbridge, Alberta.

Alberta Government (2008) 'Alberta Land-use Framework', Alberta Government, Edmonton, Alberta.

Alberta Government (2009) 'Alberta Land Stewardship Act', Alberta Government Edmonton, Alberta.

Alberta Government (2013) 'Irrigation in Alberta: A Statistical Overview – 2012', Alberta Government, Edmonton, Alberta.

AIPA, Alberta Irrigation Project Association (2002) 'South Saskatchewan River Basin – Irrigation in the 21st Century, Volume 1: Summary Report', AIPA, Lethbridge, Alberta.

Ali, M.K. and Klein, K.K. (2012) 'Water Use Efficiency and Productivity of the Irrigation Districts in Southern Alberta', Presented at the Annual Meeting of the Canadian Agricultural Economic Society. June 17–19. Niagara Falls, Canada.

AWC, Alberta Water Council (2009) 'Water for Life – Action Plan', Government of Alberta, Edmonton, Alberta.

AWRI, Alberta Water Research Institute (2009) 'Towards Sustainability: Phase 1 – Ideas and Opportunities for Improving Water Allocation and Management in Alberta', AWRI, Edmonton, Alberta.

Banks, N. and Kwasniak, A. (2005) 'The St. Mary's Irrigation District License Amendment Decisions: Irrigation District as a Law onto Themselves', *Journal of Environmental Law and Practice* Vol 16, No 1, pp. 1–18.

Bjornlund, H. (2010) 'The Competition for Water: Striking a Balance among Social, Environmental and Economic Needs'. Commentary No 302: Governance and Public Institution, C.D. Howe Institute, Toronto.

Bjornlund, H., Nicol, L. and Klein, K.K. (2007) 'Challenges in Implementing Economic Instruments to Manage Irrigation Water on Farms in Southern Alberta'. *Agricultural Water Management* Vol 92, pp. 131–141.

Bjornlund, H., Nicol, L. and Klein, K.K. (2008) 'Economic Instrument and Irrigation Water Management – A Comparative Study of Private and District Irrigators in Alberta, Canada'. In Esteve, Y.V., Brebbia, C.A. and Rico, C.P. Eds. *Sustainable Irrigation Management, Technologies and Policies II*. WIT Press, Southampton, pp. 3–14.

Bjornlund, H., Nicol, L. and Klein, K.K. (2009) 'The Adoption of Improved Irrigation Technology and Management Practices – A Study of Two Irrigation Districts in Alberta, Canada', *Journal of Agricultural Water Management* Vol 96, pp. 121–131.

Bjornlund, H., Wheeler, S. and Rossini, P. (2013a) 'Water Markets and Their Environmental, Social and Economic Impact in Australia'. In Maestu, J. Ed. *Water Trading and Global Water Scarcity: International Perspectives*. Taylor & Francis, London, pp. 68–93.

Bjornlund, H., Zuo, A., Wheeler, S., Xu, W. and Edwards, J. (2013b) 'Policy Preferences for Water Sharing in Alberta, Canada', *Water Resources and Economics* Vol 1, No 1, pp. 93–110.

Bjornlund, H., Xu, W. and Wheeler, S. (2014a) 'An Overview of Water Sharing and Participation Issues for Irrigators and Their Communities in Alberta: Implications for Water Policy', *Agricultural Water Management*, vol 145 (November) pp. 171–180.

Bjornlund, H., Zhao, X. and Xu, W. (2014b) 'Variation in the Perspective on Sharing Water – Irrigators, Their Communities and the Wider Society', *International Journal of Water Governance* Vol 2, No 1, DOI: 10.7564/13-IJWG29

Calgary Herald (2013) 'Appeals board upholds Alberta rulings denying water rights', Article in Calgary Herald on 20 September 2013, http://www.calgaryherald.com/news/alberta/Appeals+board+upholds+Alberta+rulings+denying+water+rights/8940185/story.html (accessed on 22 July 2014).

Fluker, F. (2013) 'No Public Interest Standing at the Alberta Environmental Appeals Board', The University of Calgary Faculty of Law. http://ablawg.ca/2013/02/15/no-public-interest-standing-at-the-alberta-environmental-appeals-board/ (accessed on 21 May 2013).

Harris, Cole R. (2008). *The Reluctant Land: Society, Space, and Environment in Canada before Confederation*. UBC Press. University of British Columbia. Vancouver BC.

Haie, N. and Keller, A. (2008) 'Effective Efficiency as a Tool for Sustainable Water Resource Management', *Journal of the American Water Resources Association* Vol 44, No 4, pp. 961–968.

He, L., Horbulyk, T.M., Ali, Md. K., Le Roy, D.G. and Klein, K.K. (2012) 'Proportional Water Sharing Vs. Seniority-based Allocation in the Bow River Basin of Southern Alberta', *Agricultural Water Management* Vol 104, pp. 21–31.

Huddart-Kennedy, E., Beckley, T.M., McFarlane, B.L. and Nadeau, S. (2009) 'Rural–urban Differences in Environmental Concern in Canada', *Rural Sociology* Vol 74, pp. 309–329.

Klein, K.K.., Yan, W. and Le Roy, D.G. (2012a) 'Estimating the Incremental Gross Margins Due to Irrigation Water in Southern Alberta', *Canadian Water Resources Journal* Vol 37, No 2, pp. 89–103.

Klein, K.K., Bewer, R., Ali, Md. K. and Kulshreshtha, S.N. (2012b) 'Estimating Water Use Efficiencies for Water Management Reform in Southern Alberta Irrigated Agriculture', *Water Policy* Vol 14, No 6, pp. 1015–1032.

Knafla, Louis A. (2005) 'Introduction: Laws and Societies in the Anglo-Canadian Northwest Frontier and Prairie Provinces, 1670-1940'. In Knafla, L. and Swainer, J. Eds. *Laws and Societies in the Canadian Prairie West*. UBC Press, Vancouver BC, pp. 1–56.

Lafreniere, K., Deshpande, S. and Bjornlund, H. (in press) 'Using Conjoint Analysis to Assess Irrigators' Preferences of Water Transfers in Alberta, Canada', Accepted for publication in *Society and Natural Resources*.

Lafreniere, K., Deshpande, S., Bjornlund, H. and Hunter, G. (2013) 'Extending Stakeholder Theory to Promote Resource Management Initiatives to Key Stakeholders: A Case Study of Water Transfers in Alberta, Canada', *Journal of Environmental Management* Vol 129, pp. 81–91.

Lane Miller, C., Wheeler, S., Bjornlund, H. and Connor, J. (2013) 'Acquiring Water for the Environment: Lessons from Natural Resources', *Journal of Environmental Policy and Planning* Vol 15, No 4, pp. 513–532.

MAG, Ministers Advisory Group (2009) 'Recommendations for improving Alberta's water management and allocation'. Government of Alberta, Edmonton.

Nicol, L. and Klein, K.K. (2006) 'Water Market Characteristics: Results from a Survey of Southern Alberta Irrigators', *Canadian Water Resources Journal* Vol 31, No 2, pp. 91–104.

Nicol, L., Klein, K.K. and Bjornlund, H. (2008) 'Permanent Transfers of Water Rights: A Study of the Southern Alberta Market', *Prairie Forum* Vol 33, No 2, pp. 341–356.

Nicol, L., Bjornlund, H. and Klein, K.K. (2010) 'Private Irrigators in Southern Alberta: A survey of the Adoption of Improved Irrigation Technologies and Management Practices', *Canadian Journal of Water Resources*, Vol 35, No 3, pp. 339–350.

OECD, Organisation for Economic Co-operation and Development (2001) 'Environmental Indicators for Agriculture: Methods and Results', Volume 3. OECD Paris.

Percy, D. (1977) 'Water Rights in Alberta', *Alberta Law Review* Vol 15, pp. 142–165.

Percy, D. (1996) 'Seventy-Five Years of Alberta Water Law: Maturity, Demise, and Rebirth', *Alberta Law Review* Vol 35, No 1, pp. 221–241.

Percy, D. (2005) 'Responding to Water Scarcity on Western Canada', *Texas Law Review* Vol 83, No 7, pp. 2091–2107.

Saliba, B. and Bush, D. (1987) *Water Markets in Theory and Practice: Market Transfers, Water Values, and Public Policy*, Studies in Water Policy and Management No. 12, Westview Press, Boulder, CO.

Salka, W. (2001) 'Urban–rural Conflict over Environmental Policy in the Western United States', *American Review of Public Administration* Vol 31, pp. 33–48.

Sharp, J. and Adua, L. (2009) 'The Social Basis of Agro-environmental Concern: Physical Versus Social Proximity', *Rural Sociology* Vol 74, pp. 56–85.

Taylor, L. (2009) 'Water Challenges in Oil Sand's Country: Alberta's Water for Life Strategy', *Policy Options* (July–August), pp. 44–47.

Wang, J., Bjornlund, H., Klein, K.K., Zhang, L. and Zhang, W. (2014) 'New Irrigation Technologies – The process of adoption and what influences it in Alberta, Canada', Paper presented at the joint IWREC/World Bank Conference, Washington DC, September.

Wood, S. (2013) Soil and Water Research Scientist, Irrigation and Farm Water Division, Alberta Agriculture and Rural Development, Lethbridge, Alberta Pers. Com. by e-mail on 28 November 2013.

INDEX

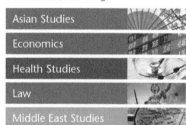